Progress in IS

More information about this series at http://www.springer.com/series/10440

James B. Pick · Avijit Sarkar

The Global Digital Divides

Explaining Change

Springer

James B. Pick
School of Business
University of Redlands
Redlands, CA
USA

Avijit Sarkar
School of Business
University of Redlands
Redlands, CA
USA

ISSN 2196-8705 ISSN 2196-8713 (electronic)
Progress in IS
ISBN 978-3-662-46601-8 ISBN 978-3-662-46602-5 (eBook)
DOI 10.1007/978-3-662-46602-5

Library of Congress Control Number: 2015933819

Springer Heidelberg New York Dordrecht London

Printed on acid-free paper

Springer-Verlag GmbH Berlin Heidelberg is part of Springer Science+Business Media
(www.springer.com)

To Dadu, a blessed presence forever in my life, my parents, and to two special women, the best part of every day, my wife Sangita and our daughter Anoushka.

—Avijit Sarkar

To my wife, Rosalyn M. Laudati, with heartfelt thanks for patience and support.

—James B. Pick

Preface

The adoption, diffusion, and utilization of information and communications technologies (ICTs) have increased steadily in many parts of the world, notably in developed countries. In other parts, especially in developing nations, ICT adoption and utilization started late, and has exhibited leapfrogging, in other words, bypassing certain parts of the adoption-and-diffusion cycle. By the end of 2013, the world was estimated to have almost 6.8 billion mobile cellular subscribers, almost equal to the world's global population, and 2.7 billion internet users; yet almost 4.4 billion people were estimated as not being online by the end of 2013 (ITU 2013). Irrespective of whether a country is highly developed, or is lagging in socioeconomic development, examining digital divides has been a continuing endeavor of policy planners and researchers in government and in other domains such as academia.

The term "digital divide" refers to the "gap between individuals, households, businesses, and geographic areas at different socio-economic levels with regard both to their opportunities to access ICTs and to their use of the Internet for a wide variety of activities" (OECD 2001). Digital divides are now at the heart of many conversations involving ICT use, socio-economic development, public policy, ICT policy, macro- and microeconomic outcomes, infrastructure development, and ethics. Digital divides have been studied from a variety of perspectives. Numerous studies have focused on quantifying digital divides, in other words, estimating the adoption and use of ICTs, and how they vary over time, place, communities, socio-economic attributes, public sector involvement versus private enterprise, and economies. Quantifying digital divides and use of ICTs have been the primary focus of organizations such as the International Telecommunications Union, which publishes its annual reports titled "Measuring the Information Society" (ITU 2013). Quantification is enabled by the expanding reach of accurate census-taking and surveys, and the creation of sophisticated indices such as the ITU's ICT Development Index. The advantage of such approaches is the longitudinal tracking of key variables that measure a country's ICT adoption status.

The literature on ICT availability, adoption, and diffusion, and the digital divide is extensive. While adoption/diffusion theory provides nuanced explanations of how technologies are adopted and diffused, its applications in the digital divide literature have been rather limited. Digital divides have been studied for diverse geographies—at the world, country, state, provincial, and city levels. Comparative digital divide studies are common between nations, but less so between developing and developed economies and for specific agglomerations such as the OECD. Digital divides have also been examined through diverse lenses such as social, economic, demographic, political, regulatory, to name a few, and in the context of specific indicators of ICTs such as internet, broadband, and mobile wireless technologies. The body of empirical literature that has attempted to understand influences of various factors on ICT adoption and utilization is vast. Sometimes this literature adopts frameworks of social reproduction (Agarwal et al. 2009) and institutional factors (King et al. 1994) as theoretical foundations; while several other studies are atheoretical and employ a variety of empirical techniques.

Most of the existing literature has two gaps. (a) While digital divide studies often compare ICT adoption between multiple nations, or focus on a single nation, a specific province, or community, systematic studies that examine digital divides at the subnational (for example, state- or provincial) level are scarce. One may wonder, why is that relevant? Subnational examination is important since there is increasing evidence in the digital divide literature that digital divides are simply not about digital haves and have-nots, the local context matters (Cecchini and Scott 2003). (b) Digital divides pertain to people and where they live—for example, in large urban agglomerations, referred to in this book as megacity or megalopolis, versus rural areas, within a state, within a province, or within a nation. The common undercurrent is geography. Surprisingly, aside from a small set of spatial scholars who have studied digital divide such as Tony Grubesic (Grubesic and Murray 2005), and Barney Warf (Warf 2013), the geography of digital divides is largely ignored. This book fills these two critical gaps.

In this book, we first examine the global digital divide at the country level, and then proceed to investigate the digital divide between the states/provinces of three of the world's most populous nations in 2013, China, India, and the United States, as well the tenth most populous nation of Japan. Together, China, India, the US, and Japan accounted for 43 % of the world population and 44 % of the World GDP in 2013 (World Bank 2014). We also examine the digital divide between countries in the African continent, which is home to more than one billion. A common theme of this book is to tease out and understand the impact of geography on ICT utilization—whether it is for the nations of Africa, or the provinces of China, using spatial statistical methods. While analyzing digital divides, most of the existing empirical research fails to account for geographic forces that might cause high technology adoption or low technology adoption nations/states/provinces to agglomerate together. If that is the case, there is a distinct possibility that the true impact of the nonspatial endogenous factors may be overestimated, sometimes leading to entirely fallacious conclusions. This book attempts to understand—does geography matter to digital divides, and if so, how?

Another consistent theme of this book is the incorporation of societal openness as it relates to the digital divide (Qureshi 2012). We live in a world today that is ravaged by conflict, civil wars, violation of national sovereignties, clampdowns on civil liberties and freedom of the press in many regions, while some other regions are largely peaceful. How does this impact ICT adoption and utilization? We attempt to answer this question within our broader empirical framework.

The book is comprised of 12 chapters, from an introduction and history of digital divides, to developing our conceptual model of ICT utilization, followed by empirical analysis of digital divides for world nations, the provinces of China, India, Japan, the US, and the countries of Africa. We present our results of comparative analysis of digital divides, along with policy recommendations, that are specific to each region/country studies, as well as generic policy implications. Finally, we outline the future of the digital divide, contextualized by our findings and the state of information and communication technology today. To elucidate our findings, to illustrate key policy recommendations, and sometimes to provide contextual information, each chapter of this book has at least one case study; a total of 19 case studies across 12 chapters. The cases are qualitative, sometimes provide secondary data, and along with our quantitative analysis of the correlates of ICT utilization offer a holistic overview of digital divides in specific contexts and specific geographies.

The case studies discussed in the book chapters are listed in Table 1. They range from case examples of ICT adoption and related policies at the country level, to specific firms that are highly regarded as enablers of innovation using ICTs, to applications of ICTs to solve socio-economic problems and foster prosperity and well-being. Also, the cases are rather evenly distributed among world regions and zoom into nations that are at various points on the human development spectrum as classified by the United Nations.

The book is intended for several audiences. It is keyed toward knowledgeable readers about telecommunications, ICTs, and the digital divide. Such knowledgeable individuals may exist in industry, government, academia, independent policy planning institutions, and other places. Each of the country-specific chapters as well as the chapter on Africa has an accompanying policy section. In addition, an entire chapter of the book is dedicated to policy. Therefore, the book is of significant interest to policymakers, notably in government, especially in the areas of telecommunications, information technology, infrastructure development, and public policy.

At a broader level, the societal implication findings are of interest to government executives, administrators, and specialists. The book is also intended for academics and students in various disciplines such as business and management, information technology and management information systems, public policy, planning, and last but not the least, geography. Professionals in the private sector, especially in technology-oriented enterprises are also going to find the book of interest. Participation by the private sector in ICT infrastructure development has often been reported in the digital divide literature, especially in developed nations. However professionals, whether in developed or developing nations, will find the findings

Table 1 Case studies in Global Digital Divide

Chapter #	Chapter title	Name/title of case study	Continent/ World region	Level of human development index (UN 2014)
1	Introduction	Singapore	Asia	Very high
		Mexico	Latin America	High
2	Historical background	Azerbaijan	Asia	High
		South Korea	Asia-Pacific	Very high
3	Theoretical model of ICT utilization	Estonia	Europe	Very high
4	Global Digital Divide	Mauritius	Africa	High
5	Digital divide in China	Alibaba	China, Asia	High
		Google China (Hong Kong)		
6	Digital divide in India	DakNet and Bhoomi	India, Asia	Medium
		Warana village project		
7	Digital divide in Japan	FTTH network in Nishiokoppe village	Japan, Asia	Very High
		Kyoto poultry traceability system		
8	Digital divide in United States	Georgia Enterprise Technology Services	United States, North America	Very high
		ICT adoption by micro-enterprises in Nebraska		
9	Digital divide in Africa	Internet diffusion in South Africa	Africa	Medium
		Mobile telephony for Nigerian micro-enterprise development	Africa	Low
10	Comparisons of countries	Lenovo	China, Asia	High
11	Roles and policies of governments	Rwanda	Africa	Low
12	Future of the digital divide	Leadership in Poland	Europe	Very high

of the book insightful as they and their organizations explore possibilities to engage in public–private partnerships toward ICT capacity development in their immediate communities, home countries, or other parts of the world. Lastly, the book is of interest to a general audience that takes interest in the global expansion in technologies.

We recognize that ICTs by their very nature are dynamic. Technology changes, and so does its impacts. It is our hope that the *Global Digital Divides: Explaining Change* book will stimulate and provoke readers, pique their interests, enable them to ask questions and propose solutions, leading to information and networked societies at global and regional levels, as well as empowered local communities.

References

Agarwal, R., Animesh, A., & Prasad, K. (2009). Social interactions and the "digital divide": Explaining variations in internet use. *Information Systems Research,20*(2), 277–294.

Cecchini, S., & Scott, C. (2003). Can information and communications technology applications contribute to poverty reduction? Lessons from rural India. *Information Technology for Development, 10*, 73–84.

Grubesic, T. H., & Murray, A. T. (2005). Geographies of imperfection in telecommunications analysis. *Telecommunications Policy, 29*, 69–94.

ITU (2013). *Measuring the Information Society*. Geneva, Switzerland: International Telecommunications Union.

King, J., Gurbaxani, V., Kraemer, K., Mcfarlan, F., Raman, K., & Yap, C. (1994). Institutional factors in information technology innovation. *Information Systems Research, 5*(2), 139–169.

OECD (2001). *Understanding the digital divide*. Report. Paris, France: Organization for Economic Co-operation and Development.

Qureshi, S. (2012). As the global digital divide narrows, who is being left behind?, *Information Technology for Development, 18*(4), 277–280.

UN (2014). Human Development Index (HDI). New York, NY: United National Development Program, United Nations. Available from http://hdr.undp.org/en/content/human-development-index-hdi. Accessed 26 November 2014.

Warf, B. (2013). *Global Geographies of the Internet*. Heidelberg: Springer.

World Bank (2014). The World Bank Data. Retrieved from http://data.worldbank.org/indicator/NV.SRV.TETC.ZS?order=wbapi_data_value_2013+wbapi_data_value+wbapi_data_value-last&sort=desc. Accessed 26 November 2014.

Acknowledgments

This project would not have been possible without the valued cooperation of many people.

First of all special appreciation to Jeremy Johnson, adjunct faculty member and MBA graduate at University of Redlands, who provided extraordinary support, GIS expertise, data management mastery, and insights on the project.

We also extend deep appreciation to Tetsushi Nishida of Nag Inc. and former MBA student and researcher at University of Redlands, who with great skill and knowledge established early quantitative and spatial support and ideas for this project.

We thank Xi Zhang for research collaboration on the China segment of the research.

Special thanks to Tommi Cahill of Department of Art at University of Redlands for her superb photographic expertise and fine support. Also we give special thanks to Vanessa Siliezar for dedicated and highly skilled staff support and MBA student Jessica Rosales for fine assistance on the index.

Thanks to chapter reviewers Bing Bai, Christopher Franklin, John Glover, Esperanza Huerta, Alicia Iriberri, Peter Keenan, Nigel Melville, Fang Ren, Namchul Shin, and David Yates, for their insightful critical comments, suggestions, and advice. Any errors that remain are the sole responsibility of the authors. We also thank anonymous reviewers of earlier journal versions and conference paper reviewers of portions of several chapters for keen insights and critical suggestions.

At various stages, several people provided ideas and support for which we are appreciative: Leo Curtis, John Glover, Keisaku Higashida, Ken Kraemer, Shintaro Nakagawa, Fang Ren, Lauren Scott, Frantz Rowe, and we thank Lee Dion of Copyright Clearance Center for support.

We thank the School of Business, University of Redlands, for research funding.

We thank Elsevier Publishing and Taylor and Francis for permission to publish some revised sections and art work from four published articles* and a book chapter* that contained earlier research related to Chaps. 4–8.

Finally, we thank Springer Senior Editor Christian Rauscher for insights, editorial advice, patience, and support.

The authors wish to acknowledge all these people and many others not mentioned who provided assistance and support.

*Publications which have some material appearing in modified form in this book, with permission of their journal or book publishers, are the following.

Nishida, T., Pick, J.B., and Sarkar, A. 2014. Japan's Prefectural Digital Divide: Multivariate and Spatial Analysis. *Telecommunications Policy*, 38, 11, 992–1010.

Pick, J.B., Nishida, T. and Zhang, X. 2013. Determinants of China's Technology Utilization and Availability 2006–2009: A Spatial Analysis. *The Information Society*, 29(1), 26–48.

Pick, J.B., Nishida, T., and Sarkar, A. 2014. Broadband Utilization in the Indian States: Socio-Economic Correlates and Geographic Aspects, in Jyoti Choudrie and Catherine Middleton (eds.), *Management of Broadband Technology Innovation*, Oxford, England: Routledge, 269–296.

Pick, J.B., Sarkar, A., and Johnson, J. 2015. United States Digital Divide: State Level Analysis of Spatial Clustering and Multivariate Determinants of ICT Utilization. *Socio-Economic Planning Sciences*, 49, 16–32.

Pick, J.B., and Nishida, T. 2015. Digital divides in the world and its regions: A spatial and multivariate analysis of technological utilization. *Technological Forecasting and Social Change*, 91, 1–17.

Contents

List of Figures

List of Tables

Chapter 1
Introduction

1.1 Introduction

Global technology change is rapid and often transformational for economies and societies. Coming from a world of fixed line telephones and micro-wave networks of the 1950s and 1960s, to today's broadband, cell phones, social media, video, and multimedia, global technology change has permeated advanced economies and ignited rapid growth in the least developed economies. The benefits of pervasive technologies include economic productivity, global collaboration, e-commerce, efficiencies and change in government and society, information richness, data-rich scientific and technological advances, political transparency, and societal knowledge.

The unfolding of technological advancement among nations worldwide has been uneven, and keyed to many permutations, notably the level of economies, democracy, human capital, infrastructural features, and innovations. Despite overall growth, there remain distinct geographic patterns of technology utilization worldwide, and continents and countries have fairly persistent relative high or low levels of information technologies. Likewise, within nations, states, and provinces, regions are uneven in geographic patterns of prevalence of information technologies. At the high end, some regions such as Silicon Valley in Northern California, Route 128 in Boston, and the Tokyo metropolitan region in Japan are known to be high-tech and innovative, while other regions are regarded as trailing digitally, such as the northeastern states in India or the western areas of China. A state or region can gain great benefits from its ICT (Information and Communications Technology) via a more productive economy, education, technologically-skilled labor force, innovation centers, creativity, entrepreneurship, and arrival of immigrants who often bring skills. Likewise, technology uses by individuals are influenced by access to knowledge, education, income, network connectivity, and social capital. Potential detriments of technology advancements are present as well, such as privacy invasion of individuals, social exclusion of Internet-deprived individuals or groups.

J.B. Pick and A. Sarkar, *The Global Digital Divides*, Progress in IS,
DOI 10.1007/978-3-662-46602-5_1

Differences in information and knowledge between nations, regions, businesses, groups, and individuals are often referred to as the *digital divide*. The digitally advanced segment of the population is separated from the digitally backward segment by a gap or divide. In most such situations, a more appropriate concept is *digital continuum*, i.e. portraying digital capability extending from high to medium to low levels, but the former term is used in this book since it is in common parlance. *Digital divide* is defined as a continuum extending from high to low digital capabilities and usage. The digital divide is geographical, since people, firms, and cities, and states have a location, and usually neighboring locations are more likely to have similar levels of technology.

1.2 Magnitude and Rapidity of Technological Change

The global diffusion of information technologies has been very rapid over the past 30 years. The internet became available for use first in the U.S. in the mid 1980s. By the late 1980s, the first U.S. backbone was established by the National Science Foundation and the U.S. Defense Department. The Internet became commercialized and widely available globally in the 1990s. Its use was further stimulated by the advent of the World Wide Web (WWW) in 1990, the first browser being developed by Tim Berners-Lee that year and the web becoming widely available in the mid 1990s. By 2013, there was an estimated 2.7 billion people using the Internet (ITU 2013). As seen in Fig. 1.1, the proportion of internet users in advanced economies such as South Korea, Japan, and the U.S. reached around 80 % of population in 2011, while middle level economies such as Brazil, China, and Mexico reached in the range of 35–45 % of population. In portions of Africa, western Asia, and the Middle East, internet use was under 10 % of population. Although the most advanced economies had very rapid growth in the decade from mid 1990s to mid-2000s, their growth rates since then have begun to taper off, while countries such as China (39 % in 2011), Mexico (37 %), and India (10 %) appear to be entering the period of their most rapid growth in the present teen's decade.

The impacts of the Internet and Web are profound on people's lives, enterprises, and governments, allowing instantaneous communication and transfer of money, assets, and business documents; enabling the cloud; providing real-time updating of multitudinous processes and data-bases; and activating an inter-connected world, which is shrinking in size in many aspects of information transfer and knowledge sharing.

The mobile phone dates from 1973, when the first bulky model appeared from Motorola. The second generation (2G) occurred in 1991, followed a decade later by the 3G generation with significantly higher bandwidth and speeds, allowing for web and internet applications to sprout on "smart phones." It is expected that the 4G generation, which commenced in 2010, will continue to grow into large market share in the late teen's decade, offering access to streaming multi-media applications at much higher speed. Mobile phone subscriptions are estimated to be 6.5 billion by 2018 (ITU 2013), which will approach the size of the human population,

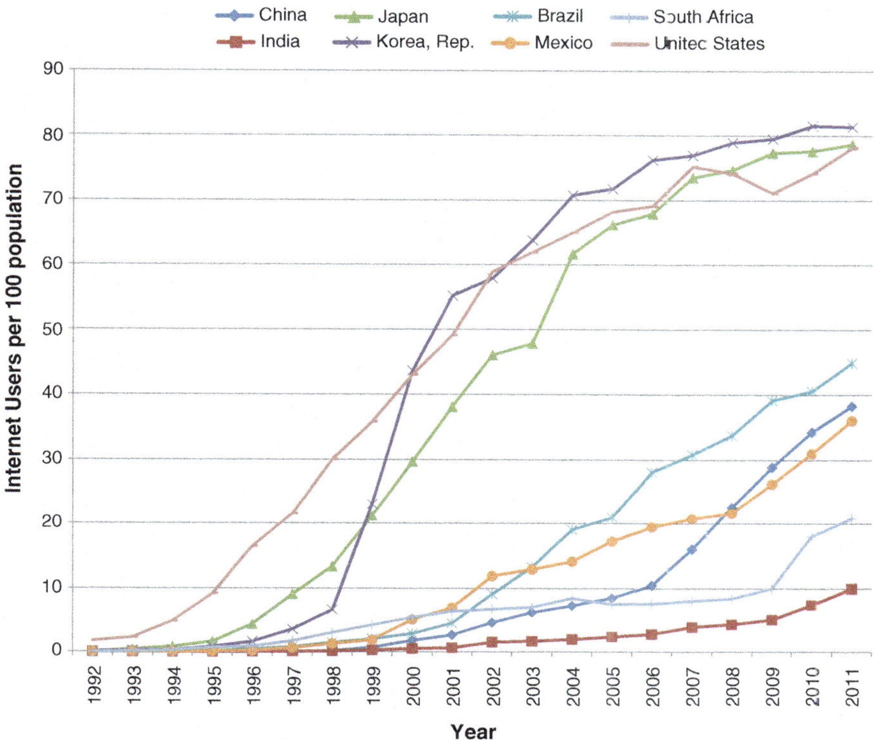

Fig. 1.1 Internet Users per 100 population, 1992–2011. *Source* World Bank (2012)

presently at 7.1 billion (Population Reference Bureau 2013). In developed nations, the active mobile phone subscriptions per 100 persons in 2013 was 74.8. Even in the economically poor continent of Africa, active mobile phones subscriptions per 100 persons was 10.9 and growing rapidly (ITU 2013). The rapidity of expansion of mobile phones, seen in Figs. 1.2 and 1.3, applies to all parts of the planet. As a world technology leader, the United States increased from 11 million mobile phone subscriptions in 1992 to 332 million in 2011 (World Bank 2012), while for China, expected in a decade or two to be the world's largest economy, mobile phone subscriptions grew from 176 thousand in 1992 to 968 million in 2011. Mobile phone users in California, U.S., are seen in Fig. 1.4. India likewise experienced rapidity of growth even faster than China over the 20 years to reach 894 million mobile phone subscribers in 2011.

The dynamic impacts of mobile phones are evident worldwide ranging from meetings of Fortune 200 businesses to basic uses by poor citizens of Africa, India, Western Asia, and South America. Even the illiterate, who can barely afford inexpensive cell phones, are still able to participate orally as part of the worldwide networks. The reason for mobile phones being perhaps the most prevalent personal

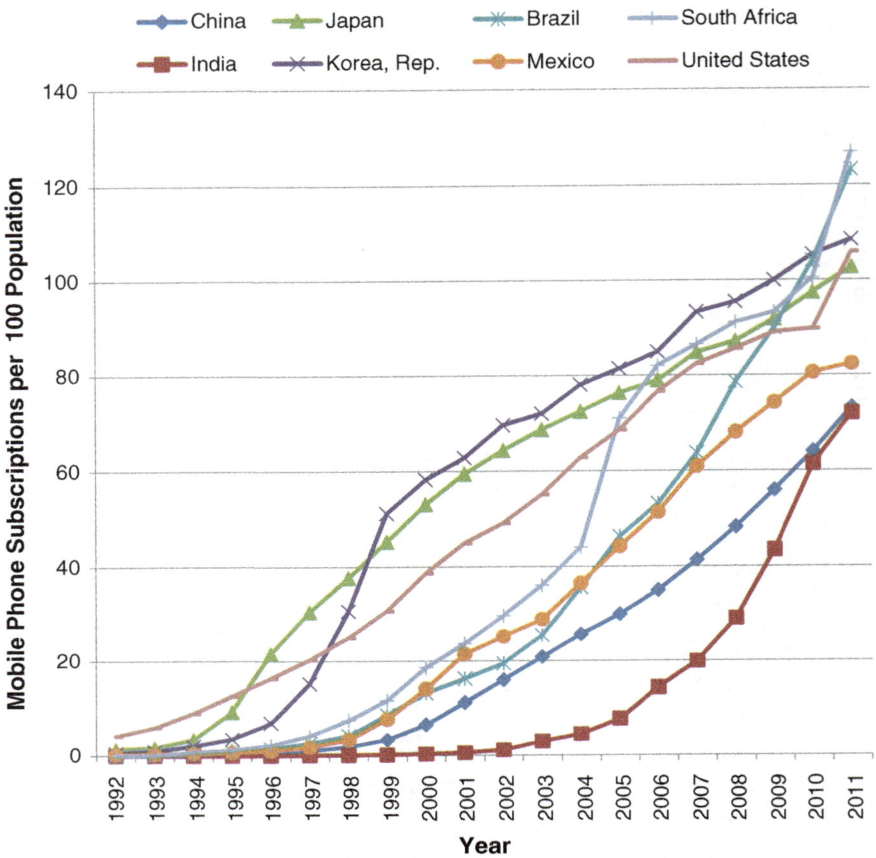

Fig. 1.2 Mobile phone subscriptions per 100 population, 1992–2011. *Source* World Bank (2012)

technology is their low cost, reliability, portability, and the relative ease for provider companies to establish networks of cell towers.

Although today information technologies are more prevalent and expanding rapidly throughout the globe, the forces motivating the remarkable increase are less well known, and the geographies of its diffusion even less so. This book addresses these lesser-known aspects.

1.3 Book's Objectives

The first goal of this book is to describe, analyze, and explain the world's rapidly changing and growing access and use of technologies, geographic patterns of technology use, and the economic, social, political, governmental, infrastructural,

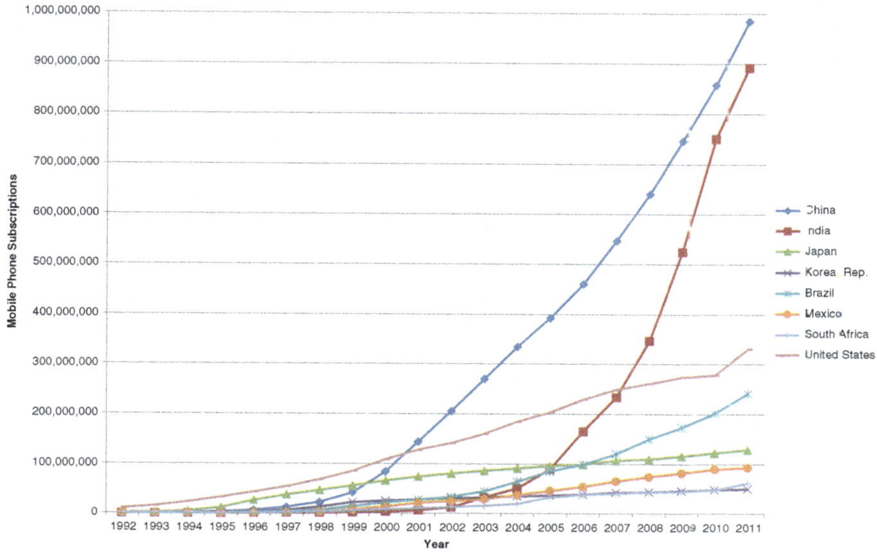

Fig. 1.3 Number of mobile phone subscriptions, 1992–2011. *Source* World Bank (2012)

Fig. 1.4 Mobile phone users in California, United States

and topographical features associated with high or low technology use. It studies not only digital divide differences among continental regions and countries worldwide, but it also drops down to smaller geographic units to analyze the distinctive utilization and geographies of technologies at the state and provincial levels. Within countries, it examines intra-country trends, patterns, and determinants for China,

India, Japan, and the United States. Their state and provincial differences have been infrequently reported on and rarely with a geographical lens. At the lower end of world economic development, it analyzes these phenomena for the African nations.

Social, economic, and political characteristics are evaluated for the extent of their influences on ICTs between countries and within them. Categories of characteristics include demographic structure, education, economy, societal openness, government support, innovation, and social capital. Some characteristics are unique to nations and regions, such as workforce in non-state-owned firms in China, cooperative society membership in India, and UN human development indices and business efficiency for African nations. Although differences between nations in these unique state/provincial characteristics do not allow the inter-country findings to be directly comparable, some data inconsistencies are unavoidable since censuses and surveys at the state/provincial level differ from country to country. This issue is often termed the modifiable areal unit problem. That limitation can only be overcome by conducting a standardized worldwide survey of states and provinces, which is beyond this book's scope.

The second goal of this volume is to examine and suggest the implications of digital divides for policymaking by governments. For instance, if innovation is related to more technology use, a government could implement or strengthen programs to foster innovation through subsidies for organizations, tax breaks, enhancement of infrastructure, and support for STEM (science, technology, engineering, mathematics) education. The policies would differ by that government's environment of economic development, culture, technological intensity, governmental capacities, national priorities, and history. Cases and examples are given of policies that have worked or failed. The policies recommended can be useful to leaders, government policymakers, economic development officials, and ICT planning officers. They may be helpful also to nonprofit organizations concerned with national or state/provincial adoption and use of technologies.

A third goal of this volume is to illustrate patterns of the planning, growth, adoption, and impacts of technologies through selected case studies, which are based on secondary sources. The cases concern specific initiatives, policies, programs, decisions, and implementations that inform the central findings of the book.

The fourth goal is to develop a robust theoretical model to explain the factors that influence the utilization of ICTs at the national and state/provincial levels worldwide. This model, presented in Chap. 3, is induced from the empirical findings drawn from extensive research literature on the digital divide. Portions of the present theoretical model draw on Adoption/Diffusion theory (Rogers 2003), Van Dijk's Theory (Van Dijk 2005), and Network Readiness Framework (Dutta and Mia 2007; Dutta and Bilbao-Osorio 2012). They are examined in Chap. 3.

1.4 Units of Analysis, Data, Statistical Research Methods

The broad empirical research framework is understood by considering units of analysis, sources of data, and the statistical techniques applied More detail is provided in Chaps. 4–9 regarding specific frameworks for study of the world as a whole, four nations, and the African continent.

The choices of units of analysis for the empirical study are: (1) the nation, and (2) the state, province or prefecture. The national unit considers the ICT levels for a nation. For instance, China in 2012 had 81.3 mobile cellular subscriptions per 100 persons and 40.9 % of households with a computer. The national unit summarizes the status of a China, but does not reflect the heterogeneous regions within it. The national unit has extensive and systematic international data available that have been the basis of most empirical studies in the digital divide literature over the past two decades. It is a unit that can apply to study of the entire planet, since data for the world's nations are collected by the World Bank, World Economic Forum, International Telecommunication Union, and United Nations.

The unit of state, province, or prefecture is a regional government entity. It is chosen in this book, for four nations, to analyze systematically the wide variation in technology use within their land extent. As seen in Table 1.1, the meaning of these governmental units is somewhat different between the four nations.

As well, the nations vary in number of units with systematic technology data available, ranging from 20 out of 36 Indian states and territories to all 50 U.S. states. Since variable definitions at this sub-national level are not standardized across countries, cross-country comparisons of state/provincial/prefectural results are not entirely consistent, yet by acknowledging the differences, cross-country comparisons can be done, as is undertaken in Chap. 10.

A variety of different sources of prefectural data were used to obtain values of dependent and independent variables. For states in the United States, among demographic independents, variables such as population and employment estimates, education, economic ones such as gross state product and income, publication estimates such as annual sales revenue of book publishers, and innovation-related variables such as research and development estimates, and patents awarded were obtained from federal government sources such as the U.S. Census Bureau's Population Census and American Community Survey (ACS), Bureau of Economic Analysis (BEA), and the National Science Foundation (NSF). Most non-governmental data were obtained for the years 2008–2010, although in several instances the sources extend back to the year 2006 to ensure completeness of data for variables such as gross state product and sales revenue estimates of newspapers and periodicals. For the dependent and independent variables, sources included the U.S. Bureau of the Census, the Federal Communications Commission (FCC), U.S. National Telecommunications and Information Administration (NTIA), and National Center for Health Statistics (NCHS), National Science Foundation, and independent sources such as Internet World Stats (www.internetworldstats.com), the DCI Group

Table 1.1 Intra-country unit of analysis for four study nations

Nation	Intra-nation unit	No.	Definition
China	Provincial level division	31[a]	Highest level Chinese administrative division for the entire nation. The secretary of the provincial committee of the Communist Party actually runs the provincial government
India	State	20[b]	Highest level Indian Administrative unit beneath the central government. The chief minister of the state has executive power and the state legislature has legislative power
Japan	Prefecture	47	A subnational jurisdistiction larger than cities and districts. The central government has the right to control prefectures. It delegates education, police, and many other powers to them. Each prefecture has an elected governor and prefectural assembly
United States	State	50	Highest level of government except for the federal government. The Constitution specifies the rights of the nation and state. Each state has an elected governor and bicameral legislature

[a]Although referred to in this volume as "provinces," 22 are provinces, 4 are municipalities (Beijing, Chongqing, Shanghai, Tianjin), and 5 are autonomous regions (Guangxi: Inner Mongolia, Ningxia, Tibet, and Xinjiang). In addition there are 2 special administrative regions (Hong Kong and Macau) and one claimed province (Taiwan)
The latter three are not considered in this research
[b]Although in 2014, there were 29 Indian states, 1 national capital region, and 7 union territories, only 20 had robust and complete data for this research

(Facebook and Twitter Users), Marcatus Center of George Mason University, and the Pew Research Center.

Sources in China were the China Information Almanac, China Statistical Yearbook, Chinese Census, and China Internet Network Information Center, while sources for India were the Indian Statistical Yearbook, Indian Department of Telecommunications, Lok Sabha (lower house of Indian Parliament), Reserve Bank of India, and India's Export Promotion Council.

Sources for Japanese prefectures were the Japanese Census, Japan Patent Office, Japan Association of Citizen Ombudsman, Japan's Telecommunications Carriers Association, and User Local Inc., while for African nations, the data came from the International Telecommunications Union, United Nations, World Bank, and World Economic Forum.

Most dependent variable estimates were from the years 2010–2012; however, in some cases, the most recent complete data were from 2008. Most variables were converted to a per capita basis (percentages or per 100 population). Descriptive statistics of all variables were calculated. Definitions and sources of the dependent and independent variables appear in tables in Chaps. 4–9.

1.5 GIS and Spatial Analysis

Dependent variables are mapped using a geographic information system (GIS). An example map of Internet users per 100 persons is shown in Fig. 1.5. Spatial analysis methods support the analysis and modeling of the relationship of objects geographically (Goodchild and Janelle 2004; Longley et al. 2011; de Smith et al. 2012).

The basic concepts of spatial techniques utilized in this study are cluster analysis with mapping, spatial autocorrelation (Moran 1950), and local indicators of spatial autocorrelation (LISA) (Anselin 1995). OLS regression analysis is applied to estimate the associations of independent variables with a dependent one. The full sequence of steps in the book's methodological approach is seen in Fig. 1.6. Each method in this diagram is covered in more detail.

K-means cluster analysis (Hair et al. 2010) is a standard statistical method that classifies sample cases into clusters, based on a set of attributes, in order to obtain homogeneity of attributes within clusters and heterogeneity of attributes between clusters (Hair et al. 2010). Tobler's Law of spatial analysis states that "Everything is related to everything else, but near things are more related than distant things" (Goodchild and Janelle 2004). Applied to cluster analysis, this law implies that cases with high similarity in attributes will also be nearer to each other geographically. Thus when a cluster is mapped, its cases tend to be neighboring each other spatially. For example, k-means cluster analysis can be performed on the eight technology variables for the sample of U.S. states. As expected, cases within each cluster tend to be near each other geographically. This procedure does not necessarily yield a cluster with totally spatially contiguous clusters.

Ordinary Least Squares (OLS) regression analysis (Hair et al. 2010) is the method used to test for the associations of independent variables on dependent ICT variables. Several types of diagnostics tests are conducted on the regressions. All independent variables are tested for multi-collinearity using the variance-inflation factor (VIF) to assure that multi-collinearity was not present. A VIF value under 5.0 indicates lack of multi-collinearity. The OLS regression diagnostic tests of Joint Wald Statistic, Koenker (BP) Statistic, and Jarque-Bera Statistic (SPSS 2011; Rosenshein et al. 2011) are applied to evaluate whether regression assumptions are met. Joint Wald Statistic is a test of the joint significance of several coefficients of individual independent variables (Wald 1943). The Koenker (BP) Statistic Test is a test for heteroscedasticity, i.e. the variance of the residuals is not constant (Lyon and Tsai 1996). The Jarque-Bera Statistic is a goodness-of-fit test of whether sample data, in this case regression residuals, have skewness and kurtosis that correspond to a normal distribution (Jarque and Bera 1980). The Moran's I test for spatial autocorrelation of residuals indicates whether or not there is spatial bias in the spatial distribution of the residuals. More details about the research methodology employed in this paper can be found in two sources (Pick et al. 2013; Nishida et al. 2014).

As a guide to the reader for the regression tables that appear throughout this volume, a valid OLS regression model, for each test, is indicated if the Joint Wald

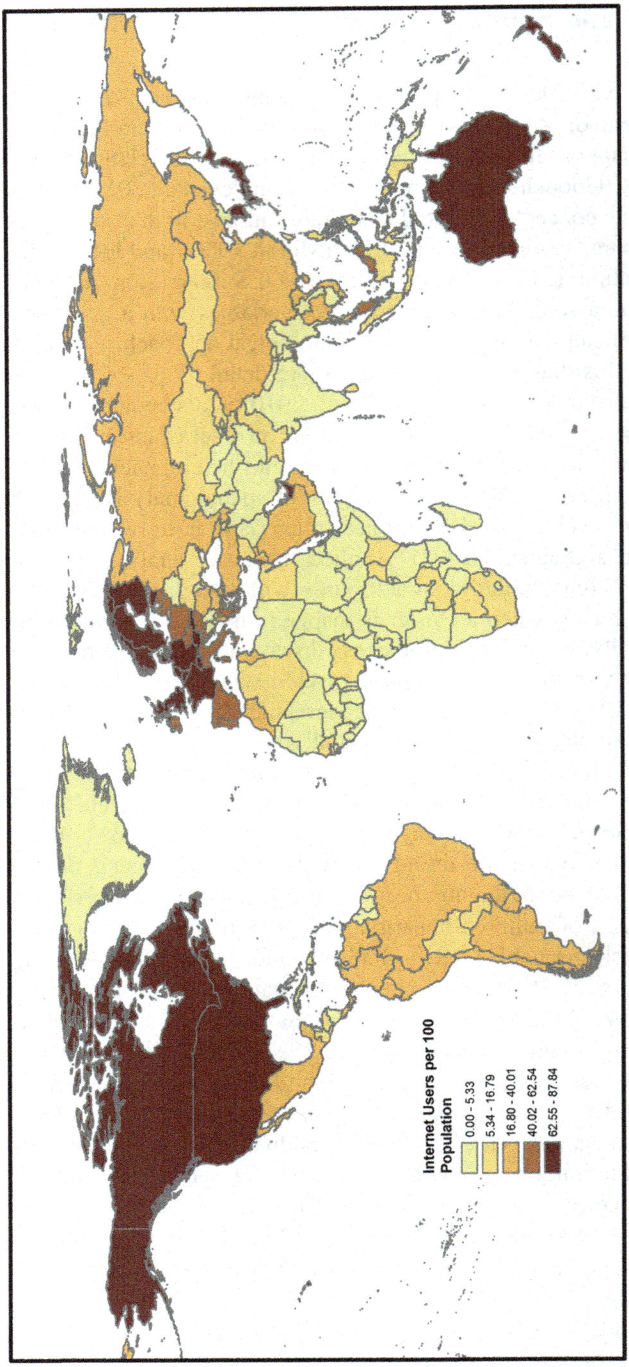

Fig. 1.5 Internet users per 100 persons worldwide, 2008

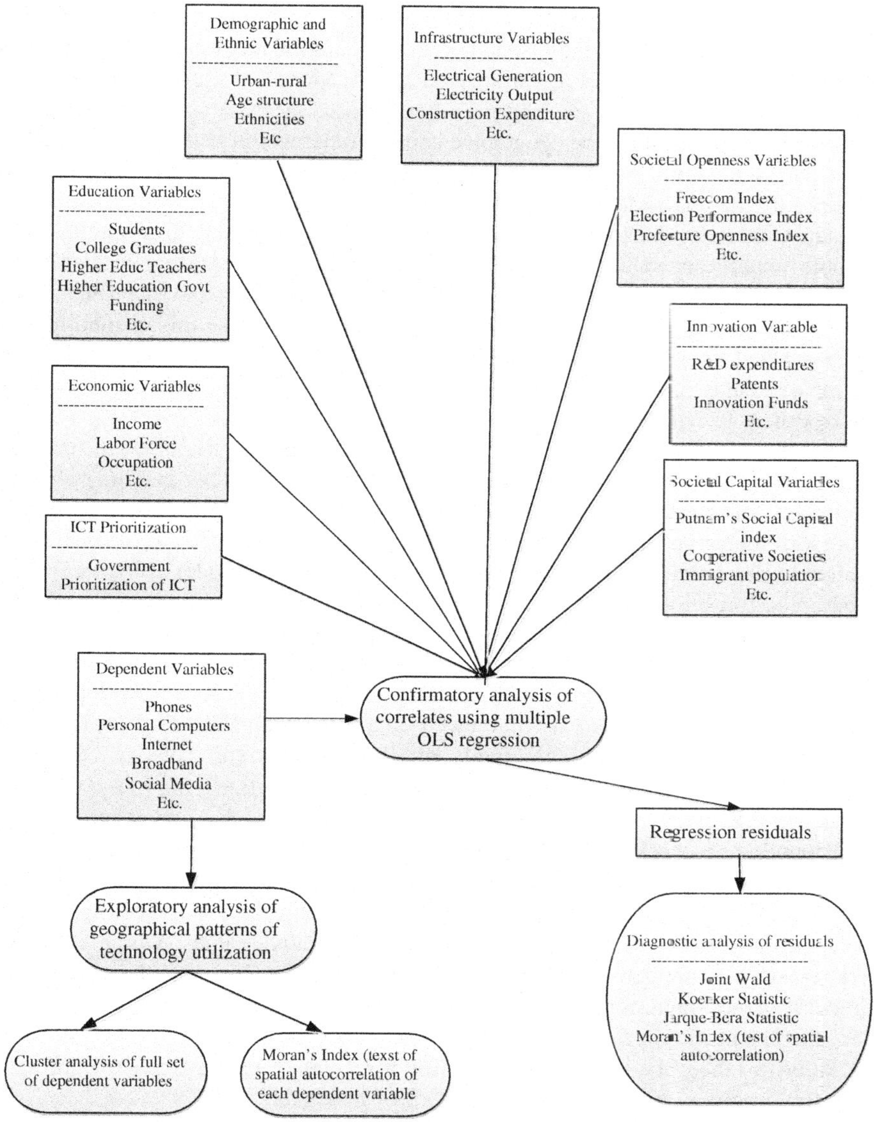

Fig. 1.6 General methodological approach utilized

Statistic is significant; if the Koenker (BP) Statistic is not significant; if the Jarque-Bera Statistic is not significant; or if Moran's Index is not significant.

Spatial autocorrelation measures similarity in the location of sample cases and similarity in the attributes of cases. Positive spatial autocorrelation means that cases with neighboring locations have similar attributes, while negative spatial autocorrelation means that cases with neighboring locations have dissimilar attributes (Longley et al. 2011). In the present study, spatial autocorrelation is measured by the Moran's I statistic (Longley et al. 2011; de Smith et al. 2012). Moran's I statistic (Moran 1950; Longley et al. 2011; Grubesic 2006) is often used to measure spatial clustering or dispersion of an attribute of the geographic units. In other words, using this statistic, we are able to identify the overall extent of clustering of states or provinces in terms of one of their attributes. The Moran's I statistic varies between 1 and −1. For the example of US states, index levels close to zero indicate a random geographic distribution of the variable in space, and values near +1 indicate like-valued states group together, while −1 indicates a high-valued state is surrounded by a low-valued one, and vice versa, i.e. a low-valued state is surrounded by a high-valued one. In applying this method for the Japan and the U.S., care was taken to resolve contiguity issues in Japan for prefectures in the major islands of Hokkaido, Honshu, Shikoku, Kyushu and Okinawa, and in the U.S. for the states of Hawaii and Alaska.

The Moran's I statistic is applied first to each dependent variable, in order to gauge whether it has random spatial distribution or not, and if not random, whether states are surrounded by similarly-valued neighbors or by differently-valued neighbors. Later on in the methodological process steps in Fig. 1.6, Moran's I is applied to the residuals of the regressions. If the regression model accounts for the geographic forces, then the Moran's I test is significantly close to 0 (i.e. random). Otherwise, the model does not capture the spatial forces.

Moran's I can be pictured on the diagram in Fig. 1.7. Moran's I varies from 1.0 (every neighbor-case is similar to its neighbor-cases) to −1.0 (every neighbor-case is dissimilar to its neighbor cases). Viewed on a map, for a Moran's I approaching 1.0, there is nearly complete contiguity of similar high-value cases and likewise nearly complete contiguity of similar low-value cases (see Fig. 1.7). In the figure, a Moran's value of −1.0 resembles a checkerboard, where each checker (prefecture) is surrounded by checkers of the opposite color (neighbors with dissimilar values). A Moran's I of 0.0 indicates a random pattern, where the neighbors for each case are random in their similarity or dissimilarity. A value of 1.0 indicates extreme spatial agglomeration of positive values and/or of negative values. Moran's I in the present study is applied to indicate, for each dependent variable, the extent and direction of non-randomness of its spatial pattern, and similarly, for each regression, the extent of non-randomness of the spatial pattern of regression residuals. In the latter instance, the goal is to have Moran's I value not significantly different from 0.0, implying a random spatial pattern of residuals.

Moran's I is a global measure for evaluating the overall spatial clustering/dispersion exhibited by a variable. However, there might be significant local clusters or outliers even if Moran's I indicates a generally random pattern. A local indicator of spatial autocorrelation (LISA) is a disaggregation of Moran's I index to show how the value of the Moran I's component of each case location contributes to the overall Moran's I for the sample (Anselin 1995; de Smith et al. 2012). Hence, for a given location, it indicates the strength or weakness of positive or negative relationships with its neighbors. It is applied in this study to indicate outlier locations,

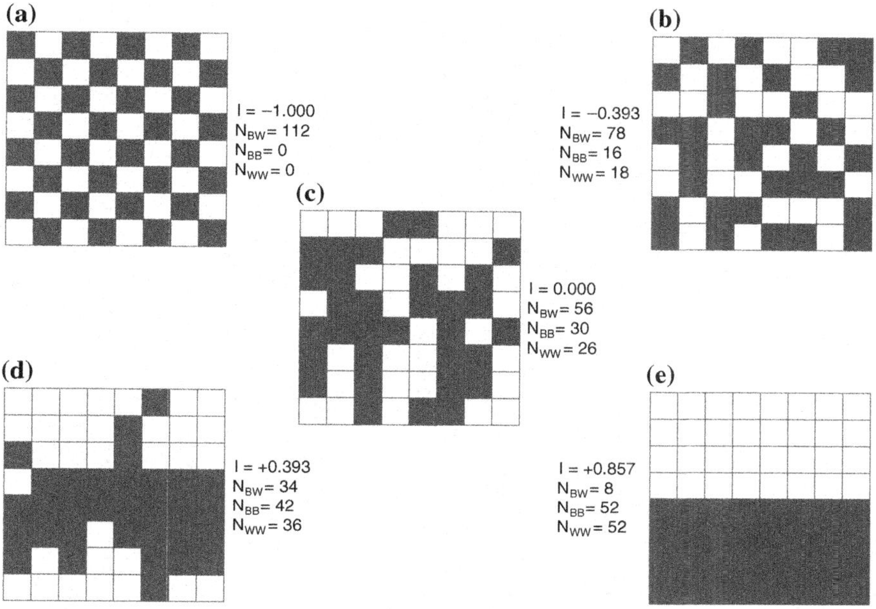

Fig. 1.7 Patterns for Moran's index, with **a** showing extreme negative value, **b** showing moderate negative value, **c** showing 0 value (random), **d** showing moderate positive value, and **e** showing extreme positive value. *Source* Longley et al. (2011)

i.e. those which contribute the most negative values in the equation for the overall Moran's I for regression residuals.

Alternative methods applied to digital divide

There are alternative methodologies that have been applied to digital divide research (see Table 1.2). Although this book utilizes the methodologies described in the last section, this section examines some of the leading alternative methods along with examples of literature studies and the reasons why that approach is not used in this volume.

Structural equation modeling (SEM) is a methodology appropriate for more complex models which demonstrates relationships between latent factors, which are estimated from underlying attributes. SEM can be applied to study dependent digital divide factors, such as technology utilization, behavioral intent to use, and e-business use. An example of a structural equation model to determine technology utilization, shown in Fig. 1.8, posited three latent factors, "government support, legal framework, and social openness," "business and technology readiness," and "socioeconomic level," which are related by pathways that ultimately influenced "technology utilization" (Pick and Azari 2011). The latent factors were formed from 16 underlying attributes.

Table 1.2 Alternative methodologies for studying the digital divide

Focus	Study	Category	Alternate analytic approach	Sample, time frame
Usage and value of e-business	Zhu and Kraemer (2005)	Structural equation modeling	PLS for sample and subsamples	10 nations, 624 surveys, 2002
Technology acceptance, influenced by cultural constructs	Srite and Karahanna (2006)	Structural equation modeling	PLS	2 surveys, 297 responses from students from 30 countries
Technology utilization construct, influenced by social, political, govt, business factors	Pick and Azari (2011)	Structural equation modeling	AMOS for sample and subsamples	110 countries, 2005
Internet access, influenced by ethnic isolation, moderated by housing and social interactions	Agarwal et al. (2009)	Probit analysis	Two-stage probit regression	U.S. Current Population Survey of 45,212 respondents, 2003
County clusters based on digital development, infrastructure, social, and economic factors	Skaletsky et al. (2013)	Visualization of component structures, geographic mapping of changes	Kohonen self-organzing component maps	179 nations, 1999 and 2007
No. of internet users, influenced by socio-economic variables	Guillén and Suárez (2005)	Economic, econometric	Panel data regressions, time series analysis	188 nations, 1997–2001
Computer, internet penetration, influenced by socio-economic variables	Chinn and Fairlie (2007)	Economic, econometric	Regression, decomposition of penetration rates across regions	161 nations, 1999–2001
Inequality in internet use, decomposed by demographic and economic variables	Lengsfeld (2011)	Economic, econometric	Decomposition of Theil Index	49,067 interviews, 2004–2005
Wireless phone diffusion, based on regional contagion theory	Kauffman and Techatassanasoontorn (2005)	Panel econometrics	Vector autoregression and variance decomposition	Time series of phone subscriber data, 24 countries, 1986–2002

Modified from Pick et al. (2015)

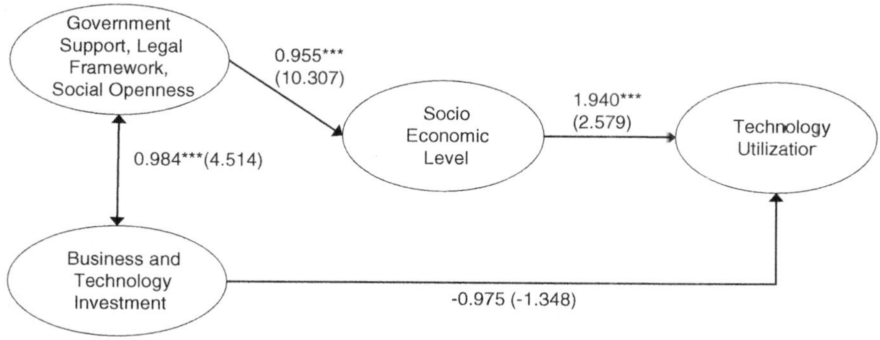

Fig. 1.8 Example of structural equation model for technology utilization. *Source* Pick and Azari (2011)

The model required a sample of at least 50, based on its having 16 underlying attributes. It was tested on a sample of 110 nations. The structure of the model provides a more complex set of relationships of factors than a multiple regression model (see Fig. 1.8). In the results given in the figure there is interaction between the two left-hand factors, and a pathway of relationships that appears at the top of the model diagram. The detailed explanation of this model is available in Pick and Azari (2011). Although the more complex relationships can be advantageous, the reason SEM is not utilized in the present study is that it does not provide residuals for the sample points, which could be geo-referenced and mapped. Hence the presence or absence of spatial bias cannot be determined. A further reason for not using SEM is that the interest in this book is to estimate and compare individual technology factors, such as mobile phone utilization or internet subscribers, rather than a combined latent factor, for which individual technologies cannot be distinguished. Lastly, it is not used because most of the sample sizes in this volume for states, provinces, and prefectures are too small to utilize with SEM.

Two other literature examples of the SEM approach are a model of complex pathways of influence of technology acceptance by individuals that combines well-known factors from the Technology Acceptance Model (TAM) with cultural factors from the Hofstede model (Srite and Karahanna 2006). A third example is a study of the influence of technological, organizational, competitive, and regulatory factors on the use of e-business by 624 firms, based on interviews in developed and developing nations (Zhu and Kraemer 2005). In these examples, sample sizes were in the hundreds. Each study validated the model as a whole, but the errors are not estimated for individual sample points, so spatial bias cannot be determined.

A two-stage *probit analysis* was conducted based on a very large individual and household sample from the U.S. Current Population Survey. Its goal was to analyze the effects of ethnic geographic proximity and other factors on internet use (Agarwal et al. 2009). Although this multi-layered study did not involve mapping and spatial statistical methods, it did include proximity as a variable. This method first estimated the proportion of internet users in a metropolitan statistical area (stage 1 of the probit analysis) and then estimated effects of 15 independent variables on that proportion (stage 2 of the probit analysis).

Geography was introduced by using instrumental variables in the two-stage model to see if causal relationships can be established on how much respondent's choice is affected by choices of geographically-proximate peers. The findings support that there exists influence of geographically-close peers on internet use. This approach benefitted by individual-level surveys, not utilized in the present research. The geography was based on peers being present together in a Metropolitan Statistical Area (MSA), but the overall geography of the sample was not studied, nor was spatial clustering or spatial bias examined. The probit method is not appropriate for studies in this volume because the present responses are quantitative rather than binary, as called for in probit analysis, and because the units of analysis are different.

Kohonen self-organizing maps is a technique that can be applied to explore how countries cluster together based on a mixture of digital, economic, and social characteristics. Kohonen analysis does pattern recognition of input signals, which can be displayed in visualization of the patterns known as self-organizing maps, which are not geographic (Kohonen 1990). This useful exploratory method has no confirmatory aspects, in contrast to the OLS regression method in this book. The Kohonen clustering process does not take account of geography but rather assesses the mathematical distance apart of vectors of attributes. In a study of a worldwide sample of countries, several geographic maps were created subsequent to a Kohonen analysis to map Kohenen-cluster changes in two regions (Skaletsky et al. 2014). The method is not appropriate for this book, since it cannot provide the geographic analysis of individual technology variables (i.e. instead it tracks Kohenen clusters), nor do confirmatory analysis.

Economic and econometric methods have been applied to understand the digital divide. An example is a panel study that used time series analysis to predict the number of internet users, based on socioeconomic variables (Guillén and Suárez 2005). The technique is based on fixed effects, using a lagged dependent variable. The generalized difference model included an auto-regressive component. The research yielded useful findings, but the method requires a sample having longitudinal data, which are not available for most of the present research. Chinn and Fairlie (2007) applied a decomposition method to panel data to explain variation in ICT uses between a pair of world regions. Such time series methods are precluded presently, because the dependent variables are rarely available longitudinally at the state/provincial levels.

In another econometric paper, inequality in internet usage by country was evaluated by the Theil Index, which in this example measures the difference between the share of the index representing Internet usage and the share representing population (Lengsfeld 2011). This index can be partitioned into its component parts. Using two samples of over 40,000 individual respondents from 22 nations, inequality was compared between countries, and by gender, age, education, occupation, income, and size of place of residence, indicating that some segments of groups were more digitally divided than others, such as strong differences for young versus old or rural versus urban residents. This approach embodies new dimensions of comparison for the digital divide. Nonetheless, the model depends on longitudinal data, precluding use with the book's mainly cross-sectional samples, and it focuses on the many ways to split the digital divide, with only slight reference to geography.

An econometric study (Kauffman and Techatassanasoontorn 2005), applied panel data in a vector autoregression model, using a variance decomposition method, to assess the influence on wireless diffusion in a nation, taking into account wireless diffusion or additional factors in other nations. Findings were compared for developing and developed countries. Its very large, longitudinal sample of individuals lacked reference to spatial location, so is not appropriate for the present research that analyzes multiple technologies and requires georeferenced locations for sample points.

1.6 Spatial Aspects of Digital Divides

The digital divide literature gives little attention to geographical aspects of information technology differences. Rather, it has been concerned with modeling and analysis of numerical digital level differences in ICT access and use between nations or between individuals, and occasionally between states or provinces. Whether those geographical units are large or small, neighbors or not, or with or without spatial associations with each other doesn't matter. The wealth of findings from these studies highlights key distinctions between characteristics of geographical units, and indicates what factors are likely to have determined the numerical differences, but largely ignores the geographical characteristics and what effect geography has. For example, is the notebook and cell phone user seen in Fig. 1.9 influenced by location? If so, the point location of the user can be georeferenced and that point can be included in spatial analysis, which provides greater potential analytic strength and precision.

This section concerns a small group of prior investigations that do recognize the geography of technological differences. They include studies of the world (Warf 2001; Takhteyev et al. 2012), U.S. (Grubesic 2004, 2006, 2010), Japan (Arai and Naganuma 2010), and China (Song 2008). They include the geography in a variety of ways, including thematic mapping of attributes, social media ties, local spatial autocorrelation (LISA analysis), and qualitative case studies of geographical settings and interventions to stimulate more technology development and use. The objective of this section is to provide background on, and examples of prior approaches to spatially examining the digital divide and to compare these alternatives to the present approach.

Worldwide spatial studies of digital divide

Global technology utilization depends partly on the underlying physical network infrastructure that supports vast volumes of information flows. Capacities for flows depend on the physical structure of the network. Other influences, social, economic,

Fig. 1.9 Notebook and cell phone user, Florida, U.S.

and behavioral, shape the actual amount of network use and flow at particular times and places.

Hence, one dimension of understanding ICT use is to analyze and map the underlying physical arrangement of fiber-optic undersea and land-based cables, broadband cables, satellites, satellite receiving and sending stations, and cell antennas and towers. This physical network has grown and changed vastly since 1969, when ARPANET was started as the first widespread network for Internet. The physical underpinning of digital divide received some attention in a paper seeking to explain why regional internet differences exist (Warf 2001). A follow-up study 12 years later examined more deeply the spatial configuration of the underlying infrastructure of the Internet (Warf 2013). For instance, satellite traffic depends on a coordinated network of satellites in geosynchronous orbit which communicate bi-directionally with about 500 public earth stations, about a sixth of which are located in the U.S. (Warf 2013). Lower speed receive-only stations number in the tens of millions. The arrangement of the earth stations can influence certain types of internet use, for example communication to remote areas of the earth lacking cables.

The world's largest bulk transmission takes place through fiber-optic cables, which are sub-oceanic or terrestrial (Warf 2013). The very expensive undersea cables are usually under joint ownership of several of the world's dominant telecommunications providers such as British Telecomm, AT&T, or Kokusai Denshin Denwa (KDD) in Japan (Warf 2013). This complex fiber-optic network is heavily concentrated in Europe, the U.S., and East Asia. As is evident in a map of the global undersea fiber-optic network (Fig. 1.10), the largest cables connect western Europe with United States, and each of them with East Asia. There is a paucity of connections with western South America, east Africa, and India. The scale and intensity of undersea cables affects the potential for terrestrial high-speed broadband in these developing regions. By traffic type, bulk corporate traffic tends to reside on fiber-optic cable, television traffic on satellites, and internet and telephone traffic on both (Warf 2013). Over time, the principal transmission platform for the Internet, has been shifting from satellite to fiber-optic cable.

Another major type of internet infrastructure is wireless, which for mobile devices depends on a backbone of cell phone towers. The density of cell phone towers is increasing in urban areas worldwide, and extending more into rural areas.

The perspective of the physical infrastructure is not the focus for this book. The reason is that physical infrastructure is increasingly prevalent and reliable, especially in the developed world, so that technology usage depends more on economic, behavioral, and social variables than on the physical network. Nevertheless, physical network infrastructure is mentioned in considering Japan in Chap. 7. The physical network becomes more important in the developing world and is mentioned concerning Africa in Chap. 9 and for the Rwanda case study in Chap. 11. For example, Central Africa has among the lowest worldwide prevalence of internet and broadband, a deficit explained by the paucity until recently of undersea fiber-optic connections to the continent and by scarcity of fiber-optic land cables connecting

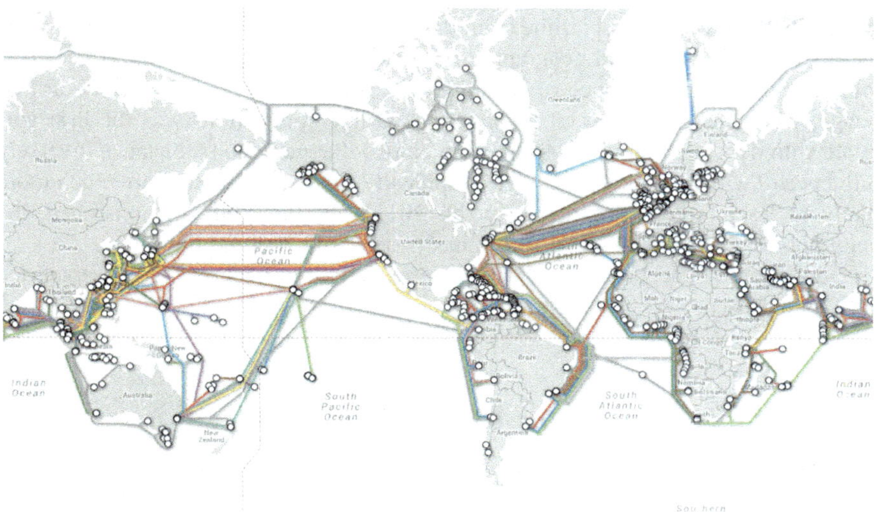

Fig. 1.10 Major fiber-optic cables worldwide, 2014. *Source* Telegeography (2014)

Table 1.3 Number of active users of social media, June 2014

Social media mode	Number of active users Worldwide, June 2014
Facebook	1,280,000,000
Qzone	644,000,000
Google+	343,000,000
LinkedIn	300,000,000
Twitter	255,000,000

Source Statista (2014)

oceanic entry points to the central interior. Likewise, in the northeast of India, there is limited fiber-optic network, which has constrained high-speed internet use. Future plans to leapfrog technological barriers and extend broadband everywhere in the developing world are examined in Chap. 12, such as the possibility of planet-wide broadband access through high altitude balloons (Google "loons").

Social media has grown in worldwide penetration as seen in Table 1.3. However, the rapid diffusion of these software services has been little studied from a geospatial viewpoint. An exception is Twitter, which has been examined for its intensity of its use worldwide (Takhteyev et al. 2012). In particular, nearly a half million Twitter messages were collected over a seven week period. They were geocoded as Twitter dyads, each dyad consisting of the "ego" i.e. the originating party and the "alter" i.e. the receiving one. Dyads were determined by the ego's "likes," one of which was randomly sampled to be the "alter." Dyads were then analyzed geographically for source and destination location. The study pointed to considerable local concentration of dyad parties, for instance, 39 % of Twitter dyad ties were at a distance from the ego of under 100 km (Takhteyev et al. 2012). The geographic, economic, and transportation factors of national boundaries, common language, and frequency of airline connections were analyzed for dyads, to see what factors resembled the dyad geography. Physical distance between dyad parties was more important than absence of separation by national boundary or common language, while the geography of frequency of airline connections was the best correlate with dyad geography. The study concluded that geographic distance inversely influences Twitter ties, i.e. parties geographically close to each other are much more likely to be in dyads (Takhteyev et al. 2012). As the study demonstrates, the geographical analysis of social media has potential to become important.

Utilization of Facebook and Twitter are examined in Chaps. 7–9 on Japan, the U.S., and Africa.

1.6.1 Spatial Studies of the U.S. Digital Divide

An early study to consider location for digital differences in the U.S. was based on spatial concepts and results, but did not utilize mapping or quantitative spatial analysis (Warf 2001). Metropolitan areas with heightened internet use were identified as Los Angeles, San Francisco, Chicago, New York, Philadelphia, Washington/Baltimore, Atlanta, Dallas, and Houston. In explaining the considerable spatial differences between states, ethnic, gender, economic, and workforce characteristics were shown to be significant. For example, laptop users in Fig. 1.11 may be influenced by their social and economic characteristics. Warf (2001) indicated that white households had one third more personal computers (PCs) at home than Blacks and nearly twice as many as Hispanics. States with substantial rural poverty had reduced internet usage. The very poor were technologically marginalized: in the late 1990s, they were "unaware of the massive economic, technological, and political changes than exclude them… from the information society" (Warf 2001).

The study (Warf 2001) accentuated the importance of physical space, in spite of predictions by some observers that virtual space would be all that mattered. Location was determined to be associated with power and wealth (Warf 2001). Another observation is that U.S. differences relate to the concepts from world systems theory of "core" and "periphery." The large U.S. metropolitan areas represent the core and provide high levels of access to internet and technological services, while the poor rural areas representing the periphery have low access and depend on the metro areas for their internet services and content. This prescient study informs the present book through its emphasis on the importance of dominant metro areas, which corresponds to our findings for China, India, and Japan in Chaps. 5–7.

Another spatial study of broadband service throughout the U.S. performed mapping at the zip code level and confirmed the greater efficiency of the broadband core (Grubesic 2010). This study included data on the number of broadband

Fig. 1.11 Laptop users in California, USA

customers and providers, as well as median income, number of financial and real estate firms, population density, education, scientific and technical firms, Black population, median age and household density (Grubesic 2010). Competition was indirectly implied by the ratio of number of broadband providers to customers.

Data envelopment analysis (DEA) was utilized, which inputs data for a geographic unit and converts them into outputs, with some parallel to regression. Using Federal Communications Commission (FCC) data, mapping of DEA efficiencies showed that "central portions of metropolitan areas in the broadband core were most efficient—with a gradual decay of efficiency as one moves toward suburban and exurban areas" (Grubesic 2010). Each metropolitan area was somewhat different in its local efficiency profile, measured on a gradient declining from its metropolitan core to the rural periphery.

This study built from an earlier one (Grubesic 2004), which considered how a dozen geodemographic lifestyle profiles selected from CACI ACORN related to the supply and demand of broadband. CACI refers to a UK company that provides multiple marketing profiles of lifestyles, which are keyed to geography. The investigation examined where in the United States each selected CACI ACORN profile is prevalent. For instance, ACORN's "Prosperous Baby Boomers" category is present in 5.8 % of the nation's broadband market. Here the spatial approach is put in a broad context: "there exists a complex and dynamic interplay between geography, socioeconomic status, market forces, and policy" (Grubesic 2004).

1.6.2 Spatial Studies of Digital Divide in China and Japan

Few spatial studies of technologies nationwide have been done in China and Japan. This is surprising, since both nations have dramatic geographic features that include mountains, deserts in China, islands in Japan, and major urban differences between remote rural areas and global megacities. The following are two examples: for China a comparison of specialized location quotients for its provinces, and in Japan a study of three broadband-deprived prefectures.

For China, a location quotient (LQ) for domain names was calculated (termed "names specialization ratio") as well as LQ for internet usage (termed "internet usage quotient") (Song 2008). A location quotient is computed as follows: (ratio of prevalence of X in a smaller geographic unit)/(ratio of X in a larger geographic unit). For instance if there are 20 cell towers per city population of 10,000, and 800 cell towers for state population of 2 million, then the LQ = (20/10,000)/(800/2,000,000) = 5. For domain names specialization ratios, two-way displays of this ratio by annual growth rate 2000–2006 are analyzed and provinces are sorted into four quadrants (see Fig. 1.12), which are labeled as "front runner," "achiever," "catch-up," and "novice" (Song 2008). A domain name is a unique name that identifies a website.

The study points to substantial technology differences: eastern provinces are dominant in internet and web uses, while those uses are weak in the far West.

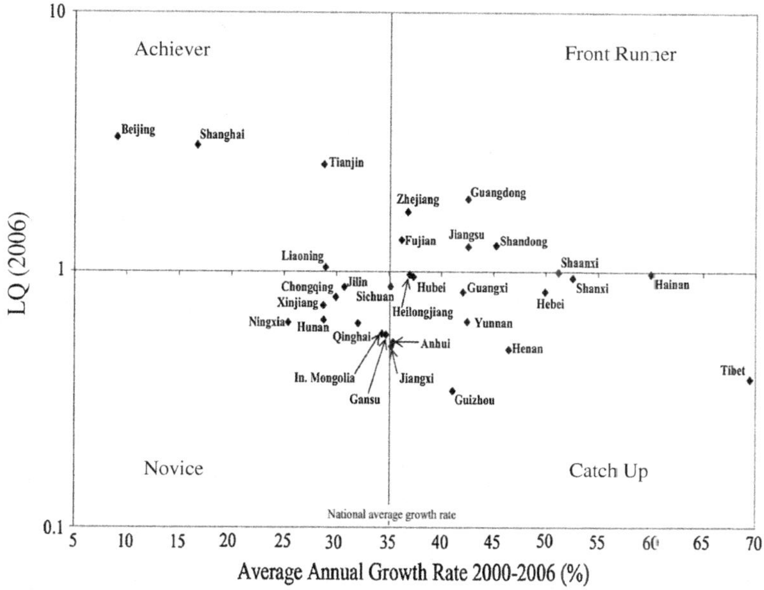

Fig. 1.12 Internet location quotients for domain names versus internet use annual growth rates, Chinese Provinces, 2000–2006. *Source* Song (2008)

Economic and educational differences are pointed out as corresponding to cyber differences, for instance PCs are viewed as a luxury in the impoverished and less educated West but a necessity in the core of rich and well educated eastern metropolitan areas. The study however performs no geographical clustering and very limited mapping analysis.

A study of Japan considered spatial aspects of the challenge of broadband in the prefectures of Hokkaido, Nagano, and Mie, which have the challenge of low levels of internet infrastructure (Arai and Naganuma 2010). Descriptive mapping was performed, that included an overview map and more detailed maps for the examples. These examples consisted of Nishiokoppe Village in Hokkaido, the Koso region in Nagano Prefecture, and areas around two small cities of the Mie Prefecture. For the latter, a mapping analysis shows the spatial distribution of cable providers for districts that include the regions of Tsu City and Matsuzaka City and the major barriers that prevented providers from extending service into adjacent remote territories. This study employed a simple geographic approach in considering the internet infrastructure and its locational challenges. For each case, it performed qualitative analysis of the history, current, and future challenges in expanding broadband access. This study (Arai and Naganuma 2010) is discussed more as a case study in Chap. 7.

1.7 Case Examples of National Evolution of Information Technologies: Mexico and Singapore

Mexico and Singapore exemplify the evolution of information technologies. They illustrate initiatives, opportunities, and issues in achieving technology use at the national level, and vary between an advanced developed nation, Singapore, and a contrasting mid-level nation, Mexico. The cases illustrate concepts introduced earlier in the chapter.

1.7.1 Singapore

Singapore was ranked by the World Economic Forum in 2012 as having the second highest Networked Readiness Index (NRI) out of 142 nations (Dutta and Bilbao-Osorio 2012). The NRI represents an average of multiple drivers of ICT readiness and usage, and is explained more in Chap. 3. For components of the NRI, this country was ranked in first place globally in economic/social impacts of IT and for its political/regulatory/business/innovation environment.

How did Singapore evolve to this high level, and what factors were essential to it?

Singapore underwent rapid economic development from 1965 to the 1990s, joining the high level of advanced economies by 2000. This was accomplished through advances in education, public safety, and political climate (Warschauer 2011). Singapore today has one of the highest GDPs per capita worldwide and is powerful as a global hub in commerce, finance, oceanic transport, and export manufacturing including electronics, chemicals, and bioscience. In spite of this economic growth, there was some repression of political freedoms, as well as some stifling of creativity and innovation. Those constraints have been detrimental to attracting world-class IT workforce to the country. There has been tight public control of information and the media, although the government was not able to exert as strong control over virtual communications.

With its steadily advancing economy, Singapore's leaders decided in 1992 on a ten-year national plan, IT2000, that had the goal to provide broadband infrastructure for public and private organizations, businesses, and citizenry across the country. As a further step in 1999, the National Computer Board and Telecommunications Authority was merged together with all other government offices of technology (Wong 2003). A second master plan iN2015, which has steered the nation from 2005–2015, not only calls for the technology base to grow more and ICT industry base to achieve world leadership, but also sets the goal to "fuel creativity and enable innovation among businesses and individuals by providing an infocomm platform that supports enterprise and talent" (IDA 2014).

Several challenges have been present over the 20+ years of strong national focus on expanding the nation's ICT (Warschauer 2011). One is the problem of media censorship, since the national government has exerted tight control of the media, including some censorship. A counter to the government's media control has been the increasing societal benefits of largely unconstrained virtual exchange of information (Warschauer 2011). Also offsetting this challenge is Singapore's effective legal system and a high rating on quality of its IT laws (ITU 2012).

A second challenge is the public educational system, which although highly rated internationally, has been criticized for its strictness and discipline which are seen to reduce students' creativity and innovation. This prompted the government to include the expansion of creative and innovative workforce and businesses in the iN2015 plan. Third, Singapore's language policy strongly emphasizes English, even though many citizens speak other native languages such as Chinese, Malay, and Indian languages. Even for English, the government has authorized moving away from local dialects sometimes termed Singlish. This also is a mixed bag for ICT use, in that English is the dominant worldwide language of the web. The language controls, inhibitive to some citizens, reflect an authoritarian government (Warschauer 2011).

Another feature of Singapore's movement into top ranks in ICT is that its internet services, technology businesses, and e-commerce are economically outwardly directed, since its domestic ICT market is relatively small. Venture capital funding has been significant and strongly favors the outward market.

The case demonstrates that a national government can prioritize strategies to grow and develop ICT, and tenaciously stick to them over two decades eventually leading to world leadership. It was able to do so by some of the very factors emphasized in this book, such as education, economic prosperity, and social capital. Although Singapore has lagged on innovation, the latest 10-year plan focuses on fostering more creative skills in technology. Its societal openness continues to lag among peer nations. However, this is moderated, because although government exerts control over some media, it is unable to control the open Web. Further, Singapore's judicial system and intellectual property protection are highly rated worldwide. Singapore in most respects represents success in bringing advanced information technologies to a highly educated, technology-conversant citizenry, with growing potential for innovation.

1.7.2 Mexico

The case of Mexico contrasts with Singapore. Its trajectory of ICT utilization has been less successful. Mexico as a developing nation has faced challenges of mediocre technical education, a small and challenged software sector, semi-monopoly dominance and high pricing in the telecommunications sector, and weak and ineffective federal planning. On the plus side has been the development of a global ICT hardware manufacturing cluster and general strides nationally in education. The 2013 Networked Readiness Index from the World Economic Forum (Dutta and Bilbao-Osorio 2012), ranks Mexico slightly below average worldwide (76 out of 142 nations). In broadband subscriber base and tariffs, it ranks about the top third but is weak on prevalence of mobile broadband, yet ranks high on mobile tariffs. A chronic problem has been inefficiency of its legal system and law-making bodies (Dutta and Bilbao-Osorio 2012).

The Mexican story starts in the 1990s with the passage of the NAFTA free trade agreement between Mexico, the U.S., and Canada. Subsequently, businesses invested more in hardware and later in the decade in newer enterprise software. During the decade, the Salinas Administration moved away from Mexico's earlier import substitution market approach to open up trade. Trade liberalization enabled multinational firms such as Hewlett Packard to locate global hardware factories in Mexico, particularly in Guadalajara, that exported to the U.S. and elsewhere in the Western Hemisphere. On the other hand, even with a federal government stimulus programs from SECOFI (Secretariat of Commerce and Industrial Development), the domestic software industry was slow to develop and mostly non-competitive globally.

Among Mexican industries, banking was the leading ICT adopter, stimulated the arrival of foreign banks that acquired most of the Mexican banking industry in the late 1990s. Wal-Mart entered the retail market, bringing modern ICT, while Volkswagen introduced world-class factory automation.

An exception to Mexico's weakness in its own software sector has been Sofftek, an outsourcing firm located in Monterrey, which was ranked 37th as an outsourcing provider worldwide by the IAOP Global Outsourcing 100 list (Softtek 2014). Sofftek has been entrepreneurial, with emphasis on process-driven IT services, just-in-time and other factory systems, insurance systems, and transaction processing systems for top-level companies in 20 countries (Softtek 2014).

Mexican citizenry were constrained in using ICTs because of a skewed national income distribution coupled with high pricing of consumer products and services (Dedrick and Kraemer 2001). By 2010, digital literacy was rising, mainly due to benefits from educational advances of the 2000s, but also through limited support from President Vicente Fox's mostly unsuccessful e-Mexico program from 2000 to 2006 to bring IT kiosks to all Mexican municipios (i.e. similar to counties). Also, digital literacy nonprofits such as Committee for Democracy in Information Technology and Centro Nacional de Comunicación Social helped at the grassroots to train and educate poorer people.

Overall, Mexico trailed other comparable countries in access to and use of internet and mobile technologies. One problem has been high cost of mobile service, about 50 % higher than for Latin America as a whole that has led to an estimated 90 % of poorer people to skip mobile phone service plans and instead utilize pre-paid cards, a pattern constraining use (DuPont 2010). Broadband rates, which were quite high in 2000 have dropped, yet are still not affordable by the preponderance of Mexicans.

In a 2012 OECD (Organisation for Economic Co-operation and Development) analysis of Mexico's telecommunication sector, concern was expressed about lack of competition, restrictions on fixed-line markets, market dominance by the telecommunications incumbent, weak regulations that are often stalled or negated, regulation tied too closely to the executive branch of the government, and deficits in provision of the communications spectrum (OECD 2012). Subsequently, President

Enrique Peña Nieto, elected in 2012, pushed and succeeded in breaking up some of the semi-monopoly hold on telecommunications of Carlos Slim's América Móvil and its cellphone operator Telcel. A new law passed in 2014 will foster more competition, improve coverage and service quality, and require provider agreements with customers to favor the consumer more (Mallén 2014).

The Peña-Nieto Administration has put forward an ambitious six-year "Agenda Digital por un México Conectado" (Digital Agenda for a Connected Mexico) that seeks to increase competitiveness, give more autonomy and weight to Mexico's regulatory bodies, guarantee transparency, apply regulations, establish courts for business disputes, and develop public infrastructure (Peña Nieto 2013). It also seeks universal access to broadband, broader information literacy, narrowing of the digital divide in the economy, and closing of digital gaps within the government. It is too early in this federal administration to see what the full outcome will be, but the opening of the telecommunication sector demonstrates progress.

The ups and downs of Mexico's ICT development points to a trajectory until recently not supported much by the federal government, but rather driven by market factors and dominant businesses resulting in the stifling of the consumer market and dependence on foreign innovation. The inability of the government to provide the legal, regulatory, technological literacy, and innovation features is being countered, at least on paper, by an ambitious 6-year plan of President Peña-Nieto.

This case relates to the research framework of this volume. Mexico experienced more of a tradeoff of pluses and minuses than for Singapore. Favoring ICT use and access are Mexico's steady rise in educational levels, a slow upward income trend, and younger age structure. Factors reducing the ICT use and access are ineffective legal and regulatory systems, high pricing stemming from the heretofore near monopoly in the telecomm industry, lack of support for the ICT domestic sector, and low extent of Mexican technological innovation. All these factors balance out for Mexico to a mediocre technology level for a mid-range economy.

References

Agarwal, R., Animesh, A., & Prasad, K. (2009). Social interactions and the 'digital divide': Explaining variations in internet use. *Information Systems Research, 20*(2), 277–294.

Anselin, L. (1995). Local indicators of spatial association—LISA. *Geographical Analysis, 27*(2), 93–115.

Arai, Y., & Naganuma, S. (2010). The geographical digital divide in broadband access and governmental policies in Japan: Three case studies. *Networks and Communication Studies, 24*, 1–2.

Chinn, M. D., & Fairlie, R. W. (2007). The determinants of the global digital divide: A cross-country analysis of computer and internet. *Oxford Economic Papers, 59*, 16–44.

Dedrick, J., & Kraemer, K. L. (2001). Impacts of liberalization and economic integration on Mexico's computer sector. *The Information Society, 17*, 119–132.

DuPont, S. (2010). *Technology in Mexican Civil Society. Report*. Washington, DC: The New Policy Institute.

Dutta, S., & Mia, I. (Eds.). (2007). *The global information technology report 2006–2007*. Geneva: World Economic Forum and INSEAD.

Dutta, S., & Bilbao-Osorio, B. (Eds.). (2012). *The global information technology report 2012*. Geneva: World Economic Forum and INSEAD.

Goodchild, M. F., & Janelle, D. G. (2004). Thinking spatially in the social sciences. In M. Goodchild & D. G. Janelle (Eds.), *Spatially integrated social science* (pp. 3–17). New York: Oxford University Press.

Grubesic, T. H. (2004). The geodemographic correlates of broadband access and availability in the United States. *Telematics and Informatics, 21*, 335–358.

Grubesic, T. H. (2006). A spatial taxonomy of broadband regions in the United States. *Information Economics and Policy, 18*, 423–448.

Grubesic, Tony H. (2010). Efficiency in broadband service provision: A spatial analysis. *Telecommunications Policy, 34*, 117–131.

Guillén, M. F., & Suárez, S. L. (2005). Explaining the global digital divide: Economic, political, and sociological drivers of cross-national Internet use. *Social Forces, 84*(2), 681–708.

Hair, J. F. Jr., Black, N. C., Baben, B., & Anderson, R. E. (2010). *Multivariate data analysis*, 7th Ed. Englewood Cliffs, NJ: Prentice Hall.

IDA. (2014). iN2015 Masterplan. Singapore: Infocomm Development Authority of Singapore. Available July 15, 2014 at www.ida.gov.sg.

ITU. (2013). *Measuring the information society*. Geneva, Switzerland: International Telecommunication Union.

Kauffman, R. J., & Techatassanasoontorn, A. (2005). Is there a digital divide in digital wireless phone technologies? *Journal of the Association for Information Systems, 6*(12), 338–382.

Kohonen, T. (1990). The self-organizing map. *Proceedings of the IEEE, 79*(9), 1464–1480.

Lengsfeld, J. H. B. (2011). An econometric analysis of the sociodemographic topology of the digital divide in Europe. *The Information Society, 27*, 141–157.

Longley, P. A., Goodchild, M. F., Maguire, D. J., & Rhind, D. W. (2011). *Geographic information systems and science*. New York: Wiley.

Lyon, J. D. & Tsai, C. -L. (1996). A comparison of tests for heteroscedasticity. *Statistician, 45*(3), 337–349.

Mallén, P. R. (2014). Mexico's telecom reform pits President Enrique Peña Nieto against billionaire tycoon Carlos Slim Helú. *International Business Times*. Available Mar 31, 2014 at http://www.ibtimes.com.

Moran, P. A. P. (1950). Notes on continuous stochastic phenomena. *Biometrika, 37*, 17–23.

Nishida, T., Pick, J. B., & Sarkar, A. (2014). Japan's prefectural digital divide: Multivariate and spatial analysis. *Telecommunications policy, 38*(11), 992–1010.

OECD. (2012). *OECD review of telecommunication policy and regulation in Mexico*. Paris, France: Organisation for Economic Co-operation and Development.

Peña Nieto, E. (2013). *Agenda digital por Un México conectado*. Mexico City, Mexico: El Gobierno de México.

Pick, J. B., & Azari, R. (2011). A global model of utilization of technology based on governmental, social, economic, and business investment factors. *Journal of Management Information Systems, 28*(1), 51–85.

Pick, J. B., Nishida, T., & Zhang, X. (2013). Determinants of China's technology utilization and availability 2006–2009: A spatial analysis. *The Information Society, 29*(1), 26–48.

Pick, J. B., Sarkar, A., & Johnson, J. (2014). United States digital divide: State level analysis of spatial clustering and multivariate determinants of ICT utilization. *Socio-Economic Planning Sciences, 49*, 16–32.

Population Reference Bureau. (2013). *World population data sheet 2013*. Washington, DC: Population Reference Bureau.

Rogers, E. M. (2003). *Diffusion of innovations*. New York: Free Press.

Rosenshein, L., Scott, L., & Pratt, M. (2011). *Finding a meaningful model* (pp. 40–45). Winter: ArcUser.

Skaletsky, M., Soremekun, O., & Galliers, R. D. (2014). The changing—and unchanging—face of the digital divide: An application of Kohonen self-organizing maps. *Information Technology for Development, 20*(3), 218–250.

Softtek. (2014). Softtek ranks once again as one of world's best providers by prestigious IAOP 2014 global outsourcing 100 list. Monterrey, Mexico: Softtek.

Song, Wei. (2008). Development of the Internet and digital divide in China: A spatial analysis. *Intercultural Communication Studies, 17*(3), 20–43.

SPSS Inc. (2011). SPSS statistical software. Chicago, IL: SPSS/IBM.

Srite, M., & Karahanna, E. (2006). The role of espoused national cultural values in technology acceptance. *MIS Quarterly, 30*(3), 679–704.

Takhteyev, Y., Gruzd, A., & Wellman, B. (2012). Geography of twitter networks. *Social Networks, 34*, 73–81.

Telegeography. (2014). Map of submarine cables. Washington: Telegeography, PriMetrica Inc.

Van Dijk, J. A. G. M. (2005). *The deepening divide: Inequality in the information society*. Thousand Oaks, CA: SAGE Publications.

Wald, A. (1943). Test of statistical hypothesis concerning several parameters when the number of observations is large. *Transactions of the American Mathematical Society, 54*, 426–482.

Warf, B. (2001). Segueways into multiple geographies of the digital divide. *Environment and Planning B: Planning and Design, 28*, 3–19.

Warf, B. (2013). *Global geographies of the internet*. Heidelberg, Germany: SpringerBriefs in Geography.

Warschauer, M. (2011). Singapore's dilemma: Control versus autonomy in IT-Led Development. *The Information Society, 17*, 305–311.

Wong, P.-K. (2003). Global and national factors affecting e-commerce diffusion in Singapore *The Information Society, 19*, 19–32.

World Bank. (2012). *World development indicators 2012*. Washington, DC: The World Bank.

Zhu, K., & Kraemer, K. L. (2005). Post-adoption variations in usage and value of e-business by organizations: Cross-country evidence from the retail industry. *Information Systems Research, 16*(1), 61–84.

Chapter 2
Historical Background

2.1 Introduction

The digital divide has been an influence in the world since the invention of computers in the mid-20th century. Early computers were restricted to small groups of highly trained users, while the general population could only read about the innovation, an early form of divide. Today, with cell phones prevalent worldwide, the concept of digital divide has changed, but remains significant and influential. This chapter focuses on how information and communications technologies evolved and expanded over the last two decades. What was the sequence of events for this growth? What drove it? How can the broader aspects of its maturation be understood?

The chapter first examines the milestones, both of technological innovations and of recognition of digital divide concepts. Stages have been proposed for the digital divide that draw on and modify Everett Roger's Adoption-Diffusion Theory (Rogers 2003), a framework that will be examined in Chap. 3. In this chapter, Roger's adoption-diffusion stages are introduced. Stages of adoption of major technologies can also be analyzed longitudinally by considering their trajectories of growth over time, which vary across nations. It will be shown that initial highly-varied levels for an innovation become less varied with time, and that the paces of convergence of usage levels for the leading ICTs tend to be offset by 5–10 years. Even if the usage levels of a technology such as mobile phones become saturated and much more even worldwide, new questions arise about variation in the technology's effectiveness, productivity, and social impacts.

Two chapter cases are presented. Azerbaijan represents a developing nation, the government of which has promoted and invested heavily in ICT, leading a trajectory of many improvements but also challenges. South Korea constitutes one of the foremost success stories in ICT worldwide stimulated through persistent central government efforts. Not resting on its ICT laurels that country diagnosed that it faced its own internal digital divide, addressed the disparities, and reduced internal

J.B. Pick and A. Sarkar, *The Global Digital Divides*, Progress in IS,
DOI 10.1007/978-3-662-46602-5_2

digital imbalances. South Korea has been exemplary in sharing the lessons of its success story by providing resources and field expertise to developing nations to improve their information technologies. The conclusion sums up these historical developments, which lead into the consideration of theories in Chap. 3.

2.2 Historical Milestones and Growth in Technologies 1990–2015

After the advent of the computer in the 1940s, during its commercial introduction in the late 1950s, computers differentially diffused into countries, states, and provinces, and individuals varied in their use of them. The computer was a technological innovation, which underwent an adoption and diffusion process, as well as changing as a technological product. The computer was initially only available to organizations rather than to individuals, due to its large size and staffing needs. With the introduction of the early commercial personal computers in the late 1970s and early 1980s, including the Apple II in 1977 and IBM PC in 1981, use of computers began to grow among individuals and households, reaching 8.2 % of U.S. households in 1984, 22.8 % in 1993, 51.0 % in 2000, and 77.2 % in 2011 (U.S. Census 2013; World Bank 2013).

As seen in Fig. 2.1, major technologies were introduced over the 25 years from 1990 to 2015, each being an innovation at the time of its market launch. This section discusses the book's major technologies of personal computers, the Internet, mobile phones, the web, broadband, and social media. The objective is to provide a broad background on what the technologies are, how they evolved, and how they are distributed by world region today. The timeline in the figure also highlights prominent digital divide initiatives that parallel the evolution of technologies. In the next section, the stages of maturation of use of technologies are discussed and detailed statistical analysis is performed for mobile phones, PCs, and broadband in the 21st century.

A key technology for this book is the mobile phone, which in subscriptions today is almost as numerous as human population. The mobile phone, or cell phone, stemmed from research at Bell Labs, beginning in the late 1940s and leading up to the original handheld cell phone introduced by Motorola in 1973. Since the early 1990s, cell phones have increasingly displaced fixed, landline phones, although the latter still remain, as evidenced at a public phones, such as those in Fig. 2.2. As seen in Figs. 1.2 and 1.3, mobile phones grew rapidly per capita as well as in gross volume during the past two decades. In the decade from 2002 to 2011, India and China experienced particularly rapid mobile phone expansion on a per capita basis, each gaining 800 million mobile phones, with China starting its spurt five years earlier than India. This volume of diffusion of a single information technology was unprecedented in human history. It strengthened and dramatically expanded personal and business information flows and communications in democratic India,

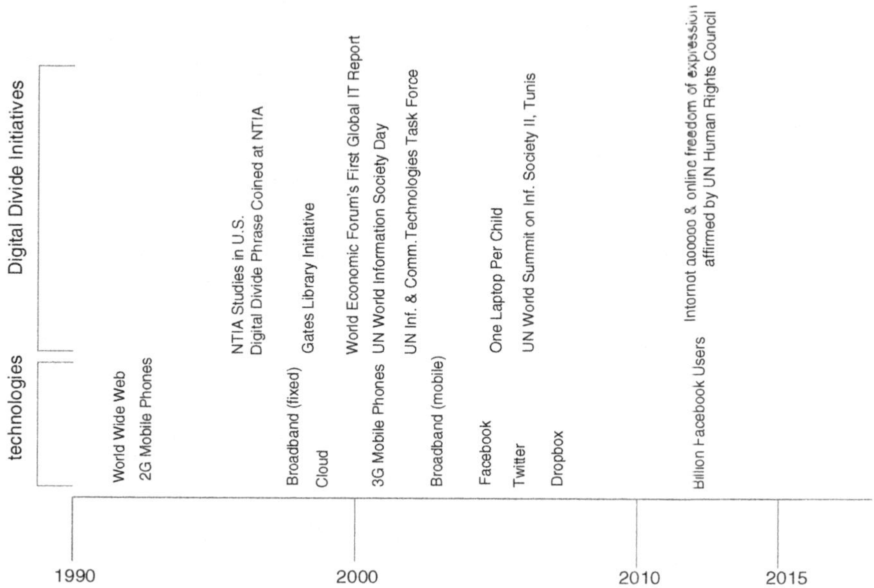

Fig. 2.1 Time line of major technologies and digital divide initiatives, 1990–2015

Fig. 2.2 Fixed, landline public phones, Illinois, USA

while radically changing the potential for citizen information flows and exchanges in China, even under the cloud of widespread censorship.

Cell phone commercial analog systems were introduced in Japan in 1979 (termed 1G–first generation systems), and the digital 2G mobile phones using the GSM protocol commenced in Finland in 1991. 3G phones, initiated in the early 2000s in Japan and South Korea, were based on broadband and packet switching; and supported internet features. Although a 3G phone is more expensive than a 2G one, there has been rapid adoption of the 3Gs especially in nations with greater internet use. Globally, by 2008, 3G (mobile broadband) was tied with fixed broadband at the level of 5 % of households, and by 2013 mobile broadband exceeded fixed broadband by threefold, at 29.5 % of households versus 9.8 % for fixed broadband (ITU 2013).

The Internet originated in 1969 with ARPANET, a research-defense network that had been started in 1966, with its first message exchange occurring at UCLA in 1969. Other internet standards such as TCP/IP were established in the early 1990s. As seen in Fig. 1.1, there was rapid increase in penetration of the Internet in advanced nations from 1996 to 2005, while their growth tapered off as levels approached 80 %. Penetration is defined as the percent of population that uses a technology. Whereas middle income nations experienced rapid internet growth from 2002 to 2011 and continue to expand rapidly, India and many lower-level developing nations were just beginning to take off in Internet use around 2008–2010.

Although broadband has grown rapidly in the past seven years, extent of use of broadband internet features by 3G phone adopters varies considerably, depending on availability and speed of broadband, education and literacy of the user, and cost of internet access. In the developing world, mobile broadband grew more rapidly than fixed phone broadband, increasing by over four-fold from 2010 to 2013, reaching 20 % of households, still far lower than its 3/4 penetration in developed nations (ITU 2013).

Another aspect of mobile broadband is its predominance as the only widespread method of broadband access in the developing world, whereas in developed nations mobile broadband is generally complementary to fixed broadband, e.g. a user might check websites on his/her cell phone and later dig in longer and more deeply on a PC. Another plus is that mobile broadband is user friendly and heavily adopted by young people, as seen in Fig. 2.3. A tradeoff is that although mobile broadband is less expensive than fixed broadband, the former tends to have volume limits on storage that constrain use (ITU 2013).

Since the data for mobile broadband are less reliable and often missing or incomplete at the state level, in this book we do not treat mobile broadband as a separate category. However, as data become more available, it will be important for country and state/provincial investigations to include it.

Fixed broadband had rapid growth initially in nations with strong fixed phone networks, and remains a stable but solid technology today mostly for advanced nations. Fixed broadband, commencing in 1997, was advantageous for countries with established fixed phone networks. A non-broadband household with a fixed

Fig. 2.3 Young user of Apple Tablet, California, USA

phone connection and DSL service available could easily and cheaply make the jump to DSL broadband. Also households with cable TV networks could readily obtain broadband if the service were offered. By contrast, in a developing nation without fixed-phone or cable networks, the options for broadband are often limited to mobile broadband, particularly in its urban areas which tend to be outfitted with cell phone towers or expensive satellite access.

As seen in Fig. 2.4, advanced nations such as South Korea, Japan, and the U.S. had extremely rapid broadband growth from 1999 to the mid-2000s, with slower growth subsequently. South Korea was particularly early and rapid in growth, the story of which appears later on in the case study. Mid-level economies of China, Brazil, and South Africa, which commenced broadband growth around 2002, have expanded at moderate rates, while the developing nations of India and Nigeria in 2013 have subscription rates under 4 %. Nigeria spurts upwards in 2012. The developing nation of Azerbaijan resembled its developing nation peers up to 2009, when it "shifted gears" into rapid exponential growth, a country that also appears as a case study later in this chapter.

The very concept of broadband has varied over time. For instance at its advent in 1997, broadband was differentiated from then prevalent dial-up lines by transmission speeds over 64 Kbps. The dividing line has continually been upgraded, so in 2013 in developed nations the accepted cutoff for broadband is greater than 2 Mbps (megabits per second) (ITU 2013). In the future, with a new standard of 4G, the broadband cutoff for developed nations might be as high as 100 Mbps. In Sweden's national broadband and ICT plan, the objective is to have 90 % of households and businesses at speeds of 100 Mbps or higher by 2020 (Pepper and Garrity 2013). The reasons for an internet service or government to assign its own threshold for broadband include the type of fiber network installed in the country, how the differences between advertised and actual speeds are resolved, and how pricing is assigned at various speeds (ITU 2013). European goals of attaining speed of 100 Mbps or higher and other broadband delivered globally in new ways are examined in Chap. 12.

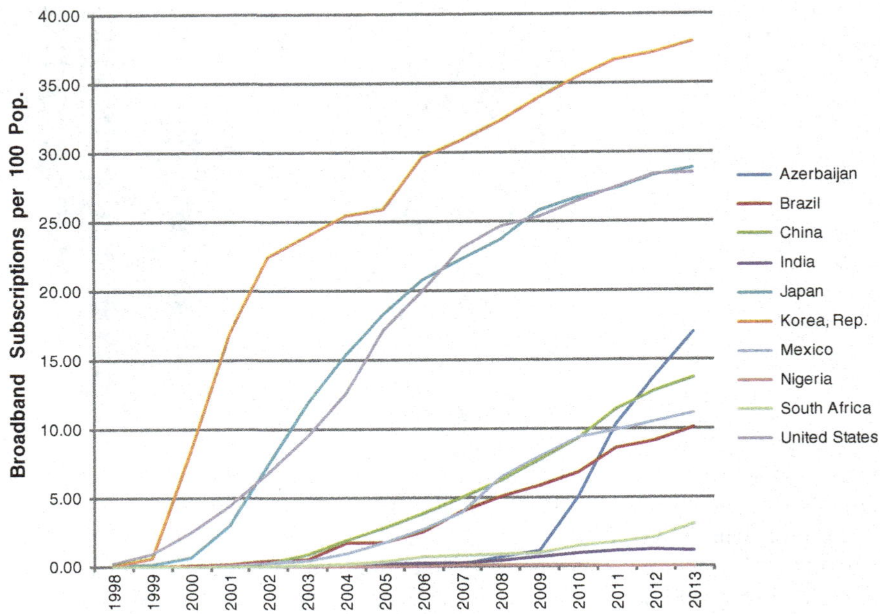

Fig. 2.4 Broadband subscriptions per 100 persons, Selected Nations, 1998–2013. *Source* World Bank (2014)

In this book, we accept the broadband definition that each data-gathering organization applies for its own datasets. The reason is that there is no systematic technique available to make accurate adjustments to multiple broadband definitions to a common standard at varied times and places. Hopefully, such a method will be developed.

The World Wide Web was introduced at the physics research center CERN in Switzerland in 2001. It has had enormous influence on Internet and broadband adoption. Data attributes for web use from samples of nations or states are not generally available from the World Bank and other leading international sources. Accordingly, we only utilize web variables in Chap. 5 for China, which has standard government variables available for number of web domain names and number of web pages.

Social media constitutes a technology that has caught on more recently. Early social networks were introduced in late 1990s and early 2000s, including Friends Reunited in 2000, Friendster in 2002, and MySpace in 2003 (Curtis 2013). Mark Zuckerberg initiated Facebook in 2004 at Harvard, while Twitter commenced in 2006 (Curtis 2013). Leading social media sites grew enormously in the past decade, with the following number of active users of the five most important social networking sites in June, 2014: Facebook (1,280,000,000), QZone (644,000,000), Google+ (343,000,000), LinkedIn (300,000,000), and Twitter (255,000,000) (Statista 2014). Qzone, popular in China and fairly similar to Facebook although in Chinese, is a service of the Chinese company Tencent. The already high, and rapidly growing,

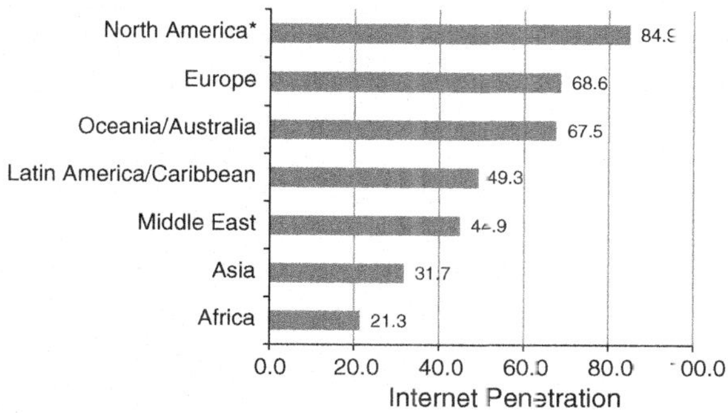

Fig. 2.5 Internet penetration by World Region, 2013. *Note* North America consists of U.S. and Canada (84.9 %). *Source* Internet World Stats (2014)

presence of social networking worldwide and its documented impacts of education, communications, and behavior justify our including it in this volume, wherever possible. However, since there is no systematic data collection on social media use worldwide, we utilize only Facebook and Twitter and restrict their use to samples for which there is reliable and systematic data collection, in particular for the U.S states, Japanese prefectures, and African nations. In the future, we expect that social networking data will become systematically available for nations from international sources, and for states/provinces from many national governments.

Not only have the major technologies in our research grown with a variety of starting points and speeds, but as mentioned in Chap. 1 they vary greatly throughout the world, as seen for internet penetration in Fig. 2.5. North America leads the world at 85 %, followed by Europe and Oceania/Australia (Internet World Stats 2014). North America in this tabulation consists 98.4 % of the U.S. and Canada, while Mexico is classified with Latin America/Caribbean. At the middle range of internet penetration is Latin America/Caribbean (49.3 %) and the Middle East (44.9 %), while Asia and Africa trail at 31.7 and 23.3 % respectively.

The elevated position of the U.S. and Canada is due to their excellent infrastructure, vigorous business environments, and educated populace, even though there are some regulatory hurdles (Dutta et al. 2011). Europe's lesser internet usage is ascribed to lower levels in Eastern Europe. In Western Europe are many of the world's leading internet nations particularly Scandinavia, the U.K., and Germany, while in Eastern Europe, there is unevenness, with moderate internet usage levels for some of the former Soviet nations now in the European Union (EU) such as the Czech Republic, Hungary, Poland, Slovak Republic, Romania, and Bulgaria, countries with moderate infrastructure and often high access costs. However, also in Eastern Europe are the nations of the Commonwealth of Independent States (CIS), most of which suffered deep drops in technology and innovation after the Soviet Union's collapse, and have only recovered somewhat (Dutta et al. 2011; Dutta and Bilbao-Osorio 2012). Two of the CIS nations with the highest internet penetration are Russia and Azerbaijan.

The Australia/Oceania region comprises the internet-leading nations of Australia and New Zealand, as well as internet-deprived ones such as Fiji and the Solomon

Islands. In Latin America and the Caribbean, although nations such as Uruguay, Chile, Argentina, and Brazil are above the world average, most countries are below average, limited by low educational levels, lack of relevant skills, marginal infrastructure, and reduced business innovation (Dutta et al. 2011, 2012).

Although Asia has some global leaders in internet such as Japan, South Korea, Hong Kong, Taiwan, and Singapore with internet penetration over 80 %, it also includes South Central Asian nations such as India, Pakistan, Bangladesh, and Myanmar which have penetration under 10 %. Africa is the most internet-deprived continent with penetration of only 21.3 %. Although this penetration is the result of rapid growth, many of its nations have rates under 10 %, including ones under 5 % in the center of the continent. The reasons for low rates in Sub-Saharan Africa include poor infrastructure, very low educational levels, relatively high internet costs, and multiple nations with sluggish and inefficient business environments (Dutta et al. 2011, 2012). The geographic contrasts in internet usage within Africa are dramatic, and have major economic ramifications, preventing viable and competitive service sectors. Chapter 9 delves deeply into reasons for Africa's stark internet disparities.

In short, this section points to technologies that have boomed worldwide in the last two decades, although following a lagged sequence of introductions and initial upward surges. Additionally there are large continental differences in penetration of ICTs, as well as wide variation within Asia and Africa.

2.3 Stages of Maturation of Technology Use

This section considers the patterns of adoption of technologies over time, in order to try to understand more thoroughly the rates at which a new technologies expand and eventually converge, and also how countries and developed versus developing regions of the world can be better understood and interpreted. Later, the progression over time and space in technologies during the period 2000–2013 is analyzed, illustrating actual sequences that the major nations have followed.

Regarded as the founder of adoption-diffusion theory, Everett Rogers formulated an extensive theory of how innovations are adopted and diffused (Rogers 2003). Although his detailed theory is discussed in Chap. 3 in relationship to the book's conceptual theory, his stages of adoption are glimpsed at in this chapter, as they relate to two subsequent digital divide stage theories. Roger posited that an innovation underwent early adoption with a few innovators; then experienced a rapid increase followed by tapering off that was rapid but slowed and became more stretched out at the end with laggard adopters. The full set of adoption-diffusion stages, seen in Fig. 2.6 consist of innovators, early adopters, early majority, late majority, and laggards (Rogers 2003). Based on this adoption curve, cumulative adoptions by percent of population over time appear as an "S-shaped curve" (see Fig. 2.7). Notice in this figure there are three "S-shaped curves" that have different rapidity of adoption.

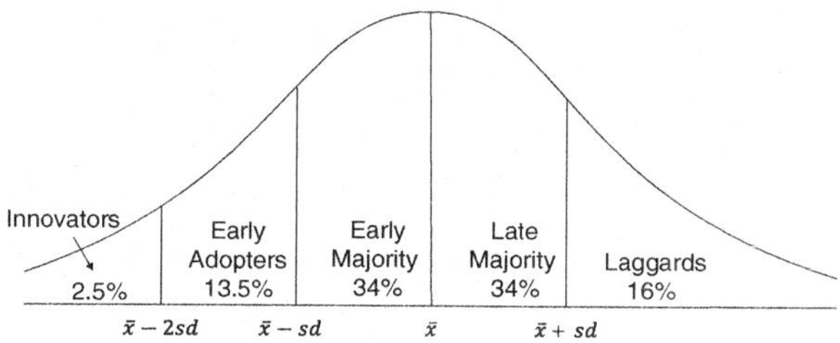

Fig. 2.6 Roger's adoption-diffusion stages. *Source* Rogers (2003)

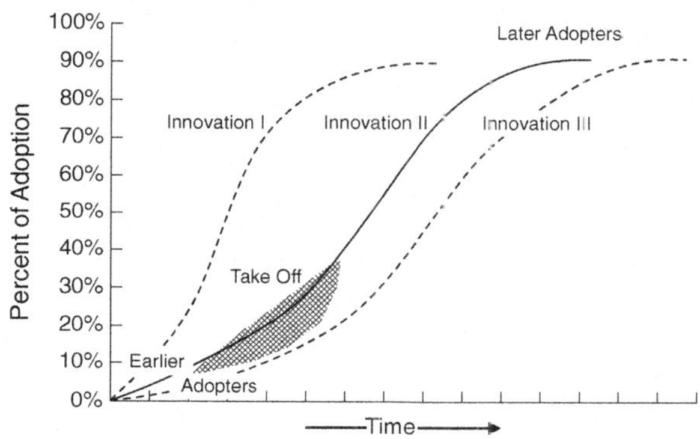

Fig. 2.7 Cumulative adoption curves, at different adoption rates. *Source* Rogers (2003)

A glance back at Fig. 1.2 shows cumulative mobile phone subscriptions over time, Korea, Japan, and Mexico follow S-shaped penetration curves, while the curves for Brazil, China, and India appear to be in the early adoption stage of rapid increase. The United States follows the S-shaped curve through the innovation and early adoption stages up to 2003, but differs in not tapering off in the early majority stage, and rather continuing to grow linearly. However, the U.S. logically would reach a limit and begin to taper at a future time point.

Based on Roger's theory, a framework for digital divide was categorized into three stages (Rueda-Sabater and Garrity 2011). We have slightly modified this framework, particularly by identifying Stage 2 with "leapfrogging," or jumping over stages.

Stage 1. First Adopters ("Early Adopters in Roger's theory)

For internet penetration their major spurt in internet use occurred between 1995 and 2005 ("early adoption stage"). By 2010, Early Adopters tended to have 75 % of population as Internet users and most households had a broadband connection (Rueda-Sabater and Garrity 2011). This group constituted mostly developed nations. For Internet they are presently levelled off at the saturation point of the S-shaped curve, such as South Korea, Japan, and the United States.

Stage 2. "Converging" Adopters ("Early Adopters in Roger's theory, but applying to countries with more recent ICT growth, often with "leapfrogging")

Nations in this category are mostly developing ones that skipped the adoption of PCs and started out in ICT adoption with rapid growth in Internet and broadband ("early adoption stage"). The adoption rate was more rapid due to extensive use of internet cafes and other cheap public venues (Rueda-Sabater and Garrity 2011). Another necessary factor is to have low-cost technologies available. Skipping adopting the PC and moving to adoption of Internet/broadband is often referred to as "leapfrogging," which implies that the nation skipped the steps of more expensive PCs and the more cumbersome infrastructure of fixed telephone networks, but rather leaped across those traditional technology steps and right to inexpensive mobile phones and Internet/broadband, such as China.

Stage 3. "Belated" Adopters ("Laggards in Roger's theory")

These countries are coming late to adopting ICTs. They have among the weakest economies and lowest incomes. In 2010, often Internet use is under 5 % and broadband under 1 % (Rueda-Sabater and Garrity 2011). These nations are "leapfroggers" even more than Stage 2's "converging adopters." However they face the barriers of the high relative cost of ICT access, low literacy and education, and poor infrastructure. However, once the cost threshold is achieved, these nations could have the most rapid rates of increase, as is being seen in some impoverished African nations today.

Another study (James 2013) expanded more on these ideas. Utilizing an extensive World Bank data set (World Bank 2008), the study indicates that although in the High income OECD countries from 1975 to 2000, 82 % of innovations reached the 50 % adoption threshold, for developing nations in that period only 9 % reached the 50 % adoption threshold (James 2013). Similar patterns occurred for innovations in earlier periods, i.e. rates in the range of 33–75 % for High income OECD countries reached the 50 % adoption threshold while only 4–11 % reached it for developing ones (James 2013). For the present research, these findings suggest that some ICTs in developing nations might stop adoption at an earlier stage and lower level and not even reach maturity.

However, James (2013) subsequently observed that for ICTs, an exception to this historical precedent is that late-starting developing nations often either equal or exceed the maturity levels of developed nations. Although starting later, the developing nations go through early adoption at much more rapid rates of increase than for developed ones, i.e. demonstrating leapfrogging similar to Stages 2 and 3 of the prior framework (Rueda-Sabater and Garrity 2011).

The two related frameworks inform our research in conceptualizing that the same innovation is likely to commence later yet grow faster in developing versus developed countries. Both are likely to reach eventually a similar maturity level, especially if costs in the developing nations became affordable.

Statistical study of sample of nations for stages of technology adoption

We explore the rates of adoption for a representative sample of 39 nations from all the continents. ICT variables for the sample are given in Table 2.1. The nations are selected as a convenience sample that includes mostly nations of large populations, and also includes the four nations studied in detail in Chaps. 5–8 as well as some African nations discussed in Chap. 9. The exploratory sample is about a third the size of the entire world sample of nations with complete data. A map of the sample, showing bar charts for mobile phone penetration in 2000, 2005, 2010, and 2013, appears in Fig. 2.8.

Before discussing the study, some background is necessary on trajectories of adoption. Cell phones and the Internet started the upward early adoption stage for advanced nations in the early 1990s, attaining late maturity in those nations about 2010 (see Figs. 1.1 and 1.2). Broadband entered early adoption for advanced nations in the late 1990s (see Fig. 2.2) and reached early maturity in about 2012–2013. However, usage of mobile phones eventually reached near saturation of population, whereas Internet use leveled off at about 80 % penetration. This difference relates to factors such as cost, infrastructure, and literacy. To use a cell phone for making calls can be done by an illiterate person at very low expense, whereas using the internet requires some literacy and is generally more expensive than a simple 2G cell phone, implying total market saturation is unlikely. By contrast broadband subscription per 100 persons, even for the most advanced usage nations such as South Korea, has reached a level of only 35–40 per 100 persons, which points to early maturity stage, with significantly higher usage expected in the future.

With this background, we analyze the usage of mobile phones, PCs, and fixed broadband, based on the sample of data from the World Bank for the period 2000 to 2011–2013. In the case of PCs, the World Bank changed its metric so that although it measured PCs per 100 population in 2000 and 2005, it ceased doing so for 2010 and beyond, but did have available percent households with a personal computer for 2011, which we used as a proxy for percent PCs. As seen in Table 2.1 and Figs. 2.8, 2.9 and 2.10, there was considerable variation between nations.

This variation reflects the varied trajectories of adoption of ICTs already noted. Consider mobile phones for twelve nations in the sample (see Fig. 2.8), It is clear that the developing nations of India and Nigeria have 2013 subscription rates of about 70 %, most of which occurred in big jumps between 2005 and 2010. By contrast, the highest levels of mobile phone use are in Finland and Saudi Arabia, both in the range of 170–175 subscriptions per 100 persons. However, Saudi Arabia, compared to Finland, started at a much smaller base in 2000 of 8 subscriptions per 100 population, and then jumped. Not only are there large differences in progression over time, but it is apparent that the starting values in 2000 had more variation than the ending values in 2013. This reflects advances worldwide in infrastructure and affordability of cell phones, leading to more convergence, as saturation gets closer.

Similar variations in progression over time are seen for PCs per 100 persons (Fig. 2.10) and fixed broadband (Fig. 2.11). For them, developing nations had

Table 2.1 Sample for analysis of longitudinal change in variation of ICT variables

	Mobile phone subscriptions/100 persons				Personal % Computers/ Households 100 persons with PC			Fixed broadband subscriptions/100 Persons			
	2000	2005	2010	2013	2000	2005	2011	2000	2005	2010	2013
Angola	0.2	9.7	48.1	61.9	0.1	0.1	6.4	0	0	0.1	0.22
Australia	44.5	89.8	100.4	106.8	46.5	68.3	82.6	NA	9.82	23.99	25.01
Bangladesh	0.2	6.3	44.9	67.1	0.2	0.1	4	0	0	0.27	0.63
Belgium	54.8	91.4	111.1	110.9	34.4	34.8	78.9	NA	0.02	17.55	29.76
Bolivia	6.9	25.9	70.7	97.7	0.2	2.3	24	NA	0.14	0.94	1.33
Brazil	13.3	46.3	100.9	135.3	4.4	10.5	45.4	0.06	1.74	6.8	10.08
China	6.7	29.8	63.2	88.7	1.6	4.1	38	0	2.83	9.29	13.63
Czech Republic	42.4	115.1	122.6	131.3	12.2	24	69.9	0.02	6.93	14.5	17.03
Dominican Republic	8.1	38.8	88.8	88.4	NA	NA	18.9	0.11	4.82	11.68	14.81
Egypt	2.1	19	90.5	121.5	2.2	3.8	36.4	0	0.2	1.86	3.26
Estonia	40.8	109.1	127.3	159.7	15.3	48.3	71.4	NA	13.52	24.02	26.54
Finland	72	100.5	156.3	171.7	39.6	48.1	85.1	0.68	22.38	28.55	30.9
France	49.1	78.3	91.4	98.5	30.4	57.5	78.2	0.33	15.41	33.74	38.79
Germany	57.7	94.6	106.5	119	33.6	54.5	86.9	0.32	12.87	31.43	34.58
Ghana	0.7	13.4	71.9	108.2	0	0.1	11.9	NA	0.01	0.21	0.27
India	0.3	8	62.4	70.8	0.4	1.6	9.5	0	0.12	0.91	1.16
Indonesia	1.8	20.9	87.8	121.5	1	1.4	12.3	0	0.05	0.95	1.3
Ireland	64.7	102.7	105.2	102.8	35.9	49.4	80.6	0	7.76	21.07	24.24
Israel	73.2	117.5	122.8	122.8	25.4	74	78.2	0	18.62	23.75	25.67
Japan	53.1	76	96.8	115.2	31.5	54.2	80	0.68	18.35	26.77	28.84
Kenya	0.4	12.9	61	70.6	0.1	0.1	8.8	NA	0.02	0.01	0.13

(continued)

Table 2.1 (continued)

	Mobile phone subscriptions/100 persons				Personal % Computers/ Households 100 persons with PC			Fixed broadband subscriptions/100 Persons			
	2000	2005	2010	2013	2000	2005	2011	2000	2005	2010	2013
Laos	0.2	11.4	62.6	66.2	0.3	1.7	7.8	NA	0.01	0.09	0.13
Lebanon	23	24.9	66	80.6	0.5	11.4	71.5	NA	3.26	6.98	9.95
Mexico	13.6	42.6	77.5	85.8	5.1	13.6	30	0.01	1.74	9.42	11.14
Netherlands	67.8	97.1	115.4	113.7	39.4	68.2	94.2	1.64	25.15	38.09	40.08
Nigeria	0	13.3	54.7	73.3	0.7	0.1	9.3	0	0	0.06	0.01
Norway	71.8	102.8	114.5	116.5	49	57.3	91	0.52	21.44	34.51	36.43
Peru	4.9	20.1	99.5	98.1	0.4	10	25.4	0	1.27	3.12	5.18
Russia	2.2	83.4	165.5	152.8	0.4	12.2	57.1	0	1.1	10.93	16.62
Saudi Arabia	6.8	57.4	189.2	176.5	0.6	35.4	62.8	0	0.27	5.49	7.33
Singapore	70.1	97.5	145.4	155.6	48.3	NA	86	1.76	14.6	24.98	25.7
South Africa	18.6	70.4	97.9	147.5	6.2	8.5	21.5	NA	0.34	1.44	3.06
South Korea	58.3	81.5	104.8	111	23.8	54.5	81.9	8.42	25.91	35.48	38.04
Spain	60.2	98.4	111.3	106.9	114.3	27.7	71.5	0.19	11.61	22.81	25.57
Turkey	25.5	64.4	85.6	93	3.8	5.2	48.5	0	2.35	9.83	11.19
United Kingdom	73.7	108.6	123.6	123.8	33.8	60	84.6	0.09	16.42	30.83	35.73
United States	38.5	68.3	91.3	95.5	58.5	76.2	77.2	2.48	17.16	26.5	28.54
Venezuela	22.3	46.8	96	101.6	4.5	8.2	19	0.02	1.32	5.6	7.31
Vietnam	1	11.3	125.3	130.9	0.9	1.3	16	NA	0.25	4.12	5.62

NA not available

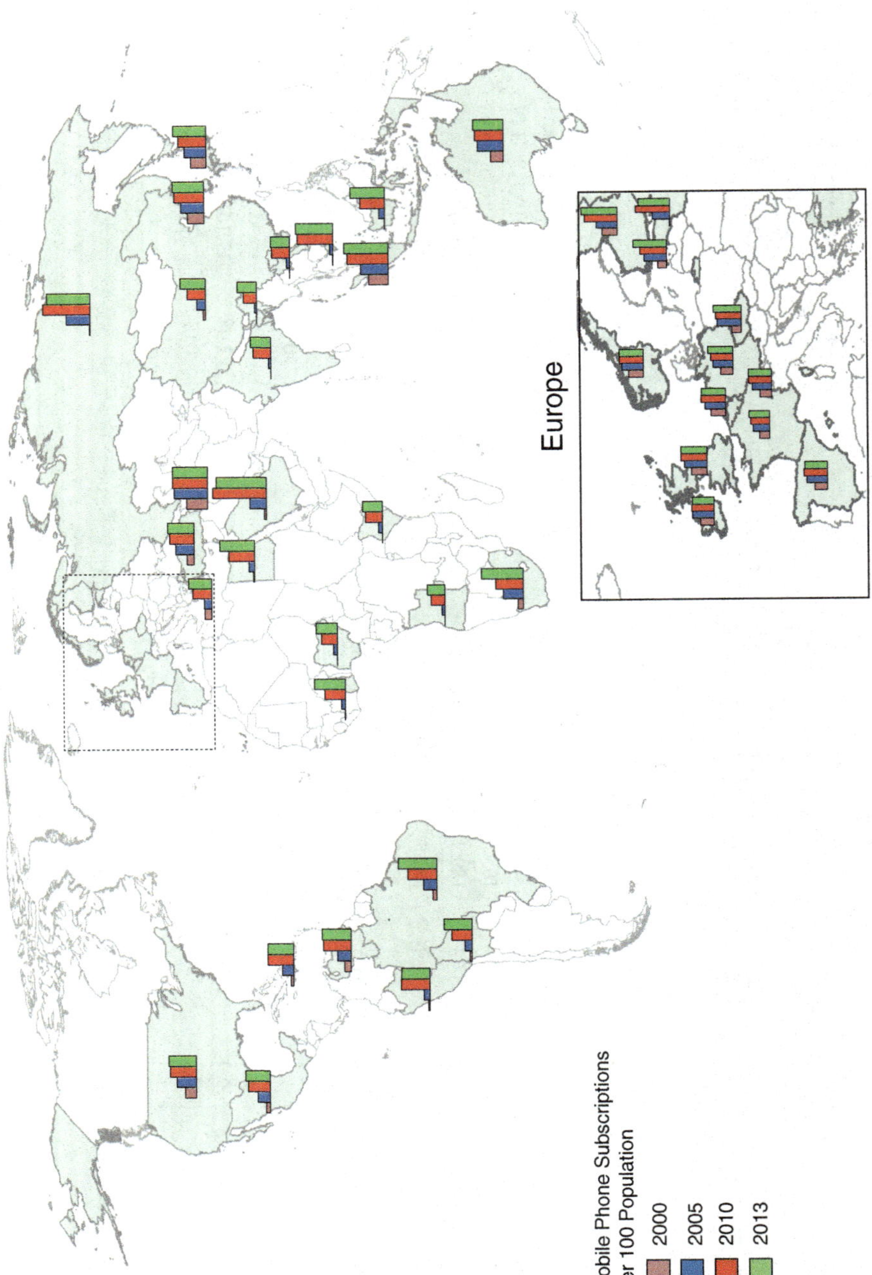

Fig. 2.8 Map of mobile phone subscriptions for 2000, 2005, 2010, and 2013, Sample of 39 Nations

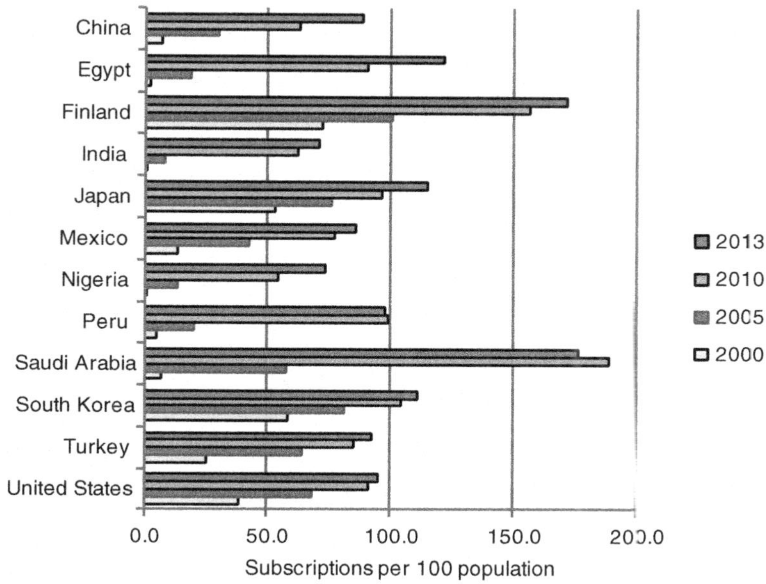

Fig. 2.9 Mobile phone subscriptions, 2000, 2005, 2010, 2013 for 12 Nations (*Data Source* World Bank 2014)

rapid growth (i.e. Roger's early adoption stage) later on than for developed nations.

Again the variation at the end dates of 2011 and 2013 is lowered, compared to the starting year of 2000. For fixed broadband, countries such as Finland, Japan, South Korea, and the U.S. grew in steady large steps from 2000 to 2013, while the poor, developing nations of India and Nigeria revealed almost no progress over the 13 years. These large contrasts relate mostly are due to lack of availability of the infrastructure for fixed broadband. India and Nigeria had late arrival times for sub-oceanic cables, and in both, access to fixed broadband was limited by the high cost of user devices and subscriptions. Another reason is the leapfrogging noted earlier, in this case leaping from no broadband to broadband-based 3G mobile phones, skipping the step of fixed broadband.

Likewise, PCs show significant gains in penetration over time, which peaks at about 80 % of households, while the low income nations of India and Nigeria only reach levels of about 10 % of individuals. Since PCs remain considerably more expensive than mobile devices, the latter countries' lack of adoption of PCs reflects cost and resultant leapfrogging of users to mobile devices such as 3G cell phones, which provide cloud-based applications that can substitute for PC applications.

To test the observation of reduction over time in variation of adoption levels, we computed the coefficient of variation of the three technologies at all the time points. The coefficient of variation for a sample is defined as 100 times the standard deviation divided by the mean. As seen in Table 2.2, each of the technologies has

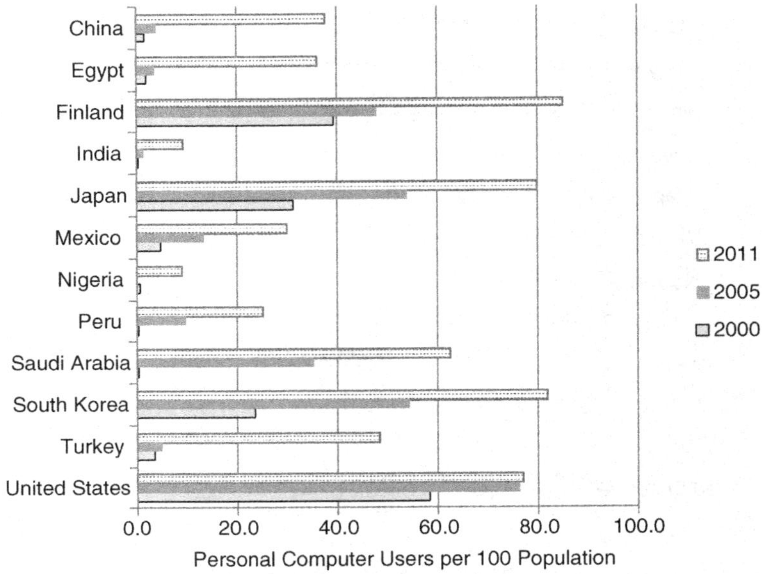

Fig. 2.10 Percent PC Users, 2000, 2005 and percent households with PCs, 2011, for 12 Nations. (*Data Source* World Bank 2014)

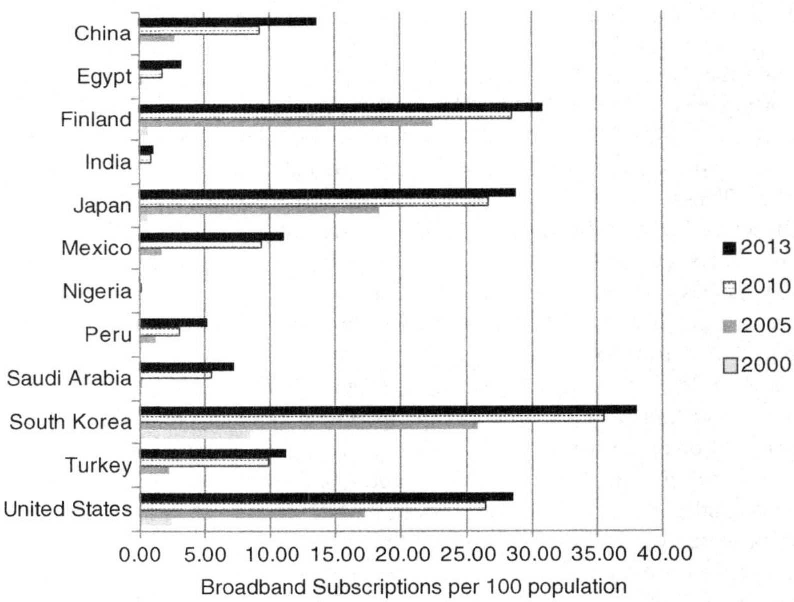

Fig. 2.11 Broadband subscriptions, 2000, 2005, 2010, 2013 for 12 Nations. (*Data Source* World Bank 2014)

Table 2.2 Longitudinal change in coefficient of variation of technology indicators

	Standard deviation	Mean	Coefficient of variation[a]
Mobile phone subscriptions 2000 per 100 pop.	27.61	29.52	93.52
Mobile phone subscriptions 2005 per 100 pop.	37.75	59.13	63.84
Mobile phone subscriptions 2010 per 100 pop.	21.92	98.89	22.17
Mobile phone subscriptions 2013 per 100 pop.	29.15	110.25	26.44
Personal computers 2000 per 100 pop.	24.27	18.56	130.73
Personal computers 2005 per 100 pop.	26.37	26.72	98.69
Percent households with personal computer 2011	31.37	50.33	62.33
Broadband subscriptions 2000 per 100 pop.	1.77	0.72	244.45
Broadband subscriptions 2005 per 100 pop.	8.43	7.17	117.60
Broadband subscriptions 2010 per 100 pop.	12.60	14.07	89.58
Broadband subscriptions 2013 per 100 pop.	13.55	16.30	83.14

[a]Coefficient of variation equals 100 times the variance divided by the mean

substantial reduction in coefficient of variation, in the range of a half to 2/3 reduction, over the entire time period. However, even after the reductions, there is substantial coefficient of variation remaining of 26 % for mobile phones in 2013, 62 % for PCs in 2011, and 83 % for broadband in 2013. This implies that mobile phones are more converged globally, followed by PCs, and broadband is least converged. Although these findings confirm that the digital divide has been reduced, they imply that substantial divide remains for PCs and broadband.

2.4 Case Study: Azerbaijan

Azerbaijan, a nation of population 9.4 million in 2013 in the South Caucasus Region, is located along the west side of the Caspian Sea, bordering Russia, Georgia, Armenia, and Iran (Population Reference Bureau 2013). A former part of the Soviet Union with majority Turkic and Muslim population, it became independent in 1991 and has prospered economically since 2000 based on its rich petroleum production sector (CIA 2014). It forms a crossroads between Asia and Europe in transportation and energy, and has been a stable nation politically (Makili-Aliyev 2013).

As Azerbaijan became prosperous, its government leaders identified ICT as a strategic sector that would provide diversity to its economy, improve internal productivity, give it regional technological leadership, and enable the nation to start up its own ICT industry. The formal start of this thrust toward ICT was the approval in 2003 of the "National Strategy on Information Communication Technologies for the sake of Development of Azerbaijan Republic, 2003–2012" (Republic of Azerbaijan 2014). This strategic plan emphasized the three "pillars" to (1) liberalize the telecommunications market, while adding regulations, (2) develop the nation's telecommunications infrastructure, and (3) foster and implement e-government and e-services (Republic of Azerbaijan 2014). A year later, the government established the Ministry of Communications and Information Technologies (MCIT), to be the lead agency to coordinate the government's efforts to stimulate businesses and non-government organizations (NGOs), and to encourage investors to deploy and expand the utilization of ICT. The government wisely provided MCIT with sufficient authority, capacity, and long-term funding to be able to achieve its goals (Aliguliyev and Gurbanov 2012).

Azerbaijan made large progress in its ICT base over the past decade, formalizing ICT advance as a national priority in 2007. A key starting factor, unusual for a developing country, was its extensive fixed phone network put in place originally in the Soviet era and developed further from the nation's ample petroleum resources, so that today all residential urban areas have fixed phone service (Aliguliyev and Gurbanov 2012). Accordingly, it was easier to implement broadband since the fixed phone service could be relatively inexpensively upgraded to DSL. The remarkable surge in broadband is evident in Fig. 2.12, starting in 2008 at about 1 % of households in 2008 and rising to 50 % level by 2013. Broadband by 2012 exceeded considerably cell phone and dial-up Internet access. Likewise, mobile phone subscriptions increased from a low-income-country level of 6 per 100 persons in 2000 to 101 per 100 persons in 2010, approaching the level for high income nations (World Bank 2002, 2012).

Simultaneously, the cost of broadband lowered by 11-fold as a percent of salary from 2009 to 2012 because of competition and government subsidies, (see Fig. 2.13), resulting in affordability similar to advanced nations.

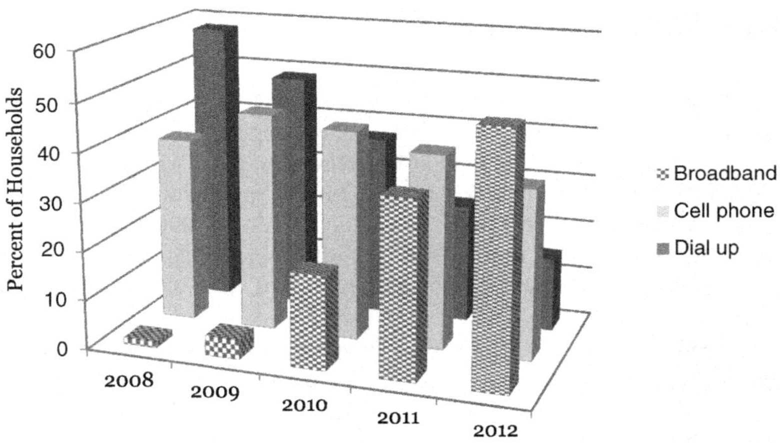

Fig. 2.12 Means of internet connectivity, Azerbaijan, 2008–2013. *Note* The other category (not shown) is less than 2.9 % of households. *Source* Republic of Azerbaijan (2012)

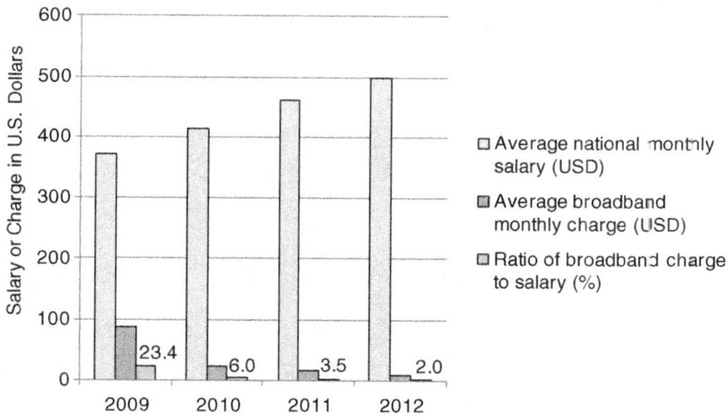

Fig. 2.13 Ratio of average broadband charges to salary (percent), 2012 (Republic of Azerbaijan 2012)

Although as noted earlier, most developing nations leapfrogged to bypass personal computers, MCIT took a different path by starting the National Personal Computer Project in 2009, which partnered with leading vendor firms such as HP and Microsoft to spread personal computers nationally at discounted prices, particularly targeting teachers and school children (Republic of Azerbaijan 2012). Hence, children who could benefit the most over their work life had much better access to PCs, albeit low by prevalence in the schools of advanced nations. Over 40,000 computers were provided this way for education (Aliguliyev and Gurbanov 2012), while the ratio of school pupils to PCs to school pupils halved from 2009 to 2012 (see Table 2.3). PCs were also used by corporations in Azerbaijan, as seen in Fig. 2.14.

Other notable government-led projects included e-government and e-services, satellites, the State Fund for Development of Information Technologies, a high technology park, and the Trans-Eurasian Super Information Highway (TASIM). In 2010–2011, several government plans and a Presidential decree started an effort to develop world-class e-government, and e-services for citizens (Republic of Azerbaijan 2012). Some highlights were the goals to achieve a state registry of citizens, e-services for taxation, e-registration of entrepreneurial enterprises, customs monitoring at a central "one-stop" website, online systems for immigration services, e-health, and the management of online availability of government information (Aliguliyev and Gurbanov 2012).

Table 2.3 Indicators of ICTs in Schools in Azerbaijan, 2009–2012

	Year			
Measure	2009	2010	2011	2012
Computer prevalence (%)	82.6	83.9	86.7	87.2
Broadband access (%)	15.8	39.1	47.4	48.3
Ratio of school pupils to PCs	35	33	27	16

Source Republic of Azerbaijan (2012)

Fig. 2.14 Corporate users of PCs in Azerbaijan. *Source* Internet Access and Training Program, Eurasia

The Trans-Eurasian Information Super Highway (TASIM) Project was started by the Azerbaijan government in 2008 to improve the connectivity of ICT infrastructure, expand the quality, and lower the cost of the Internet across the South Caucasus Region (Republic of Azerbaijan 2012). A broader and longer-term goal was to lead in linking the highly internet-developed European region through networks crossing 20 nations to link with the equally net-intensive East Asia region. Although TASIM was supported by a UN resolution in 2009, this slow-moving long-term project has so far progressed to agreements with operating firms and with cooperative governments including the construction of an undersea cable beneath the Caspian Sea connecting Azerbaijan with Kazakhstan. Ultimately, the TASIM project is intended to "contribute to create open information society in the region, to increase the speed of Internet connection and the development of Internet services, which, in turn, will lead to economic development, diversification of the economy and increased competition" (Republic of Azerbaijan 2014).

In 2010, an initiative to launch Azerbaijani space satellites commenced under the auspices of a state-owned firm, Azercosmos. This led in 2012 to successful launching of the Azerspace-1 communications satellite, with coverage of Eastern Europe, Central Asia, and North Africa (Republic of Azerbaijan 2014; Aliguliyev and Gurbanov 2012).

The government established the State Fund for Development of Information Technologies in 2012 to stimulate ICT initiatives and scientific research in state and local governments, the courts, financial and insurance enterprises, and NGOs (Republic of Azerbaijan 2014). That same year, a high technology park was established by the President in order to bring together domestic and international technology firms in a zone to encourage collaboration and exchanges.

Among the ICT challenges remaining for this country is to embed ICT in private companies, educate many more skilled ICT workers, reduce regulations on ICT, and increase openness of information (Aliguliyev and Gurbanov 2012).

This case study relates to the historical trends examined in the chapter. The country is an exception to the usual historical pattern of ICT development of a developing nation. The government took a lead role in breaking the usual developing-nation sequence, which often is reactive, and to boldly prioritize ICT as a primary national goal and pursue it concertedly for over a decade and a half. The achievement of the goals was made easier by the country's petroleum riches, which filled government coffers and made the funding of ambitious ICT plans possible. Referring to the three stages of technology development discussed earlier (Rueda-Sabater and Garrity 2011), Azerbaijan would be categorized in Stage 2-Converging Adopter. This stage was speeded up because broadband grew at very high rates, thanks to leveraging the DSL services from an existing widespread fixed phone network; because usual leapfrogging of PCs could be avoided by special targeting programs to bring PCs to schools; and because the nation's leaders were highly motivated to try to emulate a Rueda-Sabater and Garrity Stage 1-First Adopters profile, an attempt made possible in a resource-rich nation.

2.5 Case Study: South Korea

South Korea is among the most ICT-intensive of countries. In 2012, it was 12th out of 142 nations on the Networked Readiness Index (World Economic Forum 2012). In East Asia/Oceania, it ranked higher than Hong Kong (13th), Japan (18th), Australia (17th), and China (51st), although Singapore (2nd) exceeded it. In households with internet access it ranked 1st, with 97 % average access, while on broadband internet subscriptions it ranked 4th and for extent of business internet use it was in 2nd place (Dutta and Bilbao-Osorio 2012). South Korea has been careful also in planning its ICT, resulting in societal impacts in 1st place.

This was a success achieved early; in fact by 2002 South Korea had top rankings, in including 1st place worldwide for broadband penetration. Since then major indicators have continued to move upwards as seen in Table 2.4. In some cases, household and business access or use of ICT reached nearly 100 %.

How did South Korea achieve its remarkable ICT global leadership status? Does the nation have remaining issues or problems in ICT? The answers can be useful both to advanced-nation peers and to aspiring developing nations.

Table 2.4 Access and use of ICT by individuals, households, and businesses, South Korea, 2004–2011

	Year			
	2004	2007	2009	2011
Households with a computer (%)	77.8	80.4	81.4	81.9
Households with internet access (%)	86.0	94.1	95.9	97.2
Individuals using internet in last 12 months (%)	72.7	78.8	81.6	83.8
Businesses with a computer (%)	92.2	97.0	98.6	98.4
Businesses with broadband access (%)	96.3	97.7	98.8	98.7

Source Republic of Korea (2012)

Some of the explanation relates to the following factors (Choudrie et al. 2003; Choudrie and Lee 2004; Choudrie and Middleton 2014):

1. Competition. South Korean private firms competed intensively to provide the best infrastructure, encouraging continual modernization.
2. Pricing. Early pricing was targeted at the middle income person, and the competition pushed the prices lower.
3. "PC Bang." At the turn of the 21st century there was a huge sprouting of "PC bangs" (i.e. Internet cafes) around the country.
4. Evident benefits to the user. The starting base of users was targeted in the education and entertainment sectors particularly the Internet game industry which quickly perceived the benefits and value of ICT.
5. Government strategy, vision, and commitment. Strong and long-lasting commitment came from the central government.
6. Urban geography and population density. South Korea is highly urbanized into dense populated areas, enabling considerable savings on infrastructure.

The South Korean government's support for ICT development began in the mid-1990s, as the Internet started its initial upswing. The South Korean economy was growing strongly internally and its businesses had commenced to enter world markets. The central government recognized its deficiency in making available PCs and access to the Internet, and responded by starting up the Information Telecommunication Training Center (ITTC) to address the deficits in infrastructure. ITTC became the Korea Agency for Digital Opportunity and Promotion (KADO) in 1997. Its early thrusts centered on "Cyber 21," a program for housewives and other non-tech citizen segments to learn basic skills, and on the "Ten Million" program, which aspired to teach 10 million Koreans how to perform basic activities of Internet access; a goal achieved by 2002.

In 2004, the u-Korea Master Plan was introduced including strategies to develop wireless broadband, digital multimedia broadcasting, RFID, home network services, and telematics (Choudrie 2014), services that were cutting edge and ambitious at the time. Another part of the Master Plan was the Broadband Convergence Network (BCN), a public-private consortium that provided technology transfer to private firms between 2004 and 2008 (Choudrie 2014).

KADO established a remarkable program throughout the country, which refurbished old or discarded PCs for use by disadvantaged population. In all it provided 60,000 of these PCs which contributed to strengthen the underserved in the nation's ICT development.

Another objective of KADO's activities was to close South Korea's internal digital divide, helping disadvantaged citizens, such as elderly, poor, immigrant, and disabled persons. This divide in 2007 was significantly wider than for the U.S., as seen in Table 2.5; (Wong et al. 2010).

For disabled people, KADO established usage guidelines and provided software and special devices to assist them, KADO paying up to 4/5 of the costs (Woyke 2009). Another program targeted providing training and assistance to North Korean

Table 2.5 Digital inclusion in South Korea and the United States in 2007

Population groups	USA		South Korea	
Internet users (%)				
Gender				
Male	78		82.4	
Female	75		71.9	
Age				
Old	56	(Age 62–71)[a]	20.1	(Age 60+)[a]
	29	(Age 72+)[a]		
Income				
Low	61		48.4	
All	76		77.6	
Completed secondary education				
Age groups				
25–35	87		97	
55–64	87		37	

Note a = age range. *Source* Wong et al. (2010)

refugees, mostly devoid to IT skills. KADO maintained careful metrics on its detailed progress in narrowing the digital divide. Results showed steady progress, although by no means closing the gap. An example is the 7-year narrowing of the gap between all households and disadvantaged households in percent computer penetration, as seen in Fig. 2.15.

In 2009, KADO merged and became part of the National Information Society Agency (NIA), which had since 1987 been the lead government agency for implementing ICT in Korea, including support, management, and operation of many government networks and development and promotion the National Framework Plan on Informatization Promotion (Republic of Korea 2014).

Since the internal digital divide by 2009 had been narrowed considerably, NIA refocused KADO's former initiatives on external programs to assist other nations in narrowing their internal digital divides. This included offering consulting and technical assistance to nations requesting it, setting up IT Cooperation Centers, which collaborated with the partner country in providing physical centers in that country for consulting, technical advice, and sharing of technology experiences (Republic of Korea 2014). The Korea IT Learning Program convened high level meetings with technology experts from Korea and policymakers from developing nations to strengthen and improve national IT policies and planning. On request of developing nations, the Korea IT Volunteers sent volunteer groups called World Friends Korea to developing nations for 1–3 month periods, in which Korean students and IT professionals offered ICT consulting, basic training, collaborative projects, and sharing of Korean language and culture (Republic of Korea 2014).

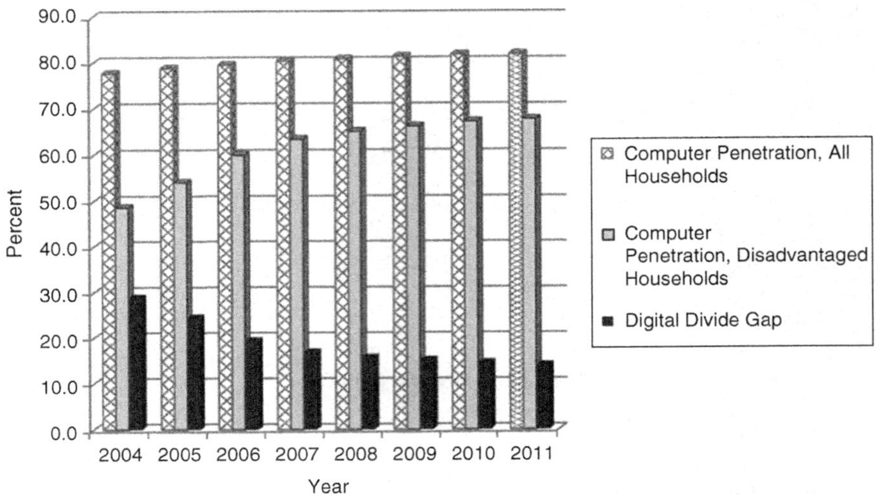

Fig. 2.15 Computer penetration and digital divide gap of disadvantaged households, South Korea, 2004–2011. *Source* Republic of Korea (2012)

During 2001–2013, the program sent 5100 Korea IT Volunteers to 70 countries, serving 100,000 people in the recipient nations.

Another initiative of the Korean central government has been to advance the development of the concept of Ubiquitous City or U-City. This is a contemporary concept that allows the users located in U-Cities to have net access anywhere/anytime and also to have access to advanced high-speed services including the Internet of Things i.e. communication with a variety of smart devices. The cities themselves take over much of the policy and planning, with several examples being U-Chongna, U-Changwon, and U-Songdo. This topic of U-Cities focusing on U-Songdo is covered in more detail in Chap. 11.

The South Korean case study demonstrates how a prosperous developed nation can be smart in planning and realizing world leading ICT levels, while simultaneously closing its own internal disparities among disadvantaged groups. However, the nation did not rest on its laurels with its success of achieving world leadership, but has increasingly extended outwards to help some of the world's poorest nations make gains in ICT knowledge, skills, and government initiatives. The case falls into the Stage One (First Adopters) in the Rueda-Sabater and Garrity model, but with the added concerns and actions to address internal and external digital divide gaps.

2.6 Conclusion

The history of the digital divide reveals rapid changes which are uneven among nations and within countries. The historical milestones caution that this history largely applies to the last two decades and that some technologies are still new and emerging. There are organizations that have recognized and worked on making the digital divide and its issues widely known.

The stages of the digital divide have a basis in the early work on adoption-diffusion of innovations (Rogers 2003) that is elaborated on in the next chapter. Stages of the divide for world regions and nations were formalized into 3-stage model (Rueda-Sabater and Garrity 2011). Another aspect of the longitudinal progression of the digital divide is that, although starting out with great variability among nations, the variation has reduced considerably over 11–13 year periods, yet substantial variation remains between nations. That remaining variation implies that the digital divide will not disappear soon, but is likely to continue to be reduced over the next several decades.

Metrics of actual behavioral and social improvements from use of ICTs reveal a divide that will continue even if technology availability is more standard. For instance, although television is fairly evenly available worldwide, differences remain in the behavioral benefits of television. The worldwide study framework of this book presently lacks the data to include such behavioral and social measures, especially at the sub-national level.

Azerbaijan represents a developing nation that might ordinarily fit into a trajectory of leapfrogging over more traditional technologies such as personal computers and fixed broadband, but instead it developed its traditional technologies as well as sprinting ahead with newer ones. This was achieved through government's strong and consistent focus on ICT as a primary priority and by an abundant petroleum base providing for citizen prosperity as well as government resources and budget.

Korea fits into the highest stage of ICT deployment, a leadership position achieved by careful government planning, implementation, and metrics. A remarkable aspect is that achieving high ICT national status was not enough, but more recently South Korea has emphasized extending its success to help multiple impoverished, developing nations worldwide close their divides.

References

Aliguliyev, R., & Gurbanov, G. (2012). Big ambitions in rapidly changing world: Azerbaijan, chapter 2.1. In S. Dutta & B. Bilbao-Osorio (Eds.), *The global information technology report 2012* (pp. 149–159). Geneva, Switzerland: World Economic Forum.

Choudrie, J. (2014). Lessons learnt from South Korea's LTE broadband development. In J. Choudrie & C. Middleton (Eds.), *Management of broadband technology innovation* (pp. 62–78). New York: Routledge.

Choudrie, J., & Lee, H. (2004). Broadband development in South Korea: Institutional and cultural factors. *European Journal of Information Systems, 13*, 103–114.

Choudrie, J., & Middleton, C. (2014). Introduction. In J. Choudrie & C. Middleton (Eds.), *Management of broadband technology innovation* (pp. 1–22). New York: Routledge.

Choudrie, J., Papazafeiropoulou, A., & Lee, H. (2003). A web of stakeholders and strategies: a case of broadband diffusion in South Korea. *Journal of Information Technology, 18*, 281–290.

CIA. (2014). *The world factbook*. Washington, D.C.: Central Intelligence Agency.

Curtis, A. (2013). *A brief history of social media. Pembroke, North Carolina, University of North Carolina at Pembroke.* Available on September 21, 2014, at http://www2.uncp.edu/home/acurtis/NewMedia/SocialMedia/SocialMediaHistory.html

Dutta, S., & Bilbao-Osorio, B. (2012). *The global information technology report 2012*. Geneva, Switzerland: World Economic Forum.

Dutta, S., Mia, I., & Geiger, T. (2011). The networked readiness index 2010–2011: Celebrating 10 years of assessing networked readiness. In S. Dutta & I. Mia (Eds.), *The global information technology report 2010–2011* (pp. 3–32). Geneva, Switzerland: World Economic Forum.

Internet World Stats. (2014). *Internet world stats*. Available September 18, 2014 at internetworldstats.com.

ITU. (2013). *Measuring the information society 2013*. Geneva, Switzerland: International Telecommunication Union.

James, J. (2013). The diffusion of IT in the historical context of innovations from developed countries. *Social Indicators Research, 111*, 175–184.

Makili-Aliyev. (2013). *Azerbaijan's foreign policy: Between East and West*. IAI Working Paper 13. Rome, Italy: Instituto Affari Internazionali.

Pepper, R., & Garrity, J. (2013). Convergent objectives, divergent strategies: a taxonomy of national broadband and ICT Plans, chapter 1.3. In B. Bilbao-Osorio, S. Dutta, & B. Lanvin (Eds.), *The global information technology report 2013* (pp. 43–52), Geneva, Switzerland, World Economic Forum.

Population Reference Bureau. (2013). *World population data sheet*. Washington, D.C.: Population Reference Bureau.

Republic of Azerbaijan. (2012). *Annual Report 2012 on Ministry of Communications and Information Technologies*. Baku, Azerbaijan: Ministry of Communications and Information Technologies, Republic of Azerbaijan.

Republic of Azerbaijan. (2014). *History of communication in Azerbaijan*. Baku, Azerbaijan: Republic of Azerbaijan. Available on September 17, 2014 at www.mincom.gov.az

Republic of Korea. (2012). *2012 Yearbook of information society statistics*. Seoul, Korea: Ministry of Public Administration and Security.

Republic of Korea. (2014). *National Framework plan on informatization promotion*. Seoul, Korea: Republic of Korea.

Rogers, E. (2003). *Diffusion of innovations*. New York: Free Press.

Rueda-Sabater, E., & Garrity, J. (2011). The emerging internet economy: Looking a decade ahead, chapter 1.2. In S. Dutta & I. Mia (Eds.), *The global information technology report 2010–2011* (pp. 33–45). Geneva, Switzerland: World Economic Forum.

Statista. (2014). *Leading social networks as of June 2014, ranked by number of active users (in millions)*. New York, NY: Statista Inc. Available on September 21, 2014 at http://www.statista.com/statistics/272014/global-social-networks-ranked-by-number-of-users

U.S. Bureau of the Census. (2013). *Computer and internet use in the United States*. Report pp. 20–569. May. Washington, D.C.: United States Bureau of the Census.

Wong, Y. C., Law, C. K., Fung, J. Y. C., & Lee, V. W. P. (2010). Digital divide and social inclusion: Policy challenge for social development in Hong Kong and South Korea. *Journal of Asian Public Policy, 3*(1), 37–52.

World Bank. (2002). *World development indicators*. Washington, D.C.: The World Bank.

World Bank. (2008). *World development indicators*. Washington, D.C.: The World Bank.

World Bank. (2012). *World development indicators*. Washington, D.C.: The World Bank.

World Bank. (2013). *World development indicators*. Washington, D.C.: The World Bank.

World Bank. (2014). *World development indicators*. Online tables. Washington, D.C.: The World Bank. Available on 10–20, 2014 at http://data.worldbank.org/products/wdi

Woyke, E. *Korea bridges digital divide. Forbes*. April 3. Available online on 10–12, 2014 at www.forbes.com

Chapter 3
Theoretical Model

3.1 Introduction

The concept of digital divide has been present since the 1980s and gained impetus in the 1990s with the advent of widespread web use. A milestone in recognizing it was the publication of a series of reports from the National Telecommunications and Information Administration (NTIA) of the U.S. Department of Commerce (NTIA 1995, 1998). They summarized and explained national representative surveys that pointed to the growth of ICTs and to variation in use of ICTs by age, education, income, and race/ethnic categories. The reports indicated for example that young, affluent, male and white or Asian persons tended to have higher rates of technology usage.

This chapter first examines the theoretical background and prior conceptual frameworks for the digital divide and then proceeds to elaborate on the book's theoretical model. Major prior theories examined are Adoption Diffusion Theory (ADT), Network Readiness Index (NRI), J.A.G.M. Van Dijk Theory (named for its originator J.A.G.M. Van Dijk), and a digital divide theory we developed in 2011 based on structural equation modeling (termed SEM-DD Theory). Although other conceptual frameworks have been put forth, we focus in these four because each bears in different respects on this volume's theoretical approach. Also these theories can inform some of the explanations of effects which appear in this volume's empirical chapters. In discussing each prior theory, strengths and weaknesses are identified, and explanation has been provided on why that theory could not be selected for the present worldwide digital divide analysis.

The chapter then turns to our proposed theoretical model for this book. That general model is applied for the world in Chap. 4, and in more detail for specific nations and for the continent of Africa in Chaps. 5–9, each chapter having a customized version of the general conceptual model. The reason we customize the models in those chapters is that the variables available are somewhat different at the state/provincial levels from nation to nation, while the African nations have a more

J.B. Pick and A. Sarkar, *The Global Digital Divides*, Progress in IS,
DOI 10.1007/978-3-662-46602-5_3

limited set of variables and a unique context of mostly low yet rapidly growing ICT use. Inconsistencies in worldwide sub-national data have been studied in other contexts such as defining urbanized areas (Forstall et al. 2009) and point to lack of worldwide standardization in defining variables and geographical units among national statistical and census agencies.

3.2 Prior Theories of the Digital Divide

3.2.1 Adoption Diffusion Theory

Adoption-Diffusion Theory (ADT) concerns how an innovation is adopted and diffused to members of a society or community over time (Rogers 2003). Rogers' concept of innovation was not limited to information and communications technologies, but included innovations as diverse as those in medicine, healthcare, management, and even freeway lane design. Although consolidated by Rogers in 2003, many parts of the theory were developed from the 1950s onwards. This section discusses the consolidated theory of Rogers (2003) and successive development of it, in the context of ICTs.

In ADT, innovations are the focus. Equipment, software, or services are innovative in the relative context in which they are introduced. For instance, Facebook as an IT service presently would be innovative in central Africa but not so in California or Hong Kong. According to ADT, innovations can be understood by their characteristics of relative advantage, compatibility, complexity, trialability, and observability.

Relative advantage refers to the improvement of an innovation compared to what served the function previously. The advantage can be technological, economic, social, or even perceived status. It also relates to the profile of the adopter. Most adopters would find a 3G mobile phone a relative advantage over a simple cell phone, but an illiterate person in say Africa might find 3G very difficult to utilize, hence not an advantage. Pricing can influence relative advantage, as improved features might be more than offset by higher price. The general conceptual model in this book does not use relative advantage because the governmental data utilized do not include variables for an individual's comparison of one technology with another. Relative advantage can however be useful in interpreting mechanisms of effects.

Compatibility represents the extent that an innovation is compatible with "existing values, past experiences, and needs of potential adopters" (Rogers 2003). For an individual, compatibility relates both to the broad environment of the person, as well as his/her job or personal needs. These types of compatibilities can be altered by the individual's education and career progression. Regarding the present model, compatibility is not directly available as a variable, but certain model variables can be explained through compatibility, for example education and/or age can connote extent of compatibility.

Complexity represents how difficult the innovation is to be understood and how challenging it is to use (Rogers 2003). Complexity is often greater in the initial adoption stage, but moderates as users become more familiar with it and as vendors tend to simplify its use. For example, some word processing software interfaces have become more friendly and simpler over time. Simplicity of use is a key success factor for rural telecenter users in India (Pick et al. 2014), and for cell phone users in central Africa, while advanced ICT users in Beijing, Tokyo, or Los Angeles are better able to handle complexity. For this research, complexity of a technology can be helpful in explaining findings and mechanisms, in varied contexts.

Trialability refers to extent an innovation can be piloted or prototyped, incurring limited effort and expense. Trialability is more important for early adopters than for late ones. For ICTs in the present study, trialability applies to new innovations in a given setting such as Facebook in Africa, rather than for diffusion of well-proven, older technologies, such as the fixed telephone.

Finally, *observability* is the extent to which an innovation is visible to people. Possession of the latest version of the iPhone is highly visible, whereas number of web pages utilized by a person is not. Another example is cell phones for youths in Japan. The young people exert special effort to have wearable cell phones, with special decorations so they are very visible, a behavior pattern referred to as "keitai." In the present research, the question of visibility refers to technology use in nations or states/provinces that are more visible. For example, there is visible technology use in parts of California, such as Silicon Valley, and in Delhi in India. Observability is important if the visibility of technology in one state or province leads to enhanced adoption and diffusing in neighboring states/provinces.

In ADT, an individual uses these perceived attributes to evaluate the pluses and minuses of the innovation, which leads in turn to the decision on whether or not to adopt it. In the process of learning about the innovation, communication channels are crucial in circulating the information about it. Having open communication channels encourages the pace of adoption and diffusion. Communication-channel factors are not available systematically from government sources as variables for the present model, but the communication and decision-making processes inform some of the examples and case studies. For instance, in an Egyptian example (Warschauer 2003), the failure in reducing the digital divide in the context of public schools can be explained by lack of communication channels with the users and by mistakes in over-weighting the attributes of the innovation for decision-making.

Another key component in ADT is the nature of behavior patterns in the social and economic system. For example, extent of use of telecenters by farmers in India is influenced by the behavior of the user support staff at the telecenter and by that of other farmers (Pick et al. 2014). In the simple setting, positive experiences by the other farmers and by cordial, motivated support staff encourage adoption. For an entire state or nation, the cumulative behavior patterns of its population represent this theoretical feature. These behaviors can apply nationwide, for example, a country's government can impose national K-12 education standards to become competent in the use of broadband. However, behavioral arguments should be used cautiously for larger geographic units, since behavior patterns, say for Facebook use in a province, might be quite varied and not generalizable.

ADT also designates categories of adopters, namely innovator, early adopter, middle adopter, late adopter, and laggard (Rogers 2003). As was seen in Fig. 2.6, these types are located at typical points of the adoption curve (Rogers 2003). Over time, adopters transition from early adopters to middle ones to late ones, with some becoming laggards. In the late stages, the rate of adoption slows down, stops, and even reverses, as seen in the cumulative adoption curves in Fig. 2.7. For most analyses in this book, a cross-section of adopter categories is represented at a single time point, rather than following them over time. However, the longitudinal perspective is useful in interpreting the progression of individuals' adoption, access, and use over time.

ADT can be modified to form a basis for understanding organizational stages in the ICT innovation process. Here simple decision making gives way to complex decisions by a hierarchy of management personnel. A sequence of group meetings, site visits, initiatives within the organization, and collaborations lead up to a decision. In the present research model, the organization is not included as a unit of analysis.

Adoption-diffusion innovation theory also is informed by a limited line of research papers concerning the geography of adoption and diffusion (Rogers 2003). As an innovation is undergoes cumulative adoption and diffusion which might follow an S-curve or other trajectory of growth over time, it would also have physical dispersion over space, since adopters are located at different geographic points from each other. Even a virtual product or service would have a geographical dispersion, with locations identified as the physical addresses of adopters. Geography can influence the spatial adoption and diffusion pattern over time through neighborhood effects, i.e. the influence of one neighbor on another; by physical geography i.e. rivers, mountains; by manmade barriers such as freeways, walls, fences, or large restricted properties; and by transportation linkages.

Early ADT geographic studies in Sweden (Hagerstrand 1952, 1953) were based on computer simulations to examine the influences of spatial distance on diffusion. Adoption probability decreased in proportion to distance from the prior adopter. Testing was done on adoption-diffusion of an agricultural innovation is a rural area, and the simulation results were compared to the actual adoption-diffusion pattern (Hagerstrand 1952, 1953).

Another early study (Whyte 1954) analyzed the adoption of window air conditioners in clusters of contiguous houses in a Philadelphia suburb. Whyte was able to identify the clusters through aerial photos, which showed the air conditioners extending out of windows, and by subsequent personal interviews on the ground. Among the findings was that although adopters did not imitate exactly the air conditioner brands of their neighbors, the brand name of air conditioners had major overall influence on whether or not a resident of the neighborhood adopted the innovation.

Another investigation considered the spatial diffusion of adopters of a loyalty program issued by a retail market, a study combining GIS with marketing theory (Allaway et al. 2003). Data from cardholders of a loyalty program were employed

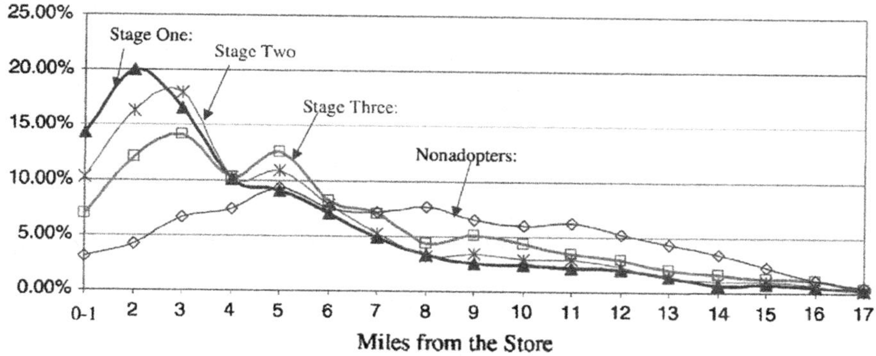

Fig. 3.1 Percent adopters of loyalty cards, by miles from central store. *Source* Allaway et al. (2003)

to follow over time customers' purchases from check-out scanners, which were linked to the customers' residential locations. At day one, there was heavy promotion of the loyalty innovation, including in-store promotions, advertising, and billboards. Adoption and neighbor influence were studied in very small areas of only 0.06 mile radius and containing 12–75 households. 18,000 respondents were allotted into Rogers' categories of Early Adopter, Middle Adopter, etc. Next, adopter categories were assigned distances from the store, for example Late Majority and Laggard categories peaked at 5 miles from the store (See Fig. 3.1).

Geographic influence was simulated as an event history model. There were effects noted for store distance, placement and presence of billboards, and neighborhood effects, all cross classified by Roger's categories. Innovation patterns were studied by examining the locations of first adopters in very small neighborhood "diffusion cells" and were found to constitute the critical "driver" for middle and late adopters to follow (see Fig. 3.2).

Although perhaps the most well-known and accepted theory for digital divide, ADT has shortcomings that have limited its selection for digital divide research. There has been limited testing of the theory for the digital divide, mainly fitting data to the S-curves of information technologies predicted by ADT using panel data for 14 years (Andrés et al. 2010; Martins and Andrés 2010). This points to a constraint that ADT applied to the digital divide depends on consistent longitudinal data, whereas most prior digital divide research has been limited to cross sectional data. Second, there is no consensus of what constitutes an innovation to be eligible for the theory. For instance, does broadband constitute an innovation for the U.S.? In Africa? What about Facebook services as innovation? Third, its emphasis on S-curves has been challenged (Van Dijk 2005). Do innovations actually follow S-curves? If the S-curve applies to a targeted sample, does it also apply to the whole population? What about innovations that fizzle out early, say after only fifty percent of early adoptions have occurred? A fourth constraint is that there is little in ADT that speaks to the economic, cultural, and political characteristics of adopters and

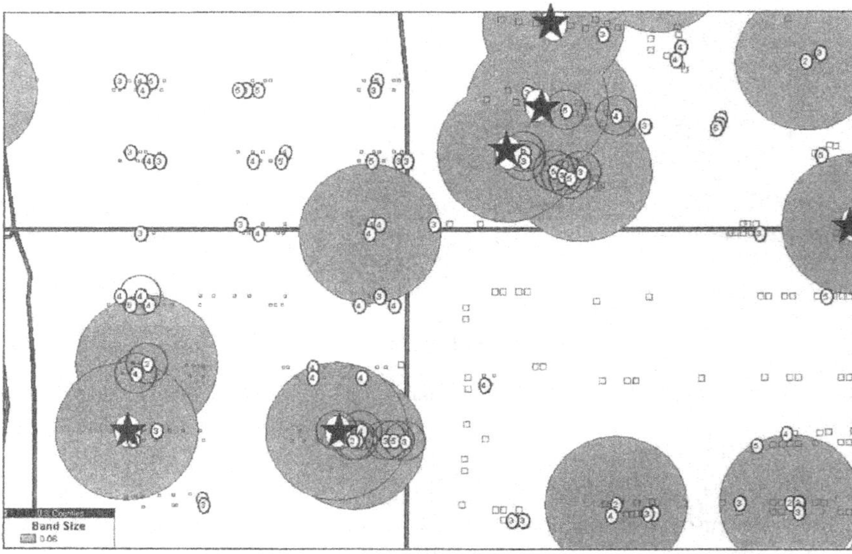

Fig. 3.2 Early innovator cells with surrounding tiny neighborhoods. *Source* Allaway et al. (2003)

non-adopters, as well as their geographic aspects. ADT has limited use in this book, because quantitative governmental data at the state and provincial levels rarely contain ADT longitudinal variables, so full ADT is not appropriate. However, its insights can be beneficial in interpreting case studies and in speculating on mechanisms for achieving higher adoption-diffusion levels, and also for interpreting changes in adoption levels for China and India in Chaps. 5 and 6, which as exceptions have data for two time points.

3.2.2 Van Dijk's Theory

Another alternative theoretical model for the digital divide is based on personal and positional characteristics, and properties of technology, which stimulate resources both material and behavioral, which lead to access of ICT (Van Dijk 2003, 2005, 2012). Access in turn impacts participation in the society, the economy, social networks, culture, politics, and institutions. Participation feeds back to influence positional characteristics forming a causal loop. The entire model (seen in Fig. 3.3), emphasizes inequalities at every major step, i.e. in positional categories, personal categories, resources, access, and participation in society. For persons in higher personal and positional categories, there is a positive, reinforcing causation from step to step, and into a positive feedback loop. For persons of lower status, the causation reinforces the low end for each of the steps and ultimately in a negative feedback loop.

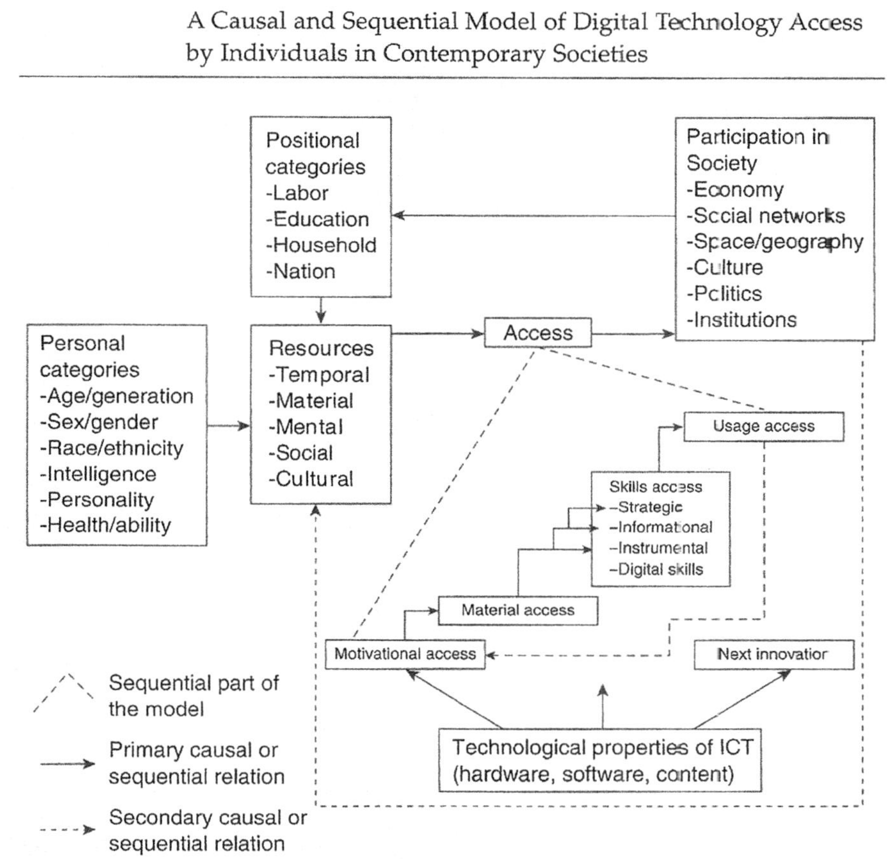

Fig. 3.3 Van Dijk's full model. *Source* Van Dijk (2005). *Note* ICT indicates information and communication technology

This theory is more elaborate than ADT about the social, behavioral, cultural, and economic factors in the digital divide. Within the broad perspective of inequalities, it provides a deeper theoretical accounting of causation than ADT. Its limited adoption among researchers partly stems from its complexity and the challenge of finding suitable empirical data.

Each step is important to the theory. *Personal categories* include demographic, social, race, ethnicity, and behavioral attributes of an individual, and the person's category level impacts his/her resources for ICT. For instance, an older person in weak health has reduced resources available for ICT. *Positional categories* relate to a spectrum of social positions that the individual holds, for instance a highly educated, British scientist, or a moderately educated Peruvian mother and part-time retail worker.

Together, personal and positional categories influence the inequalities in resources (Van Dijk 2005). Positional categories can impact resources for ICT through social exclusion, exploitation, and control (Van Dijk 2005). These are all ways for a more powerful person to eliminate resources or take them away from a weaker person.

Properties of technology (hardware, software, services, content) first of all determine if the technology is actually an innovation, and if so whether they motivate persons to access them. Facebook is considered an innovation within many lower income nations. If Facebook's innovative properties become well known in the nation, it would motivate non-users to gain access. For instance, in a poor country which tends to isolate certain people by geographical barriers, knowing about Facebook's ease in friending motivates people to gain access. Simplicity of properties is beneficial, as long as the user get what he/she seeks. By contrast, often very complex ICT properties can inhibit use by its complexity or information overload (Van Dijk 2005).

Resources to gain access to technology include availability of time for ICT, income, wealth, social capital, social networks, mental attributes, technical skills, social abilities, and knowledge. Cultural resources are advantageous for a particular culture, such as extensive guanxi (contacts) in China, cooperative society membership in India, or the town elder in an African village.

Access to ICT is a four-step process of (a) motivation to access, (b) material access, (c) skills access and (d) usage access (Van Dijk 2005). Success is not granted in today's world if the adopter only achieves the first two steps. For instance, an individual is motivated to use a 3G cellphone and is able to afford it. However that is insufficient to successfully use it, if a person hasn't learned the needed skills, or if the person has a "usage gap," i.e. underutilization of the innovation. For example, the person might squander access to broadband by using it solely for gaming with no personal or societal benefits. The example of implementing technology in Egyptian schools (Warschauer 2003) mentioned earlier in this chapter further illustrates the deficit of not moving into steps (c) and (d). Unfortunately, in the present quantitative research and preponderance of prior studies, data on (a) and (b) are readily available from public sources, but (c) and (d) are largely not present, requiring challenging and often expensive customized surveys.

Participation in Society is seen by users who get "connected," improve "connectivity," or experience transformative information, ideas and decision-making through greater access. On the other hand, exclusion from ICT participation can severely hamper a person's personal and occupational prospects in most parts of the globe.

What are the benefits and shortcomings of Van Dijk's theory. Among the pluses are that it adds details of a more elaborate theory than other theories. It spends time carefully defining concepts; it has dynamic qualities attuned to the rapidly expanding and permeating Internet and other virtual phenomena; and as a qualitative theory it can delve more deeply into individual and small group motivations and behaviors (Van Dijk 2005). On the other hand, Van Dijk's theory has the shortcomings of switching between units of analysis, only slightly including geography, and not recognizing the important digital divide constructs of societal openness and social capital. It is very complex with about 30 sub-entities, and so far has had limited empirical testing.

The Van Dijk model has some overlaps with the present one by inclusions of positional and personal attributes and in its two lower level sub-categories of access, namely motivational and material. It differs in its inclusion of its two higher level sub-categories of skills and usage access, and in including participation impacts on society as an outcome. The Van Dijk model also has a more complex set of variables and pathways than the present one. The difference is important, since the Van Dijk model does "deepen" digital divide by going beyond the outcome in most of the digital divide literature of ICT access, use, utilization, and prevalence. It goes to the next step and seeks to understand the impacts on society after gaining greater ICT access and utilization. However, even though deeply theorized, it is presently an empirical challenge to quantitatively measure, using governmental data, the model's societal impacts of access/utilization for nations and states/provinces. Although societal impact on a country can be measured and simple correlations done, the problem is to isolate what portion of the impact comes from ICT access/utilization. Van Dijk applies qualitative methods to probe this relationship at the individual or small group level, whereas the present book's simpler theoretical model stays within the practical bounds of what can be tested quantitatively with government and nonprofit data.

3.2.3 Network Readiness Index

The Network Readiness Index (NRI), introduced in 2000–2001 under the support of the World Economic Forum and INSEAD, intended to examine the relationship of technological innovations on national development and productivity, in order to gain deeper insights into competitiveness (Dutta et al. 2011). A broader intent was to enable policymakers in a country to assess its global comparative strengths and weaknesses in network readiness. In its initial form, the NRI was an exploration trying to address the complex factors and relationships that enable nations improve ICT readiness and competitiveness.

During the decade and a quarter of the NRI, ICT has expanded to almost universal presence worldwide, so the NRI's creators transitioned the Index from measuring access alone to "how to make the best use of ICT in order to improve business innovation, governance, citizens' political participation, and social cohesion" (Dutta et al. 2011). By 2012, the broadened framework of the NRI included the drivers of *ICT readiness* and *ICT usage*, which cause *economic impacts* and *social impacts* (see Fig. 3.4) (Dutta and Bilbao-Osorio 2012). The new NRI adds to the framework the outcome of economic and social impacts rather than strictly usage, which is now classified as part of the "environment."

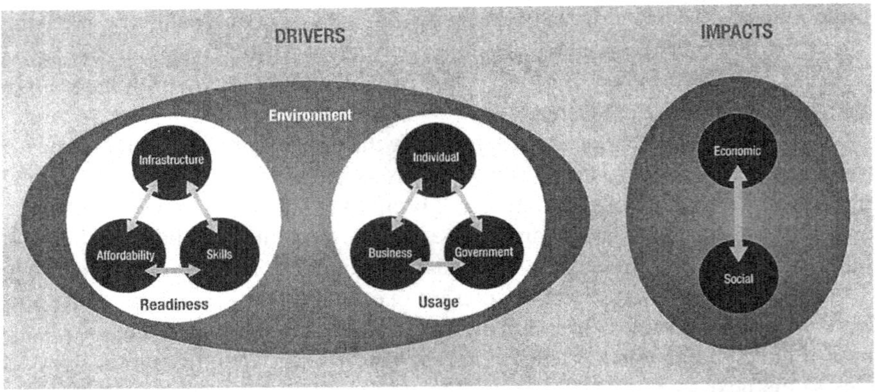

Fig. 3.4 Evolved networked readiness index. *Source* Dutta and Bilbao-Osorio (2012)

Besides adding the two impacts, NRI 2012 is designed to take into account "hyperconnectivity," defined a complex of strong relationships between people, consumers, organizations, citizens, and the state (Dutta and Bilbao-Osorio 2012). Hyperconnectivity is reflected in vast worldwide ICT connectivity, such as over 800 million active Facebook users in 2011, more than one zettabyte of data transmitted globally in 2010 (a zettabyte equals one trillion gigabytes), and widespread prevalence of large monitor screens. Hyperconnectivity has stimulated new governmental and commercial services and products and altered the way billions of people interact and communicate with each other.

With the changes, the NRI now is expressed in four sub-indexes: *readiness* (infrastructure, digital content, affordability, and skills), *environment* (political and regulatory environment and business and innovation environment), *usage* (individual, business, and government), and *impact* (economic and social) (see Fig. 3.5). Each sub-index has multiple underlying variables, for instance the readiness subindex is based on 12 variables (Dutta and Bilbao-Osorio 2012). The breadth of NRI 2012 is evident in its inclusion of 29 indicators from a customized World Economic Forum Executive Opinion Survey of individuals and 28 indicators at the national level from mostly government sources. Since 47 % of this survey's indicators is tailored for the NRI, country-level empirical data are guaranteed for half of the index.

Although the modernized NRI does provide a refined measurement of Network Readiness, it also has several drawbacks. The index itself is a single measure of readiness very widely based on 4 sub-indexes and 57 indicators, all of which are available as data. The role of the NRI itself is to serve as a summary measure that can be used for classification and comparison between nations. Relationships among all these components are not explained in the framework, so the researcher is left to evaluate them by correlation, multivariate, and other techniques, drawing on the NRI base data combined with other data sources. Second, the weighting of the variables in the NRI and its sub-indexes is not made known sufficiently for the user

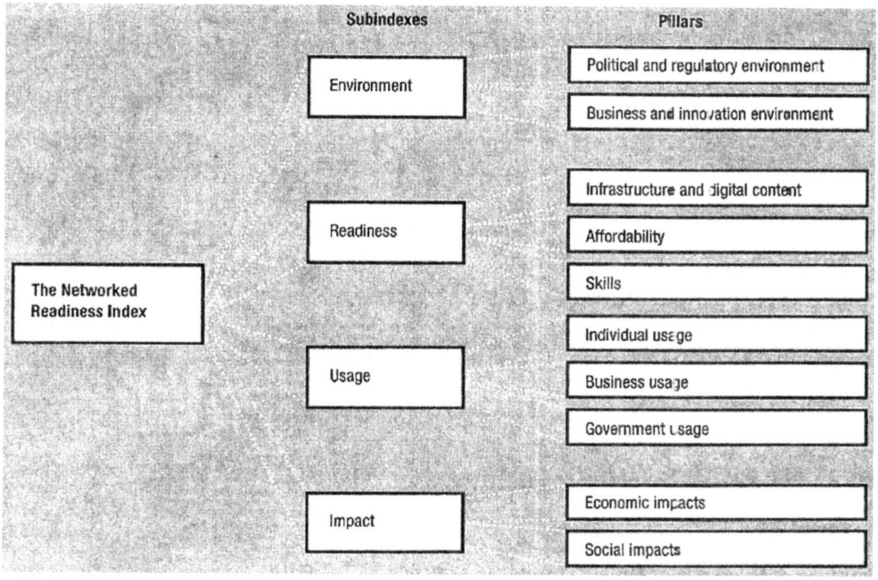

Fig. 3.5 Evolved networked readiness index structure. *Source* (Dutta et al. 2012)

of NRI, even though the NRI can be profoundly altered by different weighting schemes.

An advantage is that the earlier components in the NRI offer up to 10 years of longitudinal data, but some of the current 57 variables do not go back 5 or more years.

Why is the NRI not utilized widely in digital divide research? First of all, it does not inform many quantitative research studies because it does not posit associations between independent and dependent variables, nor the relative importance of variables. Further, it gives little attention to geographical analysis outside of displaying NRI levels by countries. Another drawback is that half of the NRI variables are collected by a special Executive Opinion Survey, which does not extend to the sub-national level of states and provinces. Although a researcher could perform his/her own sub-national survey utilizing the same survey design, it would be expensive and challenging. Some also have argued that the NRI is less suitable for developing nations, since its regulatory and governance factors are not as applicable to those countries as the IDI index from the International Telecommunications Union.

The benefit of the NRI for the present project is that its consistent variables are available for a large number of nations (up to 142 in 2012, although with some data missing). Researchers have an invaluable source to analyze digital divide across national samples, including some of the same variables utilized in this book for the world, Europe, Asia, and Africa-Latin America. On the other hand, the theoretical contributions to quantitative empirical digital divide research are limited, and mainly concern the formulation of components of ICT factors. However, NRI can inform case studies, such as the evolution of ICT in Estonia (Dutta 2007), which appears later in the chapter.

3.2.4 Structured Equation Modeling Theory

Structured equation modeling (SEM) is a methodology that frees the researcher to develop and test theories with more complex pathways and interactions than the often simple theories tested by linear multivariate models such as regression. A SEM theoretical model for the digital divide was induced from the digital divide literature (Pick and Azari 2011). The study's approach is exploratory social science, appropriate for research for which well accepted theories presently do not exist or are in tentative early form (Stebbins 2001). The researcher in such circumstance can induce factors and relationships from prior literature and combine them through reasoning into a coherent theoretical model.

For this SEM-DD Model, four constructs were induced, as well as their relationships (Pick and Azari 2011). As seen in Fig. 3.6, the constructs are business and technology investment, government support of IT/legal framework/social openness, socioeconomic level, and technology utilization (Pick and Azari 2011). Each of these factors is briefly discussed.

Business and Technology Investment

The factor consists of a blend of determinants of technology such as technology readiness, university-industry interaction, technology infrastructure, R&D, financial markets, financial investment, and foreign direct investment (FDI). All this factor's underlying variables were induced from prior research studies. For instance, FDI was included due to its strong relationship with technology use in studies of

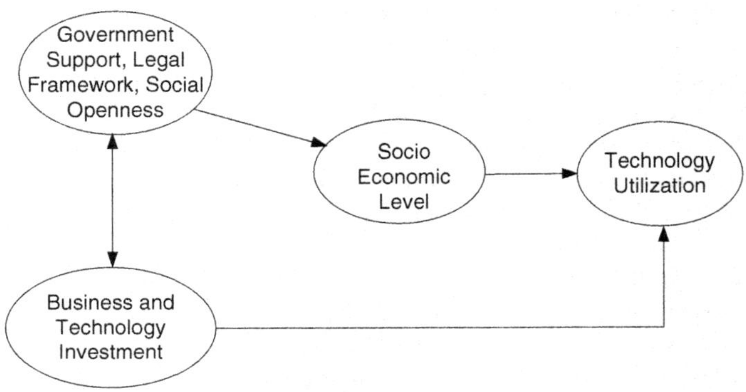

Fig. 3.6 Structural equation model for worldwide digital divide. *Source* Pick and Azari (2011)

developing countries (Baliamoune-Lutz 2003; Pick and Azari 2008). Altogether, this construct represents the business, financial, and infrastructure base to support technological development.

Government Support of IT, Legal Framework, and Social Openness (GSLO)

Some literature studies posited that the strength of a nation's legal and regulatory framework, societal openness, and government support of IT are related to ICT utilization. It is reasonable that a society that encouraging free flow of information is one in which information diffusion from the Internet would be welcome. By contrast, in an authoritarian society, free flow of information could threaten the government regime and might be regulated, censored, and/or suppressed. China somewhat typifies the latter instance; it is a mid-level nation that suppresses much information that is disdainful of the communist leadership. Government not only can suppress ICTs, but it can also stimulate ICT use through government incentives, programs, and investments. Overall, this factor concerns the position of government to try to grow or diminish ICT use in a state or nation.

Socioeconomic Level

The digital divide literature is full of examples of the influences of education and income on technology utilization. Likewise, the literature concurs that a more elderly population relates to reduced technology use. Another attribute associated with technology use is math/science education since prior research pointed to the positive influence of a STEM (science, technology, engineering, and math) training and education. Altogether, this factor comprises income, education, and age structure, and emphasizes the personal preparedness of people to make use of technology.

Technology Utilization

This factor is a straightforward set of ICT attributes that have served as dependent variables in prior research, in particular internet users, secure servers, mobile phones, and PCs.

The theory also induced the relationships shown in Fig. 3.6 between the variables. Business and Technology Investment and GSLO have a two-way relationship, while Business and Technology Investment has a direct impact on Technology Utilization. There is also a two-step pathway of GSLO influencing Socioeconomic Level and in turn Technology Utilization.

In this example, the SEM-DD approach yielded a coherent model for nations based on substantial literature support. It was tested with a sample of 110 nations and for subsamples of 55 developed and 55 developing nations (Pick and Azari 2011). Overall results indicated Business and Technology Investment and GSLO were tightly related to each other, and a robust pathway exists of GSLO influencing socioeconomic level and in turn Technology Utilization.

Overall, the SEM theoretical approach has the advantages of being grounded in existing literature variables and relationships and of supporting more complex pathways of relationships. The induced theory makes practical sense. However, potential weaknesses are that variables chosen for factors might not have sufficient internal reliability and the factors might not be sufficiently differentiated from each other enough in correlations. For SEM, large sample size is required, recommended at 100 or more, so many samples of states, provinces, or nations would be precluded as too small. Lastly, a challenge is how to incorporate geography into an SEM model. It is difficult since errors in estimation by sample points are not available. Alternatively, mapping of SEM factors, although a simple approach, can help to understand exploratory patterns.

3.3 Estonia: a Case Study to Illustrate Prior Digital Divide Models

Estonia, a small Baltic country, was spun off from the Soviet Union at its dissolution in 1992. Since then Estonia has made enormous strides in democratization and economic gain, joining the European Union in 2004. In the Soviet era, Estonia stood out as among the most highly educated and industrially advanced Soviet satellites, with special strength in military technology.

Its notable president Mart Laar for its first decade of independence established technology as a national goal and provided the leadership and government support to move the nation forward latching onto the early cutting edge of e-government and e-society applications. A multi-pronged approach was used. Strong emphasis was placed on nationwide education in computers and the internet including the Tiger Leap program of 1997 that connected all Estonian schools to the internet, provided skills training, and aspired to achieve ICT use by all school children. Another public-private program, Look @ the World, was a joint effort by the federal administration and leading companies to finance, train, and encourage use by all adults (Dutta 2007).

The government stimulated internal governmental use after independence. Importantly its Parliament in 2000 approved a constitutional right for citizens to have internet access. It also started a central government web portal, converted to all-electronic cabinet meetings, and encouraged online citizen input, many years before these became commonplace. In the trajectory towards leadership as an e-society, the federal government collaborated and partnered with Estonian technology and financial firms, which gave much more impetus to the changes (Dutta 2007). One indicator of success is that Skype originally appeared as an Estonian company, although it ended up as a Microsoft division.

As Estonia's e-society gained more attention, Scandinavian firms invested in Estonia's ICT sector, helping to modernize it. Issues remain today including raising the national income, improving STEM education, building social capital for ICT across the country, stimulating high-tech clusters of innovation, and advancing the country's technology products and services to achieve an ICT export sector.

This saga of Estonia's growth, successes, and challenges fits partially with the four theories discussed, but is supported more fully by the Van Dijk and SEM-DD models.

Fit with Van Dijk theory. In Estonia, Van Dijk's position categories were important, particularly education and status in the nation. The position of educators was crucial to the transition, while national position was crucial especially since the highest figures in the national government pushed for ICT advances. Of the model's personal categories, only age was emphasized, since children received special governmental programs, as did adults. Clearly, the Van Dijk concept of resources was important in doing so much in a little over a decade. This included providing infrastructure, social capital, and mental/cultural focus on technology. The Van Dijk (2005) 4-steps of access were accomplished by Estonia, and this theoretical stage progression is useful. Finally, there were major impacts on society, the economic, politics and institutions. Hence, the Van Dijk theory is very relevant to the case study, but at the same time challenging to operationalize and test using Estonia's national or provincial data.

Fit with the example of SEM-DD theory. The four constructs in the example of SEM-DD theory are very pertinent to the Estonia story. Starting with *business and technology investment*, technological readiness was a critical success factor. The nation at independence was already technological ready, at least a standout among the former Soviet bloc countries. The independent government leadership stressed readiness with programs for K-12 education, adults, and businesses. University collaboration with industry was emphasized during the transition, while banks participated in public-private investment programs to stimulate a nascent high-tech industry. Foreign direct investment came later on from Scandinavia, which perceived an opportune landscape to invest in. The factor of *government support, legal framework and social openness* was also of critical importance. After independence Estonia converted rapidly to a full democracy, which included open press, and even internet as a constitutional right. Also, over the two decades since independence, the Estonian federal government has given high priority to ICT and e-society. In addition, *socio-economic level* was improved during this period, through proactive education programs and economic advances. Several issues that arose in the case and are not included in this theoretical model are R&D/innovation and development of social capital. Both are components of the book's theory covered in the next section.

The two other major theories, adoption-diffusion of innovations and networked readiness index are less relevant to the case. Although adoption diffusion theory would be appropriate if several highly visible innovations were present, the case did not single out the causes of utilization of a particular innovative technology. Also, the five innovation characteristics of relative advantage, etc. were largely not examined or explained in the case. One aspect of this case which relates to ADT is the significance of communication channels to the Estonian ICT success, channels such as national leaders communicating their vision of an e-society, communication between public and private stakeholders, and communication of Estonia's success in ICT to nearby Scandinavian nations. The measurement of the Networked Readiness Index and its four sub-indices of environment, readiness, usage, and impact is one way to benchmark Estonia's ICT progress, but it doesn't speak to mechanisms and impacts of one factor on another. Finally, both the case and the four theories address very little about the geography of Estonia's rapid progress.

3.4 Theoretical Model for the Book

The theoretical model for the book emphasizes the relationship between social, economic, governmental, and economic factors and technology utilization factors. As seen in Fig. 3.7, this relationship is screened for spatial autocorrelation. If spatial autocorrelation is present, it implies that the independent variables are not able to recognize the spatial bias present. A further spatial aspect of this model is that the technology utilization factors are combined for an exploratory examination of their spatial patterns. This is very useful in visualizing the areal differences in overall technology level throughout a country or state/province.

The model given in this chapter is a generic conceptual model for the entire book. Each of Chaps. 4–9 in turn provides a specific model that is a variation of the generic model. For instance, in the next chapter a specific variant on the generic model is given for the world as whole. In Chap. 5 on China, a conceptual model variant is given for China. In all the empirical chapters, similar sets of independent factors are influencing similar sets of dependent ones. As stated earlier in the chapter, the reason for the model variants is that the variables available from census and government statistical agencies differ at the state/provincial/prefectural level from nation to nation, so most variables are somewhat differently defined between countries, and others are unique to a nation. Unique variables include, for China, staff and workers in non-state-owned units and individual investment in fixed

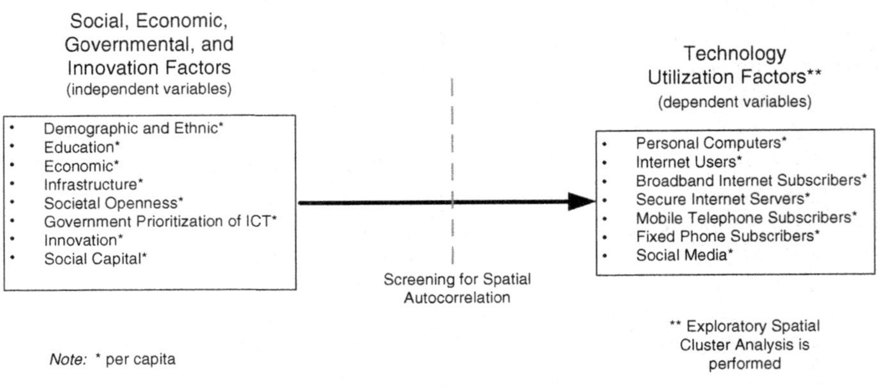

Fig. 3.7 Generic conceptual model for book

assets; for India, cooperative society membership; and for Japan, openness index for prefectures and IP phone expenditures for families. This lack of consistency is ultimately the result of lack of international standardization of state/provincial/prefectural variables and definitions.

If this book concerned only samples of nations, it would be possible to standardize all the variables because the data could be drawn from international agencies that have standard definitions worldwide, such as the UN, World Bank, World Economic Forum, and International Telecommunication Union. An alternative to the problem at the state/provincial/prefectural level would be to conduct a standardized customized survey for four nations each with 20–55 sub-national units, a survey beyond the scope of the book and requiring large staff and funding. Fortunately, the data that were used appear sound and of good quality, albeit not having worldwide standardization.

The generic model includes only variables that can be empirically tested at the state, provincial, and prefectural levels. Accordingly, several independent factors are omitted because there are not systematic variables available for states/provinces/prefectures including for example ICT governance factors, ICT affordability factors and income inequality. Likewise for the dependent factors, the distinction of separate variables for fixed broadband and mobile broadband is not included. Again the reason is lack of data at the state, provincial, and prefectural levels. It is hoped that in the future international agencies may collect systematic data for those omitted factors at the state/provincial/prefectural levels for multiple nations, enabling more factors to be added to an operational model, such as the generic one below.

When the conceptual model is tested with data, there are detailed processing steps seen in Fig. 3.8. This diagram on the left indicates the exploratory analysis of geographical patterns for technology utilization performed by two methods. As explained in Chap. 1, K-means cluster analysis measures the groupings of geographical units with the least distance between nations, based on a given k number of clusters, while Moran's Index tests to what extent, for a single variable, like units agglomerate together (positive value of the index) or are repelled from each other (negative value of the index). On the lower right of the diagram, the confirmatory analysis of the association of the independent variables with each dependent one. The difference between the regression-predicted dependent-variable values and the actual values determines the regression residuals.

The Moran's I test is applied to the regression residuals to determine if the model removes spatial bias and the residuals are random. If they are not random, then those particular model results must be treated cautiously, since the model has spatial bias. The residuals are also tested by standard OLS residual tests of Joint Wald, Koenker, and Jarque Bera Statistics.

Returning to the conceptual model, the justifications of each general factor are given along with some of the literature support, but the full literature support for specific variables is provided in Chaps. 4–9.

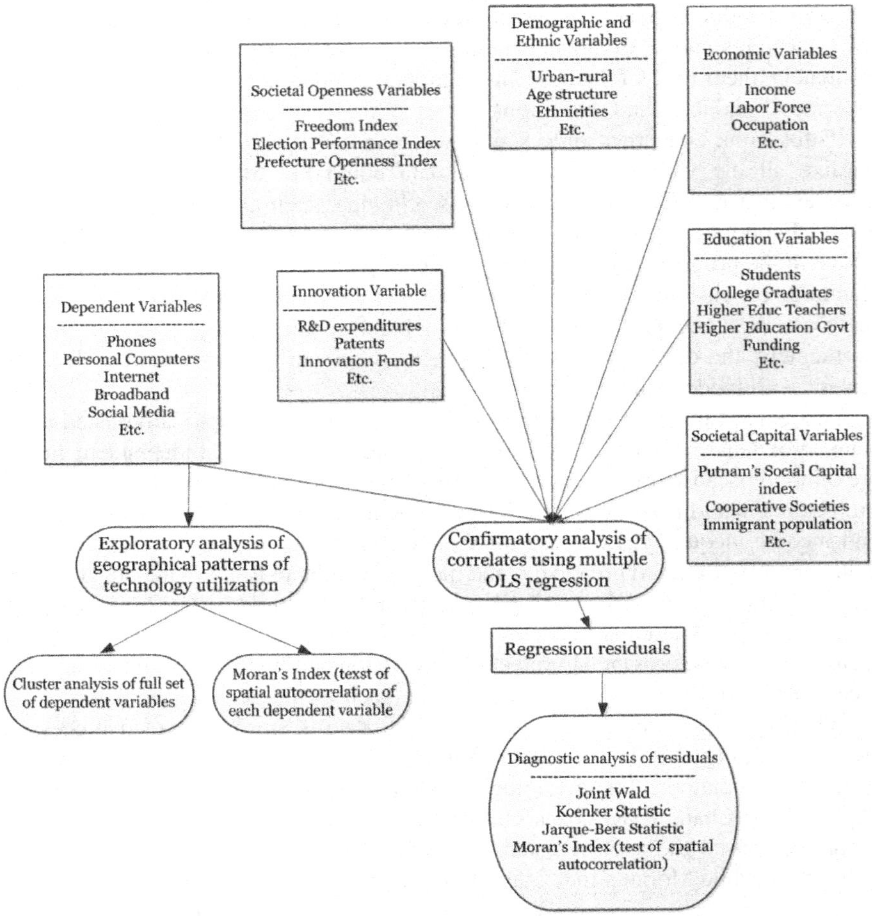

Fig. 3.8 Generic empirical testing of conceptual model

Independent Variables

Demographic and Ethnic Variables

Age structure, urban location, and ethnicities are well known in the digital divide literature to influence technology use. For nations worldwide, it is evident that children and young people are making use of computers, internet, cell phones, and social media to a greater extent than middle aged or older people. For instance, in Japan internet use decreases on the age continuum from youth to adult population (Ono and Zavodny 2005). Age and mobile phone use have become cultural in Japan, based on the concept of "keitai," which refers to a behavior pattern that includes the image of a young person, his/her attire, and expressions/behavior related to the use of cell phones (Ito et al. 2005). The mechanism may relate to ease

Fig. 3.9 Students learning and applying technologies, Midwestern University, USA

of learning at an early age. Since technology changes rapidly, older people are less able to adapt and learn well.

Urban location is also well-known to be a positive influence on ICT use. Again, in Japan urban versus rural areas tend to have improved infrastructure (Arai and Naganuma 2010) as well as higher incomes and education. The association with urban areas was also noted for China (Fong 2009) and the U.S. (Chen 2013).

Education Variables

Education is among the most well-known correlates of ICT use (Baliamoune-Lutz 2003; Yates et al. 2011; Chen 2013). There are a variety of possible mechanisms. People with more education may have been trained or educated better in technology use. In Fig. 3.9, students are learning and applying technologies. They also may seek knowledge more, with technologies as an avenue to access it. Occupationally, the more educated are more likely to have access to and need to use technologies. Lastly, education across many studies is consistently highly correlated with income. Hence a further indirect effect is that the more educated may find technology use more affordable.

Economic Variables

Economic variables are known to be important in the literature. Income has often been a correlate of ICT utilization (Baliamoune-Lutz 2003; Guillén and Suárez 2004; Lengsfeld 2011), although it is often excluded due to its usual strong correlation with education. Greater personal income makes technology more affordable, while higher national or provincial income connotes an economy having more ICT spending leading to uses by individuals, organizations, and institutions. FDI was an important determinant of Internet capacity in a study of 74 developed and developing nations (Robison and Crenshaw 2002) and a significant correlate in a study of technology availability and utilization correlates in 71 developing nations (Pick and Azari 2008), although in another study not significant for developing nations (Baliamoune-Lutz 2003). FDI has been crucial to China in developing an export-driven economy (Hale and Hale 2003; Gao and Yu 2010), especially since the 1990s in eastern coastal provinces, where the national government sought to grow the economy by linking into the global supply chain through assembly and manufacture. Other economic attributes correlated with greater technology are openness of international trade (Pick and Azari 2008), and export commodities (Pick et al. 2013). Export and trade are especially prominent in China and examined further in Chap. 5. Labor force variables included in prior studies have often lacked significance.

Infrastructure Variables

Infrastructure underlies the use of information and networking technologies. Hence, nations or states/provinces with strong infrastructure are likely to be advantaged in ICT use. An issue in the literature has been that ICT infrastructure indicators at national or state level are infrequently available. Hence infrastructure is sometimes measured indirectly by electrical production (Quibria et al. 2003), construction expenditure (Pick et al. 2013) or other underlying aspects of infrastructure.

Societal Openness Variables

Societal openness consists of indicators of freedom, democracy, open exchange of information, and well-established legal and regulatory systems. These attributes have been prominent as correlates of ICT utilization in multiple studies (Robison and Crenshaw 2002; Guillén and Suárez 2004; Yates et al. 2011). Since ICT use

Fig. 3.10 Open information exchange in a software development team. *Source* Lars Plougmann

depends on the amount and richness of content of information, open societies allow and encourage free flow that has grown rapidly based on the web, social media, locational data, the internet of things, as well as rapidly expanding use populations. In societies with laws and regulations regarding the information such as those in intellectual property and privacy, users' confidence is increased, stimulating even more use. The software team in Fig. 3.10 reflects the open exchange of information.

Government Prioritization of ICT

Government support was evident as a strong factor encouraging ICT use in the Estonia case study, as well as for Singapore in Chap. 1. It was true in the original development of the Internet in the U.S., for which ARPA in the federal government served as an incubator to develop the earliest internet applications for defense and research purposes. Government prioritization of ICT has been a positive factor the digital divide literature at the national level (Raven et al. 2008; Pick and Azari 2008, 2011). Its importance for states and provinces has been demonstrated through case studies, for example the strong furthering of ICT by the government of the State of Georgia in the U.S., as will be covered in Chap. 8. However, it is constrained for empirical work at the state and provincial levels, since systematic data for ICT prioritization are rarely available at those levels and not reported in the literature.

Innovation

Innovation has been significant in some digital divide studies. One explanation is that innovation in the present day is commonly accompanied and supported by technology. Sometimes innovation variables are entirely technological. A secondary reason is that a nation or state that has high innovation tends to encourage ICT use up and down its supply chain, and throughout the society. Innovation impacts ICT use, whether or not the innovation is technology driven. For instance in a study of ICT use worldwide, the dominant factor was per capita production of scientific and technical journal articles (Pick and Azari 2008). For China, studies consistently confirm the essential role of the Chinese government in stimulating technology availability, utilization, and innovation (Meng and Li 2002; Loo 2003; Song 2008; Gao and Yu 2010). The variable of published books serves as a proxy for innovation in China (Pick et al. 2013, and Chap. 7), since published books are beginning to appear as e-books and because the knowledge content generation from publishing is presumed to be linked to the innovation of web content generation. China's 15-year Science and Technology Plan of 2006-2021 emphasizes indigenous innovation (Gao and Yu 2010). Other studies show importance of innovation and R&D in Asia (Quibria et al. 2003), the U.S. (Azari and Pick 2005) and Japan in Chap. 7.

Social Capital

Social capital refers to the extent of linkages and ties among people in a population through social networks, physical networks, organizations that encourage human interactions and collaborations, bonding with people who have resources, and other

Fig. 3.11 Social capital: sandbagging event in Minnesota. *Source* James Postema

means (Putnam 2001; Rupasingha et al. 2006). In study of the U.S. digital divide, social capital was a leading factor that stimulated internet access and use (Chen 2013). In another study individuals' decisions to go online were impacted by internet use of peers and in communities with strong social interaction (Agarwal et al. 2009). In Chap. 6, cooperative societies in India are seen to influence uses of technologies. In the Indian context, a cooperative society serves to foster and connect firms together for cooperative work. A non-tech example of social capital, seen in Fig. 3.11, is a group of friends and neighbors interacting and working together at a sandbagging event in Minnesota.

Dependent Variables

A variety of information and computer technology variables are included, both traditional and newer ones. As dependent factors, the theoretical model does not average them together in an index, but leaves each technology factor separate. This is done because the use of each technology varies from nation to nation and within nations and needs to be put into cultural and geographic context. For instance, in northwestern Indian states, broadband has very low prevalence, so is not a major venue of use nor are its populations used to using it, while in New York or Tokyo it is highly prevalent with widespread use.

The most common variables used throughout the book are personal computers, internet users, broadband, internet servers, mobile phones, fixed phones, and social media. All of them have been included in multiple studies in the digital divide literature, with the exception of social media, which is only beginning to be studied. For specific chapters some dependent variables unique to countries are included, such as per capita number of domain names for China, and per capita IP phone expenditures per family for Japan.

3.5 Conclusion

Theories of the digital divide are varied. Four prior theories are highlighted in the chapter, namely Adoption-Diffusion Theory, Van Dijk's Theory, Networked Readiness Index, and SEM-DD Theory. Each provides different insights into the digital divide, how it has been formed, what its components are, its processes and mechanisms, and what factors influence higher or lower positions on the divide. Methodology to test the theories has varied, ranging from case studies and surveys to evaluation by quantitative models. Each theory is somewhat circumscribed by the available data, for instance quantitative models require robust and systematic data collection usually carried out by government or international agencies. Agency data provide rich and fairly complete data-sets for research, but at the same time limit the scope of digital divide factors that can be included. These data vary considerably in worldwide availability at the state/provincial/prefectural levels.

The case study of Estonia highlights a success story in reducing the digital divide for a post-Soviet nation in the decades after independence. Some critical factors were a starting base of a well educated population, a history of technology-related industry, and national leaders post-independence who viewed internet access as a constitutional right and proactively started creative programs, sometimes jointly with industry, to foster and build up both a technological and human capital base. The relevance of the four theories to the case are discussed.

Lastly, the chapter presents the book's theoretical model. It is induced from extensive literature, and is at a level that can be tested through multivariate techniques. The model also incorporates place and includes exploration of the geography of ICT use, an aspect of the digital divide that has been infrequently examined. The generic model in this chapter forms a template for specialized versions of it in Chaps. 4–9 for particular regions and nations. The generic model factors are discussed and provided with justification, which will be amplified in chapters to come.

References

Agarwal, R., Animest, A., & Prasad, K. (2009). Social interactions and the 'digital divide': Explaining variations in internet use. *Information Systems Research, 20*(2), 277–294.

Allaway, A. W., Berkowitz, D., & D'Souza, G. (2003). Spatial diffusion of a new loyalty program through a retail market. *Journal of Retailing, 79*, 137–151.

Andrés, L., Cuberes, D., Diouf, M., & Serebrisky, T. (2010). The diffusion of the Internet: a cross-country analysis. *Telecommunications Policy, 34*, 323–340.

Arai, Y., & Naganuma, S. (2010). The geographical digital divide in broadband access and governmental policies in Japan: Three case studies. *Networks and Communication Studies, NETCOM, 24*(1–2), 7–26.

Azari, R., & Pick, J. B. (2005). Technology and society: socioeconomic influences on technological sectors for United States counties. *International Journal of Information Management, 25*(1), 25–37.

Baliamoune-Lutz, M. (2003). An analysis of the determinants and effects of ICT diffusion in developing countries. *Information Technology for Development, 10*(3), 151–169.

Chen, W. (2013). The implications of social capital for the digital divides in America. *The Information Society, 29*, 13–25.

Dutta, S. (2007). Estonia: A sustainable success in networked readiness? In Dutta, S. & I. Mia (Eds.), *The global information technology report 2006–2007* (pp. 81–90). New York: Palgrave Macmillan.

Dutta, S., Mia, I., & Geiger, T. (2011). The networked readiness index 2010–2011: Celebrating 10 years of assessing network readiness. In S. Dutta & I. Mia (Eds.), *The global information technology report 2010–2011* (pp. 3–32). Geneva, Switzerland: World Economic Forum and INSEAD.

Dutta, S., & Bilbao-Osorio, B. (Eds.). (2012). *The global information technology report 2012*. Geneva: World Economic Forum and INSEAD.

Fong, M. W. I. (2009). Digital divide between urban and rural areas in China. *Electronic Journal of Information Systems in Developing Countries, 36*(6), 1–12.

Forstall, R. L., Greene, R. P., & Pick, J. B. (2009). Which are the largest? why published lists of major urban areas vary so greatly. *Journal of Economic and Social Geography (Tijdschrift Voor Economische en Sociale Geografie), 100*(3), 277–297.

Gao, P., & Yu, J. (2010). Emerging markets: Has China caught up in IT? *Communications of the ACM, 53*(8), 30–32.

Guillén, M. F., & Suárez, S. I. (2004). Explaining the global digital divide: economic, political and sociological drivers of cross-national internet use. *Social Forces, 84*(2), 681–708.

Hagerstrand, T. (1952). *The propagation of innovation waves. Lund studies in geography* 4. Lind, Sweden: University of Lund.

Hagerstrand, T. (1953). *Innovation diffusion as a spatial process*. Bulletin 15, Department of Geography, University of Lund. Lund, Sweden: University of Lund.

Hale, D., & Hale, L. H. (2003). China takes off. *Foreign Affairs, 82*(6), 681–708.

Ito, Mizuko, Okabe, D., & Matsuda, M. (Eds.). (2005). *Personal, portable, pedestrian: Mobile phones in Japanese life*. Cambridge, MA: MIT Press.

Lengsfeld, H. B. (2011). An econometric analysis of the sociodemographic topology of the digital divide in Europe. *The Information Society, 27*, 141–157.

Loo, B. P. Y. (2003). The rise of a digital community in the People's Republic of China. *Journal of Urban Technology, 10*(1), 1–21.

Martins, M. L. C., & Andrés, A. (2010). *The global digital divide revisted (2000–2004)*. Paper presented at Workshop on the Economics of ICTs, Porto, Portugal: Centro de Economia e Financas da Universidade de Porto.

Meng, Q., & Li, M. (2002). New economy and ICT development in China. *Information Economics and Policy, 14*(2), 275–295.

NTIA. (1995). *Falling through the net: A survey of the 'have nots' in rural and urban America*, Report. Washington, D.C.: National telecommunications and Information Administration, U.S. Department of Commerce.

NTIA. (1998). *Falling through the Net II: new data on the digital divide*, Report. Washington, D. C.: National telecommunications and Information Administration, U.S. Department of Commerce.

Ono, H., & Zavodny, M. (2005). Gender differences in information technology usage: A U.S.-Japan comparison. *Sociological Perspectives, 48*(1), 105–133.

Pick, J. B., & Azari, R. (2008). Global digital divide: influence of socioeconomic, governmental, and accessibility factors on information technology. *Information Technology for Development, 14*(2), 91–115.

Pick, J. B., & Azari, R. (2011). A global model of utilization of technology based on governmental, social, economic, and business investment factors. *Journal of Management Information Systems, 28*(1), 51–85.

Pick, James B., Nishida, T., & Zhang, X. (2013). Determinants of China's technology utilization and availability 2006-2009: A spatial analysis. *The Information Society, 29*(1), 26–48.

Pick, J. B., Gollakota, K., & Singh, M. (2014). Technology for development: Understanding influences on use of rural telecenters in India. *Information Technology for Development, 20*(4), 296–323.

Putnam, R. (2001). Social capital: Measurement and consequences. *Canadian Journal of Policy Research, 2*, 41–51.

Quibria, M. G., Ahmed, S. N., Tschang, T., & Reyes-Macasaquit, M.-L. (2003). Digital divide: Determinants and policies with special reference to Asia. *Journal of Asian Economics, 13*, 811–825.

Raven, V. P., Huang, X., & Kim, B. B. (2008). E-business in developing countries: A comparison of China and India. *International Journal of E-Business Research, 3*(1), 91–108.

Robison, K. K., & Crenshaw, E. M. (2002). Post-industrial transformations and cyber-space: a cross-national analysis of Internet development. *Social Science Research, 31*, 334–363.

Rupasingha, A., Goetz, S. J., & Freshwater, D. (2006). The production of social capital in U.S. counties. *The Journal of Socio-Economics, 35*, 83–101.

Rogers, E. M. (2003). *Diffusion of innovations*. New York: Free Press.

Song, W. (2008). Development of the Internet and digital divide in China: A spatial analysis. *Intercultural Communication Studies, 17*(3), 20–43.

Stebbins, R. A. (2001). *Exploratory research in the social sciences. Qualitative research methods series 48*. Thousand Oaks, CA: SAGE Publishing (SAGE University Paper).

Van Dijk, J. A. G. M. (2003). The digital divide as a complex and dynamic phenomenon. *The Information Society, 19*, 315–326.

Van Dijk, J. A. G. M. (2005). *The deepening divide: Inequality in the INFORMATION SOCIETY*. Thousand Oaks, CA: SAGE Publications.

Van Dijk, J. A. G. M. (2012). The evolution of the digital divide: the digital divide turns to inequality of skills and usage. In Bus, J. et al. (Eds.), *Digital enlightenment yearbook* (pp. 57–75). Amsterdam: IOS Press.

Warschauer, M. (2003). Dissecting the "digital divide": A case study in Egypt. *The Information Society, 19*, 297–304.

Whyte, W. H. (1954). The web of word of mouth. *Fortune, 50*(104–143), 204–212.

Yates, D. J., Gulati, G. J., & Weiss, J. W. (2011). Different paths to broadband access: the impact of governance and polity on broadband diffusion in the developed and developing worlds. In *Proceedings of 44th Hawaiian International Conference on System Sciences*, IEEE.

Chapter 4
The Global Digital Divide

4.1 Introduction

The theory, methodology, and historical perspective from the first three chapters are applied in this chapter to study the global digital dividedigital divide. A conceptual model is developed based on the general model in Chap. 3, and determinants of utilization of ICTs are examined. Additionally, the worldwide spatial patterns of ICT utilization are analyzed based on data for 110 nations.

To provide finer-grained insight, the world is divided into the developed and developing parts of the globe, that are examined separately, and their differences noted and explained. Some studies which have focused only on developing nations (Baliamoune 2003) or on developed ones (Hargittai 1999; Cuervo and Menéndez 2011; Lengsfeld 2011) add more in-depth insights for the two sides of the global divide. Other studies have compared these global regions based upon their extent of development and often found strong differences in ICT use (Kiiski and Pohjola 2002; Guillén and Suárez 2004; Pick and Azari 2008, 2011; Yates et al. 2011; Gulati and Yates 2012).

Next, the chapter disaggregates the globe even more, by examining the continental regions of Europe, Asia, and Africa-South America. Africa and South America are combined to achieve a greater sample size for regions that are predominantly developing. For the Asian region, the book will cover specialized, in-depth studies for China (Chap. 5), India (Chap. 6) and Japan (Chap. 7). Specialized analysis for Africa appears in Chap. 9.

4.2 Current Worldwide Unevenness in Access to and Use of Technologies

As seen in the first two chapters, technology uses have exploded worldwide over the past two decades with China and India leading the world starting in the mid-2000s decade in total volume of use (see Fig. 1.3 for mobile phones). In per capita

J.B. Pick and A. Sarkar, *The Global Digital Divides*, Progress in IS,
DOI 10.1007/978-3-662-46602-5_4

mobile phone use, the rising economies of South Korea and Brazil lead the way (Fig. 1.2), while for internet users per capita, Korea, Japan, and the U.S. are the leaders (Fig. 1.1). In 2013, an estimated 2.7 billion people worldwide used the Internet. However, 4.4 billion people have not yet made use of it, which points to challenges over the next 10 years to bring the Internet and other technologies to people who tend to be less literate, poorer, and/or less motivated for online use.

Broadband use in 2013 grew to reach 2.1 billion people with mobile broadband and 700 million people with fixed broadband. Mobile broadband is increasing at high rates, in particular it grew at 40 % yearly from 2010 to 2013 (ITU 2013). One industry estimate indicates mobile broadband will reach 6.5 billion subscriptions by 2018, exceeding considerably fixed broadband (Ericsson, cited in ITU 2013). The reasons for this disparity include that wireless mobile broadband depends on less expensive infrastructure investments largely in urban areas, compared to the high investments necessary for country-wide fixed fiber networks. Although it is cheaper, mobile broadband presently has limitations on bandwidth that reduce its usefulness compared to fixed broadband. Rural population in the developing world is somewhat deprived of fixed broadband, since fixed fiber networks are usually not available outside of major urban areas (ITU 2013). Overall, in 2013, about a billion people, 15 % of the world's population, had access to high-speed broadband. Future 4-G mobile systems will expand the bandwidth and partly remedy the disparity present in poor, rural areas. Another novel but uncertain alternative is a stratospheric-based broadband network service, such as Google loons discussed in Chap. 12.

The next section intends to look more closely at the differences between the developed and developing world in use of broadband, mobile phones and other ICTs. Later in the chapter, the determinants of the ICT use for the world, and developed/developing parts of it, and its continental regions are analyzed and interpreted.

4.2.1 Differences in ICT Levels in the Developed and Developing World

The world's technology level grew during the 2000s, on a per capita basis, in mobile phones, internet use, and broadband use, while lowering somewhat in fixed phone use (Fig. 4.1). Looking closely, total cell phone subscriptions in 2013 are nearly on parity with the world's population, while active mobile-broadband subscriptions, are growing the most rapidly (ITU 2013). This reflects the increasing prevalence of 3-G cell phones and their popularity as substitutes for other means of internet access, such as laptops. This is especially significant in the developing world (ITU 2013), as seen in Fig. 4.2. Even though mobile-broadband prevalence is nearly four times more prevalent in the developed versus the developing world, the rates of growth are much faster for the developing world. A similar pattern holds for

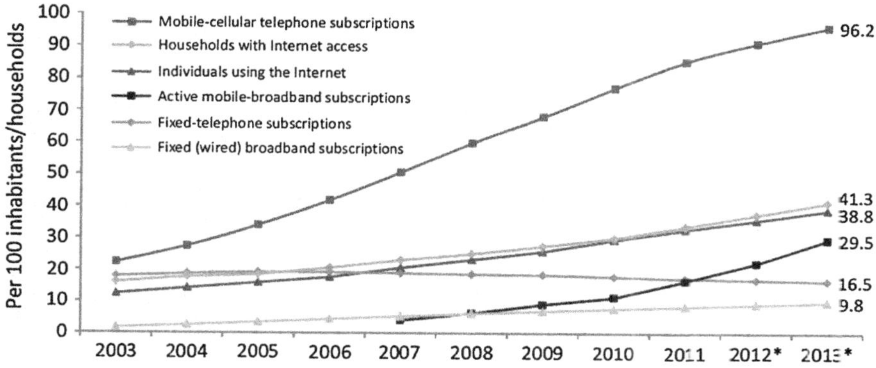

Fig. 4.1 Penetration of seven mobile and internet technologies worldwide, 2003–2013. *Note* * estimate. *Source* ITU (2013) world telecommunication/ICT indicators database

Fig. 4.2 Penetration of active mobile-broadband subscriptions worldwide and by development level, 2003–2013. *Note* * estimate. *Source* ITU (2013)

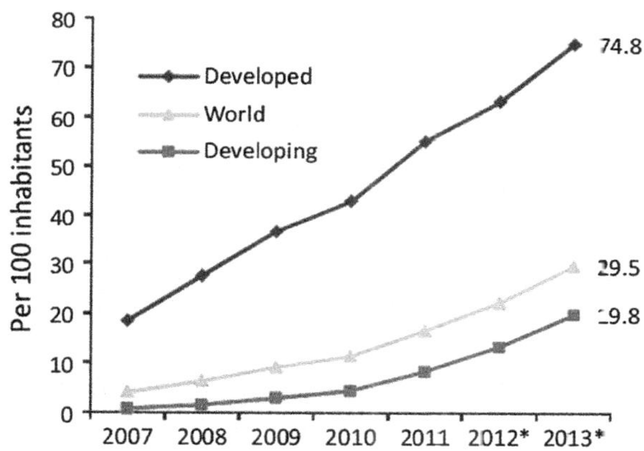

conventional fixed wire broadband (ITU 2013) i.e. the developed part of the world has 4.5 times the prevalence of the developing one. For fixed broadband penetration, the differences by world regions are dramatic, ranging from 27 per 100 persons in Europe to 3.3 per 100 in the Arab states and only 0.3 per 100 in Africa (ITU 2013).

The extraordinary disparity in ICT penetration rates between regions provides justification for disaggregating the world and studying portions of it separately. The analysis later in this chapter disaggregates the world into developed and developing parts and also into the world regions of Europe, Asia, and Africa-Latin America combined, a division not as detailed as the one shown in this Fig. 4.3.

Some of the historical reasons for the disparities by region in the phone levels were covered in Chap. 2, including differences in economic development, government encouragement, competitiveness of communication provider firms, cost, and

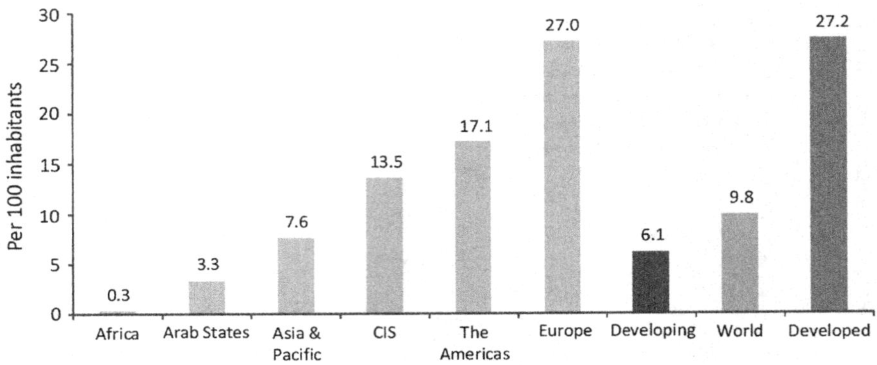

Fig. 4.3 Penetration of fixed broadband subscriptions by region and development level, 2013. *Note* CIS is the commonwealth of independent states, former Soviet Bloc Countries. *Source* ITU (2013)

regulations. Prior studies that compare determinants of ICT levels by development level have sought to explain the mechanisms that differentiate ICT penetration for the world, developed and developing nations, and/or world regions (Hargittai 1999; Kiiski and Pohjola 2002; Baliamoune-Lutz 2003; Quibria et al. 2003; Cuervo and Menéndez 2006; Pick and Azari 2008; Yates et al. 2011; Pick and Azari 2011; Gulati and Yates 2012). These prior studies are brought up in interpreting the chapter findings.

Gender Unevenness in Digital Divide

In addition to the gap in ICT penetration rate by geographical region, a digital divide is also observed in ICT use by gender. A gender gap in ICT usage is persistent worldwide, although more accentuated for developing nations including in Africa (Deen-Swarray et al. 2012). It is closely associated with income and education gaps. Women worldwide tend to have lower incomes than men, which leads to lower their ICT use, due in part to costs of hardware, software, bandwidth, and services. Also, in many nations, women have lower average educational levels than men. As a result, women may have less educational exposure, training and education in technology and associated skills.

The gender gap can be examined by studying the sex ratio, which is the number of males divided by number of females. In 2013, the global population sex ratio was 1.07, implying that globally there were somewhat more males than females (Population Reference Bureau 2013). However, the global sex ratio of online users of the Internet was 1.11 (ITU 2013), which implies a 4 % reduction in online use for women, controlling for the population sex ratio. For developed nations the online sex ratio is 1.02, which implies a 2 % online advantage for men, while for developing nations it was 1.16, i.e.

indicating a 9 % advantage online for men. The dramatic online disadvantage for women in developing nations may be due to women's lesser education and income in those countries and to less opportunity for internet access in the workplace (ITU 2013). The gender gap is not included in the digital divide analyses in this volume, because the online-use data by gender for nations and states/provinces are presently not systematically and reliably collected. Some governments and organizations are aware of the gender gap and are working on a variety of programs and initiatives to reduce or eliminate this aspect of digital divide.

4.3 Model for Understanding the World and World Regional Digital Divides

Based on the book's conceptual model (Fig. 3.8), a specific conceptual model for the world digital divide at the national level is presented in Fig. 4.4. The model has nine independent factors that relate to five dependent ones. The model performs the screening for spatial autocorrelation, in order to evaluate whether or not spatial bias has been eliminated. For the independent factors, all of the categories listed in Fig. 3.8 are represented in Fig. 4.4 except for infrastructure and social capital. The five dependent ICT factors are personal computers, internet users, broadband, secure internet servers, and mobile phones. Definitions and sources of the world variables are provided in Table 4.1.

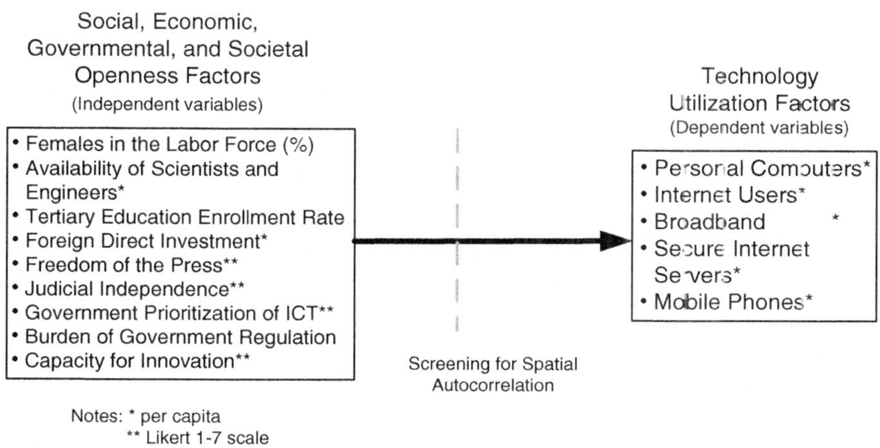

Fig. 4.4 Specific conceptual model for digital divide in world, developed/developing regions, and continental regions

Table 4.1 Definitions of world variables

Variable	Dependent or independent	Definition	Year of data collection	Data source	Measurement	Transformation
PCs	Dependent	Personal computers per 100 population	2008	WEF	Numerical	LN*
Broadband internet subscribers	Dependent	Broadband internet subscribers per 100 population	2008	WEF	Numerical	LN*
Internet users	Dependent	Internet users per 100 population	2008	WEF	Numerical	LN*
Secure internet servers per million pop.	Dependent	Secure internet servers per million population	2008	WEF	Numerical	LN*
Mobile telephone subscribers	Dependent	Mobile telephone subscribers per 100 population	2008	ITU	Numerical	
Tertiary education enrollment rate	Independent	Gross tertiary education enrollment rate	2007	WEF	Numerical	
Percent of females in the labor force	Independent	Percent females age 15–64 in the labor force	2008	WB	Numerical	
Foreign direct investment	Independent	Foreign direct investment equity flows (in millions of US dollars) per 1,000 pop.	2008	WB	Numerical	LN
Capacity for innovation	Independent	Extent of conduct of formal research and pioneering in products and services	2009	WEF	Likert 1–7 Scale	
Availability of scientists and engineers	Independent	Extent to which scientists and engineers are available	2009	WEF	Likert 1–7 Scale	
Freedom of the press	Independent	Extent the press is free	2009	WEF	Likert 1–7 Scale	

(continued)

Table 4.1 (continued)

Variable	Dependent or independent	Definition	Year of data collection	Data source	Measurement	Transformation
Judicial independence	Independent	Extent to which the judiciary in your country independent from influences of members of government, citizens, or firms?)	2008	WEF	Likert 1–7 Scale	
Government prioritization of information and communications technologies	Independent	How priority the government in your country places on information and communication technologies?	2009	WEF	Likert 1–7 Scale	
Burden of government regulation	Independent	Extent of burden for business to comply with government administrative requirements (regulations, permits, reporting)	2009	WEF	Likert 1–7 Scale	

LN natural logarithm
*for developing countries
ITU International Telecommunication Union
WEF World Economic Forum
WB World Bank

These variables were mostly discussed and justified in Chap. 3. Two of the independent variables, females in the labor force and availability of scientists and engineers, were not included in the sub-section on independent variables in Chap. 3, so their justification is given here.

Females in the labor force. This factor is infrequently included in digital divide studies. The reason for including it in the current study is that working represents a way for women to achieve greater exposure to, training in, and use of technologies in the workplace (see Box). This variable is consistently available for national samples from the World Bank, so is available for the investigation in this chapter that focuses on data at the national level. However, in Chaps. 5–8 which analyze sub-national data for states, provinces, and prefectures, the variable is mostly not available. Therefore this variable is excluded in these analyses. In an earlier study, female labor force participation was a significant determinant of internet hosts per capita, ICT expenditure as percent of GDP and Index of ICT Infrastructure Quality for developed nations but not developing ones (Pick and Azari 2008). This variable is not included for the sample of African nations in Chap. 9 because of errors introduced by the high proportion of informal labor force in that continent, estimated at 48 % in North Africa and 72 % in sub-Saharan Africa.

Availability of scientists and engineers. This variable is included because prior research results have shown a highly technical and scientific workforce to be a significant correlate of ICT use, in particular it was influential on information systems/data processing receipts and on payrolls for U.S. counties (Azari and Pick 2005) and on internet expenditures per family for Japanese prefectures (see Chap. 7). The mechanism may be that ICT use increases in a geographic unit with high levels of scientists, engineers, and technical employees due to the strong technology skills of these workers, and by externalities of their presence, such as support workers, technical services, and nearness to universities emphasizing computing and information technology.

In other words scientists and engineers tend to simply incorporate high levels of technology use in all aspects of their lives (work and personal—office and residence) and therefore significantly influence everything around them. Further, their presence affects adoption of technology and use by others living near them in the "geographic unit," in this case the nation.

4.3.1 The Present World Sample and Differences in the Developed and Developing Regions

The present investigation utilizes data from a sample of 110 nations compiled by the World Bank (2010), World Economic Forum (2010), and International Telecommunications Union (2009). The average values of the ICT variables (dependent) and socio-economic variables (independent) are presented in Table 4.2.

Table 4.2 Average values of world sample variables, by development level, 2010

Variable	Dependent or independent	Averages for samples			Ratio—developed to developing nations
		World	Developed nations	Developing nations	
PCs per 100 pop.	Dependent	21.472	36.205	5.517	6.6
Broadband internet subscribers per 100 pop.	Dependent	8.777	15.779	0.927	17.0
Internet users per 100 pop.	Dependent	32.573	50.865	11.568	4.4
Secure internet servers per million pop.	Dependent	152.863	286.489	3.240	88.4
Mobile telephone subscribers per 100 pop.	Dependent	87.552	111.121	58.202	1.9
Females in the labor force (%)	Independent	41.044	42.014	41.021	1.0
Availability of scientists and engineers (1 = lowest, 7 = highest)	Independent	4.174	4.447	3.840	1.2
Tertiary education enrollment rate	Independent	36.406	53.393	18.139	2.9
Foreign direct investment per capita	Independent	0.386	0.690	0.047	14.7
Freedom of the press (1 = lowest, 7 = highest)	Independent	5.115	5.519	4.665	1.2
Judicial independence (1 = lowest, 7 = highest)	Independent	4.053	4.659	3.318	1.4
Government prioritization of information and communications technologies (1 = lowest, 7 = highest)	Independent	4.667	4.803	4.458	1.1
Burden of government regulation (1 = lowest, 7 = highest)	Independent	3.283	3.195	3.333	1.0
Capacity for innovation (1 = lowest, 7 = highest)	Independent	3.292	3.756	2.785	1.3

Sources World Bank (2010), World Economic Forum (2010), International Telecommunication Union (2009)

It is clear from Table 4.2 that for every variable except Burden of Government Regulation, developed nations have a higher value. The ratio is especially high for secure internet servers, reflecting that internet servers can be located anywhere and serve remote users. It makes sense to locate servers in advanced economies which would tend to be safer and more easily maintained by skilled staff. Broadband subscribers are also heavily weighted towards developed nations, a phenomenon explained by broadband's reduced availability, lower reliability, and higher cost in developing countries. When the dependent variables are displayed by world region

(Figs. 4.5 and 4.6), all three continental regions differ from each other, with Asia having the highest penetration of cell phones and PCs, while Europe has the highest internet and broadband penetration rates.

Regarding the independent variables, the largest disparity occurs for foreign direct investment per capita, which reflects the relative size of economies between developed and developing nations. Higher education is about 3 times more prevalent in developed nations. Percent of women in the labor force is nearly even between developed and developing nations, due in part to widespread presence of women in the informal labor force of developing nations. Perception of government regulation is slightly greater in developing nations, which is exemplified by India's extensive regulation and by strong government regulatory controls in China. Innovation capacity, i.e. the extent of conduct of formal research and pioneering in products and services, is perceived as about 30 % stronger in developed countries, which connotes the much higher level of innovation in those nations especially the highly advanced economies such as the U.S., Japan, and countries in Western Europe. Democracy and openness attributes are higher in developed nations.

4.3.2 Spatial Distributions of Global Technology Use

The worldwide distribution of fixed broadband, seen in Fig. 4.5, reveals that fixed broadband subscriptions are highest in the United States and Canada, western and northern Europe, South Korea, Japan, Australia, New Zealand and Chile. These are among the world's most advanced nations economically and educationally. In the middle range are the BRICS nations of Brazil, Russia, India, China, and South Africa, with India the lowest of the BRICS, plus Eastern Europe, and several Central and South American nations including Mexico, Columbia, Venezuela, Peru, Uruguay, and Argentina. At the low end is Sub Saharan Africa, except for South Africa. Although the Middle East is missing data, studies have shown it to be mostly at the low end, with the exceptions of Saudi Arabia, Israel, and the Gulf states (World Economic Forum 2012).

The worldwide pattern for mobile phone subscribers (Fig. 4.6) somewhat resembles that of fixed broadband, but differs in the very high prevalence of mobile phone subscriptions in Western and Eastern Europe and extending into Russia, Chile, Argentina, New Zealand, and Saudi Arabia, while moderate range includes the United States, France, Brazil, Australia, north Africa, and South Africa, with Canada, China, Mexico, Peru and Ecuador at the low moderate level. Fairly consistent with fixed broadbandbroadband, the lowest levels are in central Africa, most of the Middle East, and India, which has limited telecommunication network infrastructure to serve its vast rural population.

The global disparities in ICT use seen in these two examples underscore the difficulties in generalizing worldwide on the use of ICTs. The next section turns to a detailed analysis of the determinants of ICT use.

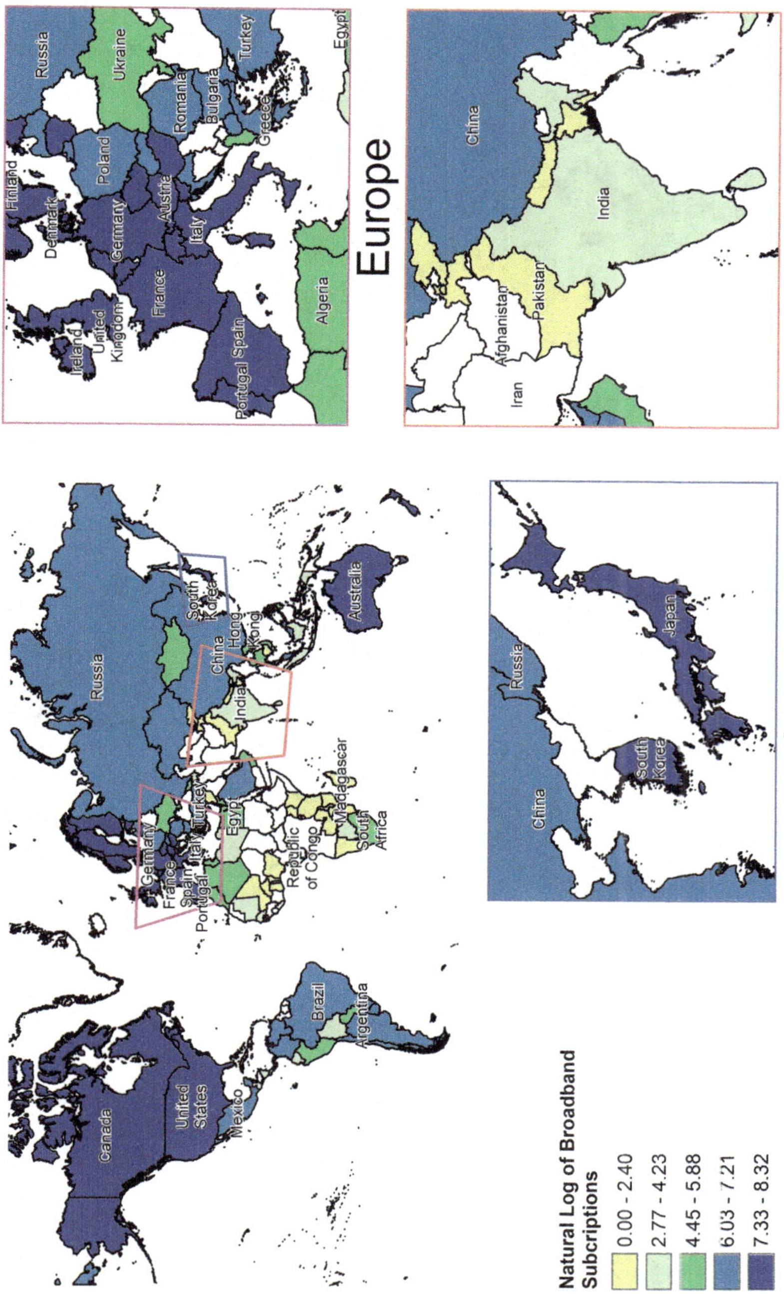

Fig. 4.5 Global distribution of fixed broadband, 2010. *Source* World Bank (2010)

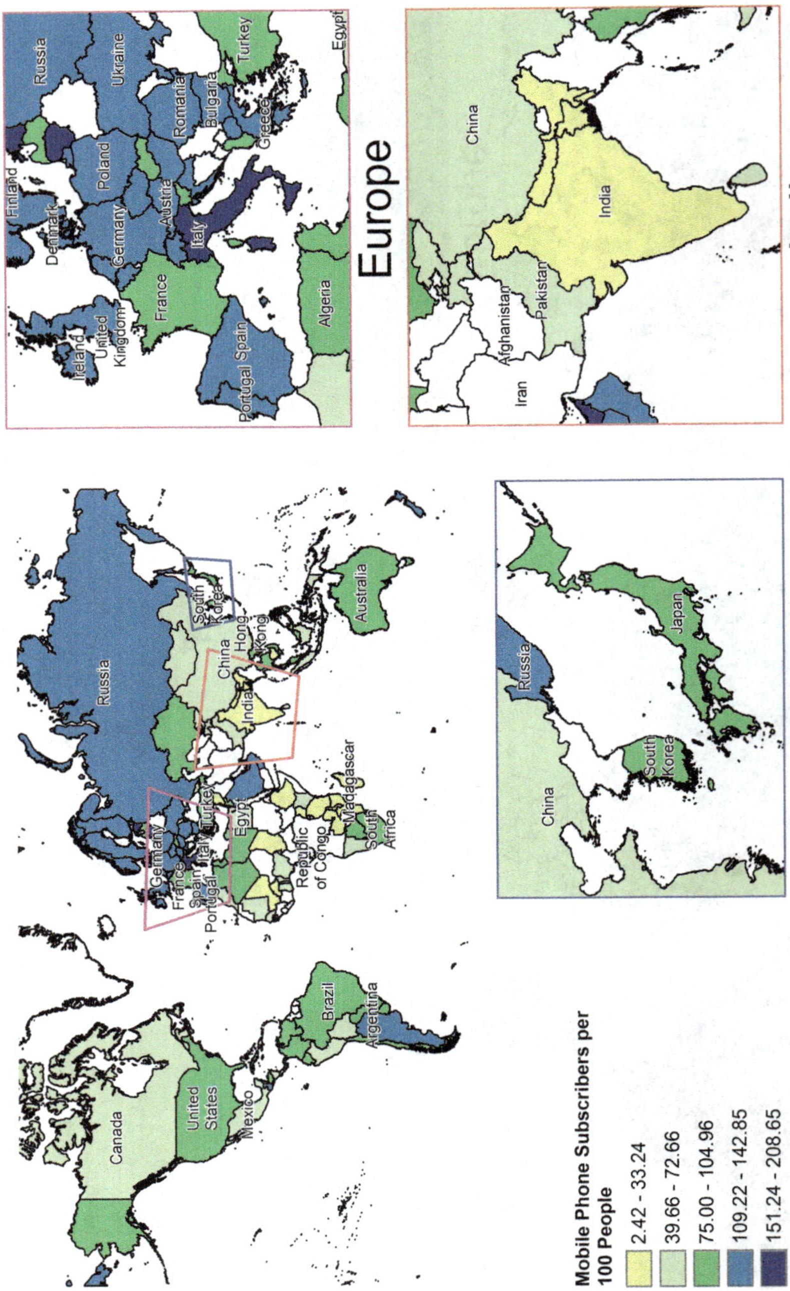

Fig. 4.6 Global distribution of mobile phone subscribers, 2010. *Source* World Bank (2010)

4.4 Determinants of Technology Use

The OLS regression and spatial autocorrelation methodology presented in Chap. 1 and summarized in Fig. 1.6 are followed for this chapter, with the exception that global cluster analysis is not performed. However, cluster analyses are performed later in Chaps. 5–9 for states in four major nations and for the nations in the continent of Africa.

OLS Regression Findings

OLS regression analysis is first conducted for the world as a whole, with results indicating that the most important determinants of technology use are tertiary education, followed by foreign direct investment (FDI), judicial independence, and capacity for innovation (Table 4.3). Freedom of the press is significant for broadband and internet servers, while women's labor force participation is inversely related to broadband and internet use.

The OLS regression diagnostic tests of Joint Wald, Koenker, and Jarque-Bera indicate the OLS regression model is sound. Please note that the tests confirm that the model meets its regression assumptions when Joint Wald is significant and when the other tests are non-significant. The three tests are explained in Chap. 1.

However, the Moran's I test for spatial autocorrelation of regression residuals is significant for four of the five regressions, with only broadband having insignificant spatial autocorrelation. Moran's I is positive, which indicates for the four regressions that the independent variables in the regression model do not account for spatial proximity effects, in other words there is spatial bias present without knowing what it stems from. The spatial agglomeration that characterizes positive, statistically significant, Moran's I can be observed the map of global mobile phone subscribers in Fig. 4.6, in which similar levels of mobile phone subscriptions are evident and agglomerated within western and northern Europe, for most of Asia, and for most of Africa. South America is mixed, with a variety of levels of mobile phone subscribers.

One way to attempt to reduce this spatial bias problem would be to modify or replace the present set of independent variables. Another approach, and the one adopted here, is to disaggregate the sample of nations into continental regions, within which nations are roughly at the same level of broadband, and then re-run separate regressions for each continental region, testing again to determine if the spatial bias is reduced or eliminated. Accordingly, we disaggregate the world into European nations, Asian nations, and African-South American nations combined, based on the UN definition of these continental regions (UN 2011). We also analyze the results for samples of the developed countries and the developing countries based on gross national product per capita. Developed nations are defined as ones having gross national income per capita greater than or equal to $4400, while developing nations are ones with gross national income per capita less than $4400. The dividing point of $4400 is determined by a 60:40 weighted average of the

Table 4.3 Standardized regression results for dependent variables, 2008–2009, for world nations

	LN of PCs per 100 pop.	LN of broadband internet subscribers per 100 pop.	LN of internet users per 100 pop.	LN of secure internet servers per 100 pop.	Mobile telephone subscribers per 100 pop.
Tertiary education enrollment rate	0.496***	0.623***	0.692***	0.534***	0.682***
Availability of scientists and engineers				−0.168*	
Capacity for innovation	0.211*	0.203***		0.259**	
Burden of government regulation					
Freedom of the press		0.145*		0.198***	
Judicial independence	0.281***		0.264***	0.239***	
Government prioritization of information technology					
Percent of females in the labor force		−0.117*	−0.156**		
LN of foreign direct investment		0.147**		0.096*	0.168**
Regression adjusted R square and significance level	0.666***	0.710***	0.656***	0.796***	0.544***
Sample size (N)	106	108	110	109	110
OLS regression tests					
Joint wald statistic	315.1***	392.0***	278.5***	663.6***	130.9***
Koenker (BP) statistic	6.161	9.828	5.661	10.249	3.469
Jarque-Bera statistic	2.663	1.213	2.681	0.36	2.943
Test of spatial autocorrelation of residuals					
Moran's index	0.204	0.072	0.164	0.201	0.193
p value	0.005**	0.306	0.023*	0.006**	0.008**
Z score	2.783	1.024	2.272	2.732	2.662

*Signif. at 0.05
**Signif. at 0.01
***Signif. at 0.001

means for lower-middle-income and upper-middle-income nations in the 2010 World Development Indicators (World Bank 2010). The specific developed/developing and continental findings are as follows.

Developed and Developing Countries

For developed nations, the dominant variables are higher education, capacity for innovation, and judicial independence (see Table 4.4). The spatial bias, as measured by Moran's I, is controlled by the model. These findings are consistent with the results for Europe and Asia, so will be deferred to the discussion of those continental regions. For developing countries, the major correlates are higher education and foreign direct investment, both of which are significant in regressions for all five ICT variables. Those findings are also reflected in the results for the combined continental regions of Africa and South America, so discussion is deferred.

Europe

The findings for Europe indicate that the most important determinant is judicial independence, followed by capacity for innovation. The regression diagnostics support an unbiased model, including Moran's I indicating lack of spatial bias, as seen in Table 4.5.

In concert with the findings on judicial independence, the European Commission has emphasized internet and information rights (European Commission 2014) including data protection, copyright, and related rights. It has instituted legislation based on rules from the World Intellectual Property Organization and established copyright protection of information products without reducing innovation and creativity. The findings are consistent with a prior study of broadband in developed nations worldwide, which found greater broadband use is related to a governance index that includes regulatory quality, control of corruption, and free flow of information, all of which relate to judicial independence (Yates et al. 2011).

Capacity for innovation is associated with use of broadband and PCs. Since innovation is defined broadly, a European innovative nation is likely to have innovation clusters that utilize considerable technology such as technology parks and medical research complexes, as well as attract scientific, engineering, and technical workforce who tend to have high computer usage levels. An example is Finland, which has been innovative in mobile and network technologies and has substantial scientific and technical workforce, who utilize technologies intensively. Another example is Estonia, the case profiled in the last chapter, which fostered multiple technology start-ups that have spawned innovations such as Skype, advances helped by the country's highly skilled workforce that heavily utilizes ICT, a labor force assisted by proactive government training programs. Although Eastern Europe has been overshadowed by Western Europe in innovation, the East is catching up, as seen in a tech conference held in Kiev, Ukraine (Fig. 4.7).

Table 4.4 Standardized regression results for dependent variables, 2008–2009, for developed countries and developing countries

	Developed countries				
	PCs per 100 pop.	Broadband internet subscribers per 100 pop.	Internet users per 100 pop.	Secure internet servers per 100 pop.	Mobile telephone subscribers per 100 pop.
Tertiary education enrollment rate		0.333***	0.347***	0.296***	0.333**
Availability of scientists and engineers					
Capacity for innovation	0.613***	0.454***	0.450***	0.280*	
Burden of government regulation					0.244
Freedom of the press					
Judicial independence	0.277*	0.242***	0.267*	0.498***	
Government prioritization of information technology					
Percent of females in the labor force		0.142*			
LN of foreign direct investment		0.148***			
Regression adjusted R square and significance level	0.665***	0.830***	0.692***	0.692***	0.136**
Sample size (N)	54	57	57	57	57
OLS regression tests					
Joint Wald statistic	129.2***	0.356.0***	194.3***	104.4***	13.5**
Koenker (BP) statistic	4.900	5.169	6.166	12.403**	2.531
Jarque-Bera statistic	3.054	1.946	0.484	3.300	0.315
Test of spatial autocorrelation of residuals					
Moran's index	0.008	−0.035	0.182	0.131	0.063
p value	0.824	0.885	0.083	0.202	0.485
Z score	0.223	−0.148	1.732	1.276	0.698

(continued)

Table 4.4 (continued)

	Developing countries				
	LN_PCs per 100 pop.	LN of broadband internet subscribers per 100 pop.	LN of internet users per 100 pop.	LN of secure internet servers per 100 pop.	Mobile telephone subscribers per 100 pop.
Tertiary education enrollment rate	0.271*	0.527***	0.369***	0.424***	0.515***
Availability of scientists and engineers					
Capacity for innovation			0.248*		
Burden of government regulation					
Freedom of the press				0.225*	
Judicial independence					
Government prioritization of information technology					
Percent of females in the labor force			−0.269**		
LN of foreign direct investment	0.535***	0.299*	0.279*	0.431***	0.264*
Regression adjusted R square	0.511***	0.514***	0.515***	0.560***	0.476***
Sample size (N)	52	49	52	50	52
OLS regression tests					
Joint Wald statistic	46.8***	73.1***	65.9***	45.9***	67.0***
Koenker (BP) statistic	3.283	2	9.570*	12.403**	0.391
Jarque-Bera statistic	1.438	0.84	0.27	2.046	2.787
Test of spatial autocorrelation of residuals					
Moran's index	−0.101	0.126	0.006	0.129	0.311
p value	0.546	0.302	0.054	0.284	0.014
Z score	−0.603	1.032	0.184	1.07	2.446

*Signif. at 0.05
**Signif. at 0.01
***Signif. at 0.001

Table 4.5 Standardized regression results for dependent variables, 2008–2009, for nations in Europe

	Europe				
	PCs per 100 pop.	Broadband internet subscribers per 100 pop.	Internet users per 100 pop.	Secure internet servers per 100 pop.	Mobile telephone subscribers per 100 pop.
Tertiary education enrollment rate					N.S.
Availability of scientists and Engineers					N.S.
Capacity for innovation	0.450*	0.436**			N.S.
Burden of government regulation					N.S.
Freedom of the press					N.S.
Judicial independence	0.421*	0.534***	0.886***	0.883***	N.S.
Government prioritization of information technology					N.S.
Percent of females in the labor force					N.S.
LN of foreign direct Investment					N.S.
Regression adjusted R square	0.676***	0.854***	0.778***	0.773***	N.S.
Sample size (N)	31	31	30	31	31
OLS regression tests					
Joint Wald statistic	69.9***	172.2***	132.8***	87.9***	
Koenker (BP) statistic	1.657	4.455	2.078	7.478**	
Jarque-Bera statistic	1.15	1.805	0.679	1.634	
Test of spatial autocorrelation of residuals					
Moran's index	−0.127	−0.077	−0.023	−0.081	
p value	0.407	0.692	0.947	0.669	
Z score	−0.827	−0.396	−0.066	−0.428	

N.S. Not significant
*Signif. at 0.05
**Signif. at 0.01
***Signif. at 0.001

Fig. 4.7 Speaker at Tech Conference in Kiev, Ukraine. *Source* Internet Technologies and Innovations

Asia

In Asia, the major determinants of ICT use are higher education, capacity for innovation, and to a lesser extent, foreign direct investment and judicial independence. The Asian regressions meet all diagnostic tests, including lack of spatial bias as indicated by Moran's I (see Table 4.6). Education has been strongly supported over the past decade by governments in many Asian nations, including China, Australia, South Korea, and Vietnam. Education has been identified as an important determinant of ICT use in Asia (Quibria et al. 2003; Ono and Zavodny 2006; Pick et al. 2014). Quibria found that higher education was significantly related to internet and PC use for samples of 56–68 nations, controlling for Asian nations. He explained that education is complementary to internet use, as well as enhancing a country's capacity for industry change and innovation, which in turn encourage IT use (Quibria et al. 2003). The campus of Tsinghua University, eminent in Technology and Innovation, is seen in Fig. 4.8.

Asian capacity for innovation is explained by justifications similar to those for Europe. Innovators are themselves computer users and stimulate many others to be users (Bass 1969). For instance, the Chinese government emphasized innovation in the 2000s by sponsoring innovation centers in dozens of cities, seeking with incentives the return of Chinese research innovators who went overseas, and starting universities that emphasize innovation such as the Hong Kong Institute of Science and Technology (patterned after MIT in the U.S.). In Japan after World War II, some consider that innovation was sparked by the visit of quality engineer, William Edwards Deming. Subsequently, the Japanese government for decades supported industrial innovation, most recently by an ambitious strategic plan for research and innovation released by the Council for Science and Technology Policy which reports to Prime Minister Shinzo Abe (Cyranoski 2013).

The association of foreign direct investment (FDI) with PCs, broadband subscribers, and mobile phone subscribers has a variety of explanations. FDI can lead to manufacturing facilities which directly influence the recipient country's domestic

Table 4.6 Standardized regression results for dependent variables, 2008–2009, for nations in Asia

	Asia				
	PCs per 100 pop.	Broadband internet subscribers per 100 pop.	Internet users per 100 pop.	Secure internet servers per 100 pop.	Mobile telephone subscribers per 100 pop.
Tertiary education enrollment rate	0.404***	0.459***	0.447***	0.621***	0.409**
Availability of scientists and engineers					
Capacity for innovation	0.433***	0.409***	0.280*		
Burden of government regulation					
Freedom of the press					
Judicial independence			0.355**	0.391**	
Government prioritization of information Technology					
Percent of females in the labor force					
LN of foreign direct investment	0.534***	0.397***			0.790***
Regression adjusted R square	0.577***	0.878***	0.781***	0.732***	0.611***
Sample size (N)	27	30	30	30	30
OLS regression tests					
Joint Wald statistic	50.9***	204.7***	183.9***	42.6***	43.2***
Koenker (BP) statistic	1.479	5.088	4.358	8.658	2.013
Jarque-Bera statistic	8.562*	1.237	0.844	1.141	0.512
Test of spatial autocorrelation of residuals					
Moran's index	−0.99	0.018	−0.184	0.111	0.19
p value	0.859	0.759	0.32	0.371	0.188
Z score	0.178	0.306	−0.994	0.894	1.316

*Signif. at 0.05
**Signif. at 0.01
***Signif. at 0.001

Fig. 4.8 Campus of renowned Tsinghua University, Beijing, China

supply of ICT devices and infrastructure. As detailed in Chap. 5, foreign direct investment to China was crucial in building the country's export economy in the 2000s to a world-class level that interlinks with the global supply chain (Hale and Hale 2003; Gao and Yu 2010) and has required substantial build-up and uses of ICT. A similar investment stimulus to ICT use occurred earlier in Japan and South Korea, and is in process in some other Asian nations that are becoming more export-oriented, such as Malaysia and Vietnam. In contrast to Europe, judicial independence is unimportant as a factor for Asian ICT use, which might relate to less prevalence of democratic governments in Asia than in Europe.

Africa and South America

In Africa and South America, the most significant determinant is press freedom, followed by higher education, which is correlated to uses of four ICTs, and foreign direct investment, important for PCs and internet servers. The regression diagnostics are all favorable including the Moran's I test (see Table 4.7).

Freedom of the press has not been previously included in digital divide studies for Africa and South America. For developing nations, its importance was not supported by several prior studies that included democracy and societal openness factors (Onyeiwu 2002; Guillén and Suárez 2004; Baliamoune-Lutz 2003; Yates et al. 2011). By contrast, another study of developing and developed nations found a strong effect on internet diffusion from a democracy index (Robison and Crenshaw 2002). Since the data in the cited studies are one to two decades old, the effect noted here might reflect that some African nations and many South American ones became less authoritarian in the 2000s, enabling more widespread and open use of ICTs. Press freedom presumably reflects the more open societies.

Education was seen from Chap. 3 to be one of the most widely supported general factors influencing ICT use. However in a study of 39 developing nations in the period 1999–2001, the quality of math and science education was significant although not as much as the factors of scientific publications and FDI (Pick and

Table 4.7 Standardized regression results for dependent variables, 2008–2009, for nations in combined Africa and Latin America

	Africa and Latin America				
	LN of PCs per 100 pop.	LN of broadband internet subscribers per 100 pop.	LN of internet users per 100 pop.	LN of secure internet servers per 100 pop.	Mobile telephone subscribers per 100 pop.
Tertiary education enrollment rate		0.716***	0.650***	0.330**	0.687***
Availability of scientists and engineers					
Capacity for innovation			0.217*		
Burden of government regulation					
Freedom of the press	0.248*	0.315**	0.279**	0.348***	0.298**
Judicial independence					
Government prioritization of information technology					
Percent of females in the labor force					
LN of foreign direct investment	0.538***			0.427***	
Regression adjusted R square	0.425***	0.591***	0.582***	0.667***	0.541***
sample size (N)	45	42	45	44	45
OLS regression tests					
Joint Wald statistic	37.4***	111.4***	98.5***	139.9***	62.9***
Koenker (BP) statistic	5.398	0.000	0.655	6.022	4.414
Jarque-Bera statistic	0.639	1.427	1.908	0.386	1.404
Test of spatial autocorrelation of residuals					
Moran's index	0.243	0.105	0.116	0.128	0.207
p value	2.217	0.306	0.248	1.244	0.056
Z score	0.026	1.024	1.155	0.213	1.903

*Signif. at 0.05
**Signif. at 0.01
***Signif. at 0.001

Azari 2008). Another earlier study based on 47 developing nations likewise did not find the factors of adult literacy and index of education to be significant on diffusion of internet use or personal computer diffusion 1998–2000 (Baliamoune-Lutz 2003). Also in the book's present more extensive study in Chap. 9 on Africa in 2010–2012, literacy was not significant. One difference of these studies from our result is our education variable of higher education, which in Africa and Latin America may have more impact on ICT use than lower levels of education or literacy. We can say that the influence of higher education does not seem generalizable to other levels of education, and that various educational levels need to be subjected to more intensive future research. The explanation of FDI's importance on ICT in the poorer economies of Africa and Latin America is due both to the influence of foreign investments for building of more modern infrastructure and to ICT skills training of local citizens employed in foreign-owned firms that receive FDI.

4.5 Revised Conceptual Model

The conceptual model for the world digital divide was tested using national data from 2008 to 2009, and it was shown that spatial bias is present for 4 of the 5 dependent variables in the world model as seen in Fig. 4.9. To eliminate this bias, the model is disaggregated into developed and developing nations, as well as into continental regions. Spatial bias disappeared in all the disaggregated samples. This suggests that the conceptual model should be disaggregated into separate sets of independent factors for developed and developing countries, each set influencing the dependent variables. As seen in Fig. 4.9, factors that have been shown to be

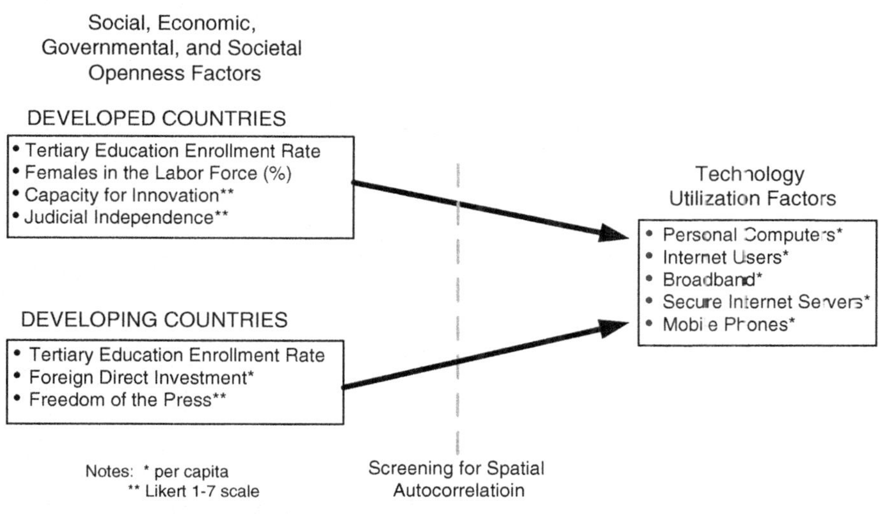

Fig. 4.9 Revised conceptual model for digital divide worldwide

more influential in developed countries and developing countries are partitioned in the revised conceptual model, with some factors shared in common. With more empirical testing, this model could be developed to include an expanded set of factors in each partition on the left.

4.6 Mauritius Case Study

Mauritius, an African island nation located in the middle of the Indian Ocean, exemplifies the progress possible for a nation in the developing world while also providing an example to demonstrate this chapter's model and its findings. Mauritius has had a long history. Settled by the Dutch in the 17th century, it was taken over by the French in 1715, and subsequently taken by the British in 1810, who held it until the country's independence in 1968. The island's Indian Ocean location lent military importance to it under the British rule, after which it emerged in 1968 as an independent state based on a rural agricultural economy. Since independence, it has undergone remarkable economic transformation so that presently it has a middle level, service-oriented economy with among the highest per capita income in Africa. Today Mauritius has 1.33 million people and a 2013 GDP of $21 billion (U.S. Central Intelligence Agency 2014). The nation has about half rural population, which has tended to perpetuate unevenness in socio-economic levels.

Early in the nearly half century since independence, the country adopted a national goal to embrace technologies, including ICT. This leadership has come from the nation's top leaders, who in 1997 created the Ministry of Information and Communication Technology, one of the world's first federal government departments of ICT, followed a year later by issuance of its National ICT Strategic Plan, the first of a series of strategic ICT plans that continue to the present. In 1999 the Cyber City Project commenced with support from the Indian government in the form of a $100 million credit line, half of which was allocated to build a Cyber City to incubate and host ICT companies and business process outsourcing (BPO) firms. Since the latter firms would partly serve the Indian market, this was a form of foreign direct investment. FDI was also provided to connect Mauritius to an undersea fiber-optic cable that interlinks South Africa and India and subsequently to build land-based fiber-optic networks, which today are widespread and even offer fiber-to-the-home, a first for Africa.

In 2003, the country's telecommunication sector was completely liberalized, and in the 2000s, a business process outsourcing (BPO) sector developed, mostly to serve South Africa, India, and a few European nations. By 2011, the country's ICT sector accounted for 6.5 %of GDP and employed 15,000 employees (Oolun et al. 2012).

How were all these ICT milestones accomplished in a small island nation? One of the critical success factors was a strong legal and regulatory framework put in place after independence, along with government encouragement of an environment of broad democratic and open access to ICT and its information. A sequence of laws were passed over the past 20 years that provide basic freedoms, privacy protections, and statutes for computer misuse and crime, including for example, the Copyright Act, Data Protection Act, Independent Broadcasting Act, and Computer Misuse and Cybercrime Act (Oolun 2012). Furthermore, the government created initiatives for ICT literacy outreach, a universal ICT education program, and community empowerment program. There was even provision for three cyber caravans to travel daily around Mauritius providing ICT access and assistance in remote, underserved areas (Oolun 2012).

A challenge for the nation's goals of ICT is that the country's higher education system is small, with only 8,000 students entering it yearly. To attain the world class standards of outsourcing the nation aspires to, there needs to be a larger professional ICT workforce than this system is able to produce. To try to solve this gap the government collaborated with private industry to establish the ICT Academy as a school of the further training of high school graduates in technology industry practices including offering widely known certifications such as those of Oracle, Microsoft, and SAP (Oolun et al. 2012). Training was also given by a nonprofit organization DotConnectAfrica that provides socio-technological advancement throughout Africa, as seen in Fig. 4.10.

The Ministry of Information and Communication Technology (MICT) has been responsible for planning and implementing ICT in the country, results of which have been mostly positive in spite of some challenges. The Ministry's National Information and Communication Strategic Plan 2011–2014 sought to position ICT as one of the nation's leading economic sectors and to establish Mauritius as a global ICT hub (Republic of Mauritius 2012a, 2014). The plan's 15 programs and 124 projects are highly ambitious, but are only partly realized due to workforce capacity issues of a small country as well as high cost of bandwidth. The high cost contrasts with the lowering of rates throughout continental Africa due to increased competition, whereas Mauritius's bandwidth market is non-competitive, dominated by one sub-oceanic fiber provider and a single incumbent telecommunications firm,

Fig. 4.10 Training session by DotConnect Africa, Mauritius. *Source* http://DotConnectAfrica.Org

Mauritius Telecom (Republic of Mauritius 2014). The solution aspired to but not achieved was to mandate a competitive market and open access by competing companies to the oceanic cable and the incumbent's land-based infrastructure, which have led to cost reductions and use of greater bandwidth. An extensive new National Broadband Policy 2012–2020 (Republic of Mauritius 2012b) now seeks to rectify these issues.

Another problem in the strategic plan is the limited progress to implement the goal of comprehensive e-government. E-government is slowed down due to limited buyoff by government stakeholders and resistance of many workers. This issue is targeted to be resolved in the MICT's plan for e-Government Strategy 2013–2017 (Republic of Mauritius 2013).

The Mauritius story of extraordinary ICT progress for a developing nation albeit with some obstacles can be compared to this chapter's global ICT model and results. In concert with the model's regression findings, the Mauritius case accentuates the critical role of higher education to produce skilled ICT workforce, as well as the essential provision of FDI by the Indian government to initially stimulate the development of an ICT/BPO sector.

4.7 Policy Implications

As technology uses diffuse and explode into higher levels of ICT use worldwide, world organizations and nations confront challenges in understanding its growth, geographical patterning and factors driving the expansion. The model in this chapter has proven useful in probing these change processes and relationships. Independent factors from the general conceptual model in Chap. 3 were posited as having influence on five technology utilization factors. This section briefly discusses the direct implications from the model and empirical research in this chapter. There are similar sections in Chaps. 5–9 for four major countries and for Africa, as well as full discussion of the broad roles and policies of governments related to digital divide in Chap. 11.

On testing the model, there was a problem of spatial bias when the model was applied worldwide, so the model was disaggregated into separate independent determinants for developed and developing regions. The revised conceptual model reflects these differences. Consequently, a government or world organization trying to understand and create policy for the digital divide needs to position its country along the economic development continuum, and put forward policies appropriate for that growth stage. Based on the stage of economic development, different policies can be adopted.

Among the most consistent and foremost implications of this chapter's findings is the importance of higher education for policymakers in national governments and world organizations. Higher education stands out as a major determinant of use for the world and all the world divisions analyzed except for Europe. For non-European nations and organizations, policies should be developed to strengthen and broaden higher education in the country or region.

For European nations, government policymakers should consider strengthening further judicial independence and legal protections of private information. Another step for European planners and policymakers is to foster innovation through funding, tax incentives, government sponsorship and other measures. There are several success examples in Europe that encouraged technology innovation through federal government incentives, for instance Ireland in the late 1990s and early 2000s and the Estonian government, which was seen in Chap. 3 to have supported R&D in ICT to stimulate technology start-ups.

Turning to Asia, government policymakers need to emphasize higher education, as well as provide policies and planning for innovation and judicial independence, steps which China is presently struggling with and only slowly moving forward on. Policies that encourage FDI are recommended. India for instance has encouraged FDI in the form of "captive centers," i.e. India-based offices of major international outsourcers such as General Electric (Oshri et al. 2011).

In Africa and Latin America, government policies that support and build up higher education are recommended, even if these strategies are long-term in nature and costly. Another policy suggested is to encourage legislation providing guarantees and rights of freedom of the press as well as other citizen rights. The implementation of these policies may be a hard struggle, because many of the poorest nations are autocracies that historically have suppressed press and personal freedoms.

4.8 Conclusion

The planet is undergoing rapid transformation into a vastly more digital and hyperconnected world. ICTs are becoming commonplace for many billions of people, yet true success requires that the requisite skills and government support also be available to stimulate ICT use. This chapter examined the geographic distribution of ICTs worldwide, finding that some large continental regions such as Western Europe are much more advanced in ICT use compared to central Africa or parts of South America, which have low to very low ICT use. A conceptual model is developed similar to the general model in Chap. 3, based on induction from the literature on the global and world-regional digital divides.

An analysis of determinants of ICT use worldwide reveals that the world as a whole is spatially agglomerated and that the model of ICT determinants at the world level is mostly spatially biased. Moving to the levels of developed/developing nations, and major continental regions, there are distinctive differences in determinants of usage, without spatial bias.

The conceptual model is illustrated by the example of Mauritius over the five decades since its independence from Great Britain in 1968. The Mauritius case qualitatively validates the developing world model, in its emphasis on education and foreign direct investment.

The detailed findings are translated into recommendations for national, regional and global policymakers, to help them customize approaches suitable for their income levels and geographical locations.

References

Azari, R., & Pick, J. B. (2005). Technology and society: socioeconomic influences on technological sectors for United States counties. *International Journal of Information Management, 25*(1), 25–37.

Baliamoune-Lutz, M. (2003). An analysis of the determinants and effects of ICT diffusion in developing countries. *Information Technology for Development, 10*(3), 151–169.

Bass, F. M. (1969). A new product growth for model consumer durables. *Management Science, 15*, 215–227.

Cuervo, M. R. V., & Menéndez, A. J. L. (2006). A multivariate framework for the analysis of the digital divide: Evidence for the European Union-15. *Information & Management, 43*, 756–766.

Cuervo, M. R. V., & Menéndez, A. J. L. (2011). Assessing the regional digital divide across the European Union-27. *Telecommunications Policy, 35*, 220–237.

Cyranoski, D. (2013). Japan aims high for growth: Innovation in science is at the heart of government plans to boost the economy, May 29. *Nature, 497*(7451).

Deen-Swarray, M., Gillwald, A., & Morrell, R. (2012). Lifting the veil on ICT gender indicators in Africa. Evidence for ICT policy action, policy paper 13. *Capetown, Africa*: http://www.researchictafrica.net/.

European Commission. (2014). *Information society*. Brussels, Belgium: European Commission. Available on August 21, 2014 at http://europa.edu/legislation_summaries/information_society.

Gao, P., & Yu, J. (2010). Emerging markets: Has China caught up in IT? *Communications of the ACM, 53*(8), 30–32.

Guillén, M. F., & Suárez, S. I. (2004). Explaining the global digital divide: economic, political and sociological drivers of cross-national internet use. *Social Forces, 84*(2), 681–708.

Gulati, G. J., & Yates, D. J. (2012). Different paths to universal access: The impact of policy and regulation on broadband diffusion in the developed and developing worlds. *Telecommunications Policy, 36*, 749–761.

Hale, D., & Hale, L. H. (2003). China takes off. *Foreign Affairs, 82*(6), 36–53.

Hargittai, E. (1999). Weaving the Western Web: Explaining differences in Internet connectivity among OECD countries. *Telecommunications Policy, 23*, 701–718.

ITU. (2009). *Yearbook of statistics*. Geneva, Switzerland: International Telecommunications Union.

ITU. (2013). *Measuring the information society*. Report. Geneva, Switzerland: International Telecommunication Union.

Kiiski, S., & Pohjola, M. (2002). Cross-country diffusion of the internet. *Information Economics and Policy, 14*, 297–310.

Lengsfeld, H. B. (2011). An econometric analysis of the sociodemographic topology of the digital divide in Europe. *The Information Society, 27*, 141–157.

Ono, H., & Zavodny, M. (2006). Digital inequality: A five country comparison using microdata. *Social Science Research, 36*(2006), 1135–1155.

Onyeiwu, S. (2002). Inter-country variations in digital technology in Africa. Discussion Paper No. 2002/72. World Institute for Development Economics Research, United Nations University, Tokyo, Japan.

Oolun, K., Ramgolam, S., & Dorasami, V. (2012). The making of a digital nations: Toward i-Mauritius, Chap. 2.2. In: S. Dutta & B. Bilbao-Osorio (Eds.), *The global information technology report 2012*, Geneva, Switzerland, World Economic Forum and INSEAD (pp. 161–168).

Oshri, I., Kotlarsky, J., & Willcocks, L. P. (2011). *The handbook of global outsourcing and offshoring* (2nd ed). Basingstoke, UK: Palgrave Macmillan.

Pick, J. B., & Azari, R. (2008). Global digital divide: Influence of socioeconomic, governmental, and accessibility factors on information technology. *Information Technology for Development, 14*(2), 91–115.

Pick, J. B., & Azari, R. (2011). A global model of utilization of technology based on governmental, social, economic, and business investment factors. *Journal of Management Information Systems, 28*(1), 51–85.

Pick, J. B., Nishida, T., & Sarkar, A. (2014). Broadband utilization in the Indian states: Socio-economic correlates and geographic aspects. In J. Choudrie & C. Middleton (Eds.), *Management of broadband technology innovation* (pp. 269–296). Oxford, England: Routledge.

Population Reference Bureau. (2013). *World population data sheet*. Washington, DC: Population Reference Bureau.

Quibria, M. G., Ahmed, S. N., Tschang, T., & Reyes-Macasaquit, M.-L. (2003). Digital divide: Determinants and policies with special reference to Asia. *Journal of Asian Economics, 13*, 811–825.

Republic of Mauritius. (2012a). *National information and communication technology strategic plan 2011–2014: Towards i-Mauritius*. Port Louis, Mauritius: Ministry of Information and Communication Technology, Republic of Mauritius.

Republic of Mauritius. (2012b). *National broadband policy 2012–2020*. Port Louis, Mauritius: Ministry of Information and Communication Technology, Republic of Mauritius.

Republic of Mauritius. (2013). *e-Government strategy 2013–2017*. Port Louis, Mauritius: Ministry of Information and Communication Technology, Republic of Mauritius.

Republic of Mauritius. (2014). National computer board. Port Louis, Mauritius: Ministry of Information and Communication Technology, Republic of Mauritius. Available on August 19, 2014 at http://www.ncb.mu/English.

Robison, K. K., & Crenshaw, E. M. (2002). Post-industrial transformations and cyber-space: A cross-national analysis of Internet development. *Social Science Research, 31*, 334–363.

United Nations. (2011). Composition of macro geographical (continental) regions, geographical sub-regions, and selected economic and other groupings. New York: United Nations. Available on February 18, 2011 at http://unstats.un.org/unsd/methods/m49/m49regin.htm.

U.S. Central Intelligence Agency. (2014). Africa: Mauritius. *The World Factbook*. Washington, DC: Central Intelligence Agency.

World Bank. (2010). *World development indicators*. Washington, DC: World Bank.

World Economic Forum. (2010). *Global information technology report*. Geneva, Switzerland: International Telecommunications Union.

World Economic Forum. (2012). *Global information technology report*. Geneva, Switzerland: International Telecommunications Union.

Yates, D. J., Gulati, G. J., & Weiss, J. W. (2011). Different paths to broadband access: the impact of governance and polity on broadband diffusion in the developed and developing worlds. In: *Proceedings of 44th Hawaiian International Conference on System Sciences*. IEEE.

Chapter 5
China's Digital Divide

5.1 Introduction

China has the largest volume of information technology users and continues to grow its digital stature. In 2012 China had 864 million mobile telephone subscriptions, 463 million Internet users, and 127 million broadband subscribers, which compares to respective totals in the United States of 282 million mobile telephone subscribers, 232 million Internet users, and 87 million broadband subscribers (ITU 2012; Population Reference Bureau 2012). China has moved into first place among nations in its volume of ICT use, an advance that predominantly took place since the turn of the 21st century.

However, on the World Economic Forum's Network Readiness Index (NRI), China's world rank was only 58th in 2013 (World Economic Forum 2013). Mainland China is also very diverse regionally in technology, with high levels of ICT use in its two largest metropolises of Beijing and Shanghai and in its southeast manufacturing and export region, which contrasts with relatively low levels of ITC availability and utilization in its vast western and southwestern regions.

China's unprecedented economic expansion in this century along with its government's quest for leadership in innovation and technology have been motivating forces for its ICT expansion. This chapter begins by examining the history of China's technology expansion, which has had major accomplishments along with setbacks such as its inability to develop ICT in its western region at the rate of the rest of the nation and the ambiguities in the government control policies for certain types of information.

Next, the chapter discusses in more depth China's regional differences, concentrating on spatial distribution of technologies in two marker years of rapid growth, 2006 and 2009. Next, it presents the specific conceptual model for technology determinants in China, based on the book's general model in Chap. 3. In justifying the Chinese conceptual model, the chapter gives special attention to independent variables that were not available for the general model, in particular

J.B. Pick and A. Sarkar, *The Global Digital Divides*, Progress in IS,
DOI 10.1007/978-3-662-46602-5_5

employment in non-state-owned units, export commodities value, individual's fixed asset investments, published books, public libraries, and expenses in government departments of education, science, public health and culture. The major determinants of use of ICTs in 2006 and 2009 are analyzed and discussed. China's remarkable growth can be viewed as part of the "second wave" of the global Internet to middle-level nations in the 2000s, after the "first wave" occurred in economically advanced countries in the 1990s. China now seeks to leap beyond emulation of foreign technology leaders and generate indigenous innovation and world competitiveness in selected industries, an approach sometimes referred to as "techno-nationalism" (Economist 2011; Gao and Yu 2010).

Later in the chapter, the case study of Alibaba reveals that firm's exceptional growth in ICT and e-commerce, reinforcing the empirical results for our conceptual model. The Google case illustrates government curtailment of Internet content, filling in for a deficit of democracy and societal openness attributes in our formal model. Lastly, the chapter examines technology policy implications for China, before concluding.

5.2 Historical Background

In 1979, as Chairman Deng Xiaoping led in opening up the Chinese economy, computers were very sparse, only present in the military but rarely in the state owned enterprises that almost entirely comprised the business sector. This stands in large contrast to the United States, which had developed its computer industry and business systems starting in the late 1950s, and India, which had a vibrant early computer industry and skilled workers (McFarlan et al. 2012). Similarly, a large disparity existed between China and then British Hong Kong; for example in 1980 China averaged 4 phones per 1000 people, while Hong Kong had 460 phones per 1000. In the 1980s, as China's economy strengthened, low cost manufacturing was emphasized and foreign direct investment encouraged, often leading to joint ventures and acquisition of foreign intellectual property (World Bank 2013).

With the early backing of the Chinese central government, ICT progressed fairly slowly in the 1980s, but was spurred by the Chinese government's establishment in 1980 of Special Economic Zones (SEZs), initially in the East and Southeast. SEZs were a success serving as growth foci of surrounding regions, including the development of parts of Guangdong Province into the world's manufacturing center.

After the advent of the Internet in China in 1992, Chinese ICT went through three phases that correspond to adoption-diffusion theory from Chap. 3 (Guo and Chen 2005). In the initial phase (1993–1995), the first national internet backbone, called the National Computing and Networking Facility, was built and e-mail was initiated into use. In the next few years, other national and most provincial networks were added to the backbone, as well as connections to some pioneering private enterprises. In 1994, China's National Information Infrastructure Steering Committee was created, which functioned for over a decade to provide nationwide platforms including for the Internet, e-commence, and credit card payments (Guo and Chen 2005). The speed of this phase was due to the then openness to foreign technology companies, to government support, and to large demand for technology (Guo and Chen 2005).

In the contagion phase (1995–2000), networks multiplied and e-commerce started up and expanded rapidly. This phase abruptly ended with China's version of Dot Com Bust in 2000, a decline bankrupting many companies and slowing overall progress. During the cooling phase (2000–2004), there was emphasis on controlling unchecked technology growth (Guo and Chen 2005), while the Internet continued to diffuse across the nation and into secondary level provinces, although not to remote areas (Song 2008). The nation's five year plans now formally included ICT expansion, and China's IT services and software industries started to grow, with government support (McFarlan et al. 2012). Based on the plans the government has continued to make significant investments in the development of the Internet (Raven et al. 2007).

In the decade that began in 2004, termed the permeation phase (Guo and Chen 2005), ICT has become a standard feature of all medium and large sized Chinese enterprises. The ICT industry grew much larger although with a domestic orientation, while several Chinese firms were able to break into worldwide leadership, in particular Huawei in telecommunications and networking; Lenovo, which in 2005 purchased IBM's PC division and in 2013 became the world's largest PC manufacturing, pushing aside Hewlett Packard (McFarlan et al. 2012, and see Chap. 10 case study); and Alibaba covered later as a case study. Recent Chinese five year policies and plans have stressed technology and innovation as national goals. Chinese policy and planning experts are seen in Fig. 5.1 attending a Beijing seminar on Big Data.

In volume, the Chinese ICT development and advances during the past 15 years have been the largest in world history, resulting in the country's world leadership in consumer ICT usage, ICT manufacturing, and e-commerce. Some weaknesses that remain include moderate levels of domestic technology innovation, minimal ICT bandwidth and usage in remote regions, limited Chinese international technology presence, and persistence of government programs to control and censor information content for consumers.

The analyses in this chapter concern the 2006–2009 mid period in this century's vast Chinese ICT technology expansion.

5.3 Current Dimensions of Access and Use of Technologies in China

This section analyzes China's ICT usage and geographical distribution, and determinants of use during 2006 and 2009. The study is based on the nation's 31 administrative units, which consist of 22 provinces, 5 autonomous regions, and the 4 centrally administered municipalities of Beijing, Shanghai, Tianjin, and Chongqing. In this book, the 31 units will be referred to as provinces.

Fig. 5.1 Chinese policy and planning experts attending a big data seminar, Beijing, China

5.3.1 China's East and Southeast: Global Factory

China's Southeast and East have been the most prosperous and ICT-intensive areas in recent Chinese history, and continue to be so today. The most prominent early centers for ICT in China in the 1990s were Beijing, Shanghai, and Guangdong, and they remain so. Beijing, the capital and academic and cultural center of the nation, received early government support and has developed a famous high-tech zone of Zhongguancun. Shanghai, as the financial and business center of mainland China, developed early and rapidly in ICT. In the economic boom of the first decade of the 21st century, the southeastern and eastern provinces expanded ICT concomitant with their roles as export-oriented manufacturing areas.

In Guangdong Province, the Pearl River Delta region that extends from Shenzhen south to Hong Kong and east to Guangzhou is today among the world's greatest manufacturing complexes, so much of the stimulus for the region's IT stems from its crucial role in production facilities and networks. For example, the city of Shenzhen grew very rapidly from 1.2 million to 10.5 million population from 1990 to 2012, with its metropolitan economy growing to $237 billion, and centered on technology-related industries such as pharmaceuticals, medical equipment, electronics, and telecommunications. Its ICT prominence is reflected in having the world's largest Apple manufacturing plant, as part of Foxconn's megafactory of up to 250,000 workers. Shenzhen's human resources benefit from the city's presence of satellite

Fig. 5.2 Worker at monitor screen factory in Shenzhen, China. *Source* International Labor Organization

campuses of Beijing and Tsinghua Universities, as well as of local universities. A Shenzhen factory producing monitors is seen in Fig. 5.2.

Also in Guangdong Province is the city of Dongguan, which is a city of 9 million about halfway between Shenzhen and Guangdong. Many transnational corporations (TNCs) located there since the mid 1990s. PC manufacturers from Taiwan have built a cross-border industrial complex in this city, which has stimulated ICT infrastructure and utilization (Yang and Liao 2008). Its complex of PC companies has strong production chain relationships with suppliers in Taiwan, which together produce world-class networked technology. The TNCs have had weak relationships with local Chinese suppliers; The city government aspires to strengthen those ties in the future, which would require even more ICT upgrading for the city and region (Yang and Liao 2008). Since the urban zone of Dongguan is only a small part of a vast export manufacturing region, the example points to FDI as only one of the drivers of ICT growth throughout the east-southeast of the nation.

Also in the Southeast are the two Chinese Special Administrative Regions (SARS) of Hong Kong and Macau, each of which broke off from colonial rule—Hong Kong from Great Britain in 1997 and Macau from Portugal in 1999. They are not included in this chapter's analysis, because their government data do not match that of mainland China, but rather they conduct their own censuses.

Both Hong Kong and Macau are among the world's most advanced locations for ICT. The World Economic Forum ranked the Hong Kong SAR as 14th worldwide in its 2013 Networked Readiness Index, compared to a rank of 58th for China (WEF 2013), while in mobile phone subscriptions Hong Kong is in first place worldwide at 2.15 subscriptions per person, compared to China in 115th place at 0.73 per person (WEF 2013). Although not in the WEF data-base, Macau also has over 2 mobile phone subscriptions per person (ITU 2013). In the Asia Pacific region, Macau has the lowest fixed broadband prices as percent of national income per capita, with Hong Kong in 3rd place.

The reasons they are at extraordinary ICT levels stems from their very high per capita income, Hong Kong's role as world financial center, presence of leading universities, and government initiatives before and after sovereignty transfer to

China. For instance, Hong Kong SAR connected to its eighth high-speed undersea cable in 2012 and more connections are planned. Consequently, Hong Kong has 185 broadband internet service providers, with 87 % of households connected to broadband internet (ITU 2013).

In China's East and Southeast, as seen for 2009 in Fig. 5.3, broadband rates are consistently high averaging about 15 subscribers per 100, which although much lower than the two SARs, is very high for the mainland.

The central region of China, in between the extremes of the East/Southeast and West, has benefitted by moderate educational and income levels, and trickle-down effects from the leading ICT areas in the East/Southeast. Because the Central area of China is more populated and prosperous, and less remote than most of the West, it has moderate ICT infrastructure, including fiber-optic networks, with less dependence on the mobile networking serving the remote areas of the West.

5.3.1.1 The Trailing West of China

In late 1990s, in response to the economic disparity between regions, the Chinese government brought forward the "Develop the West" policy. Although the West comprises twelve provinces and administrative units, it covers 73 % of the nation's land area and 28 % of population, and in 2009 had 20 % of GDP, 21 % of fixed asset investment, and 8 % of foreign direct investment (Ma and Summers 2009). In general it is much poorer than the other regions (Moxley 2009) and slower in moving to a market-based and export-oriented economy. On the other hand, it contains most of China's major natural resources, in particular 60 % of reserves of coal, 70 % of natural gas reserves, 80 % of water resources, and the preponderance of rare earth minerals (Ma and Summers 2009). The "West" concept also has come to connote a somewhat-mistaken frontier and backward image of this part of China (Ma and Summers 2009).

The multi-faceted Develop the West policy sought to develop the region through improvement in human resources, expansion in technology and science, infrastructure improvement, development of energy and natural resources, reduction in state-owned enterprises, growth of the market economy, and expansion of foreign direct investment (Ma and Summers 2009; Moxley 2009). China's 10th 5-Year Plan starting in 2001 included many stimuli for the West, including taxation, targeted expenditures by the provincial and national government, and improved governance (Ma and Summers 2009). Consequently, the West experienced higher economic growth rates during the 2000s than the nation as whole. For instance, its foreign direct investment (FDI) increased by seven fold during 2000–2009; however, this FDI was targeted at resource extraction and industry that serves the domestic market, rather than to developing export-oriented linkage with global supply chains as in the east (Ma and Summers 2009). Another issue in the West has been its lower level of human development (i.e. education, life expectancy, and economic development) compared to the central and eastern regions (Ma and Summers 2009). Not only does this gap limit potential development of technological workforce, but it also reduces the lifestyle incentives for educated, prosperous ICT workers to migrate to the West from the East.

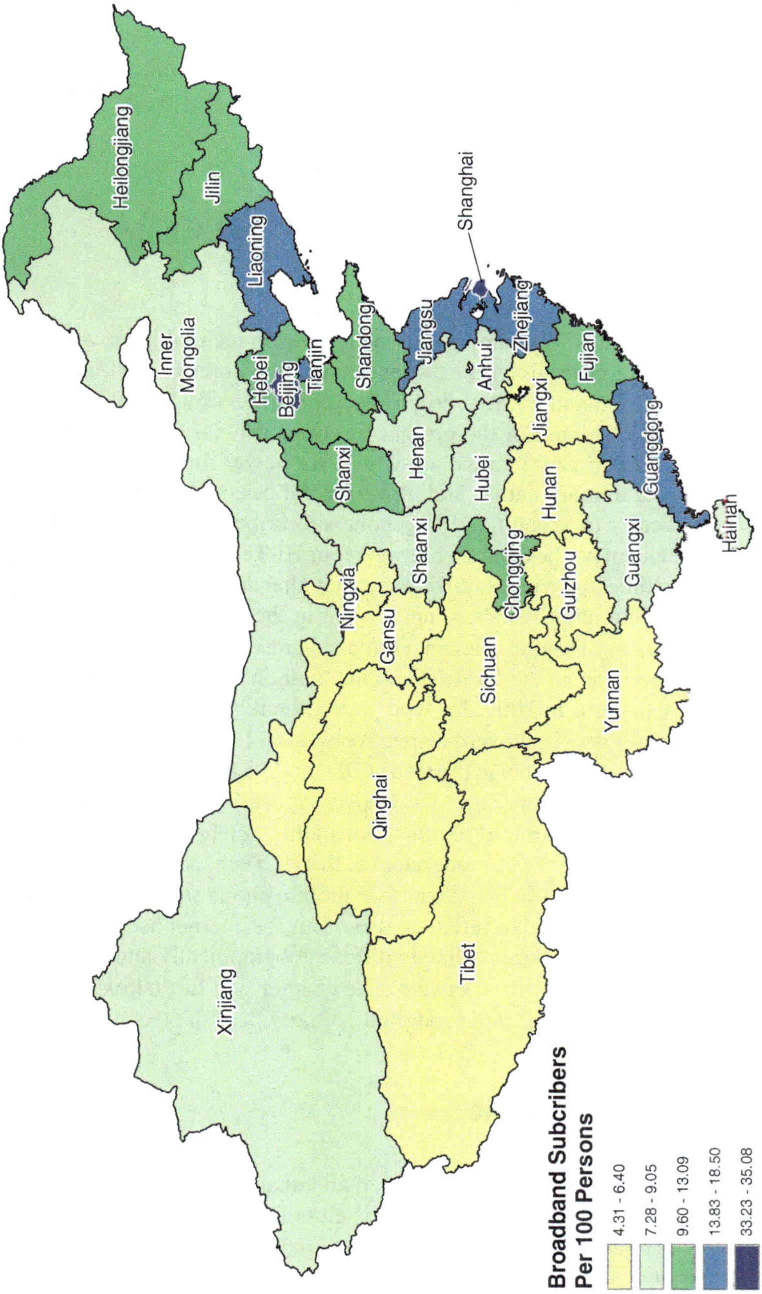

Fig. 5.3 Broadband subscribers per 100 persons in Chinese Provinces, 2009

Government programs have promoted Internet infrastructure and e-commerce in the West, installing fiber-optic trunk lines, introducing wireless capability in remote and gradually bringing in financial networks and platforms to the region (Davison et al. 2005). However, compared to the East, high-tech investments have been quite limited, and some attempts at large technology investments in the West, such as in Xinjiang in the late 1990s, failed (Davison et al. 2005).

5.4 Spatial Arrangement of Technologies Within China, 2006 and 2009

Within this context of great ICT advances but unevenness in technology availability and utilization across different regions, the present analysis explores ICT diffusion and spatial distributions in 2006 and 2009, followed by application of the book's general ICT utilization model to China. For the present recent provincial data, the regional differences, all stated in per capita terms, are widest for newer forms of access particularly web pages and domain names and also for rural pcs, and narrowest for the traditional usage indicator of urban fixed telephone subscribers The regional differences are strikingly evident by the contrast between four Hi-Tech provinces, the major metropolitan municipalities/provinces of Beijing, Shanghai and Guangdong (with its large cities of Guangdong and Shenzhen) and Fujian in the southeast; five Middle-Tech provinces in the central region (Henan, Human, Jiangxi, Shanxi, and Sichuan); and five Low-Tech provinces in the far west and far south, namely Gansu, Xinjiang, Tibet, Yunnan, and Guizhou. In 2009, for web pages, the regional averages for Hi-Tech, Middle-Tech, and Low-Tech were respectively 28.9, 1.1, and 0.45 web pages per capita, with Beijing and Shanghai in the lead at 70.7 and 35.8 web pages per capita. This points to the high outlier status the two leaders, as well as the vast difference between the High-Tech and other two regions. For mobile telephone subscriptions, regional differences were present but moderate, i.e. the Hi-Tech, Medium-Tech, and Low-Tech regions had respectively, 77, 25, and 25 mobile phone subscribers per 100 population. The moderate regional differences also occur for internet users and urban pcs. These technologies have had more time to diffuse geographically and for mobile phones and internet are relatively inexpensive. The chapter will further examine the regional contrasts in the context of rapid longitudinal growth in the technologies.

5.5 Clusters of Similar Provinces

In this section, clusters of provinces are identified and analyzed for 2006 and 2009. As explained in Chap. 1, k-means cluster analysis classifies states or provinces into clusters, based on the full set of dependent ICT variables, with the objective to achieve the most homogeneity with clusters as possible. For this chapter, six clusters were designated by K-means cluster analysis for 2006 and six clusters for 2009. The provinces included in respective clusters can be seen in Table 5.1. This analysis is

Table 5.1 Cluster characteristics, China, showing policy areas for 2006 and 2009

Year 2006

Dependent variable	Clusters 2006						Ratio of high to low clusters	Ratio of high to low clusters, excluding Beijing and Shanghai
	Policy area 1		Policy area 3	Policy area 2	Policy area 4			
	Cluster 1	Cluster 2	Cluster 3	Cluster 4	Cluster 5	Cluster 6		
Pop. (in 1000s)	15,810	18,150	93,040	199,710	792,231	172,372		
Broadband	0.2120	0.2321	0.0780	0.0806	0.0359	0.0223	6.5	2.2
Domain names	0.0497	0.0208	0.0069	0.0051	0.0013	0.0007	73	10.1
Internet user	0.2960	0.2810	0.1968	0.1502	0.0880	0.0535	5.5	3.7
Mobile subscribers	0.9937	0.8868	0.7650	0.4552	0.2926	0.2054	4.8	3.7
Urban fixed telephone	0.5093	0.6061	0.3051	0.2876	0.1551	0.1166	5.2	2.6
Web pages	70.6626	35.8387	4.6896	4.8151	1.2242	0.7412	95.3	6.5
	Beijing	Shanghai	Guangdong	Chongqing	Guangxi	Anhui		
				Fujian	Hainan	Gansu		
				Jiangsu	Hebei	Guizhou		
				Tianjin	Heilongjiang	Tibet		
				Zhejiang	Henan	Yunnan		
					Hubei			
					Hunan			
					Inner Mongolia			
					Jiangxi			
					Jilin			
					Liaoning			
					Ningxia			

(continued)

Table 5.1 (continued)

Year 2006

Dependent variable	Clusters 2006						Ratio of high to low clusters	Ratio of high to low clusters, excluding Beijing and Shanghai
	Policy area 1		Policy area 3	Policy area 2	Policy area 4			
	Cluster 1	Cluster 2	Cluster 3	Cluster 4	Cluster 5	Cluster 6		
					Qinghai			
					Shaanxi			
					Shandong			
					Shanxi			
					Sichuan			
					Xinjiang			

Year 2009

Dependent variable	Clusters 2009						Ratio of high to low clusters	Ratio of high to low clusters, excluding Beijing and Shanghai	Ratio of averages area 1 in 2009–2006
	Policy area 1		Policy area 3	Policy area 2	Policy area 4				
	Cluster 1	Cluster 2	Cluster 3	Cluster 4	Cluster 5	Cluster 6			
Pop. (in 1000s)	17,550	19,210	132,650	64,082	558,340	524,776			
Broadband	0.3209	0.3448	0.1662	0.1766	0.1040	0.0661	5.2	2.7	1.5
Domain names	0.1687	0.0755	0.0213	0.0270	0.0072	0.0037	20.4	7.3	3.5
Internet user	0.6285	0.6096	0.4892	0.4706	0.2869	0.2022	3.1	2.4	2.1
Mobile subscribers	1.0402	1.1001	0.8716	0.8503	0.5702	0.4173	2.6	2.1	1.1

(continued)

Table 5.1 (continued)

Year 2009									
Dependent variable	Clusters 2009						Ratio of high to low clusters	Ratio of high to low clusters, excluding Beijing and Shanghai	Ratio of averages area 1 in 2009–2006
	Policy area 1		Policy area 3	Policy area 2	Policy area 4				
	Cluster 1	Cluster 2	Cluster 3	Cluster 4	Cluster 5	Cluster 6			
Urban fixed telephone	0.3956	0.4807	0.2569	0.2665	0.1526	0.1131	4.2	2.4	0.8
Web pages	549.8488	171.0206	48.7393	72.6208	10.9862	6.5049	84.5	11.2	6.8
	Beijing	Shanghai	Fujian	Tianjin	Chongqing	Anhui			
			Guangdong	Zhejiang	Guangxi	Gansu			
					Hainan	Guizhou			
					Hebei	Heilongjiang			
					Hubei	Henan			
					Inner Mongolia	Hunan			
					Jiangsu	Jiangxi			
					Jilin	Qinghai			
					Liaoning	Sichuan			
					Ningxia	Tibet			
					Shaanxi	Xinjiang			
					Shandong	Yunnan			
					Shanxi				

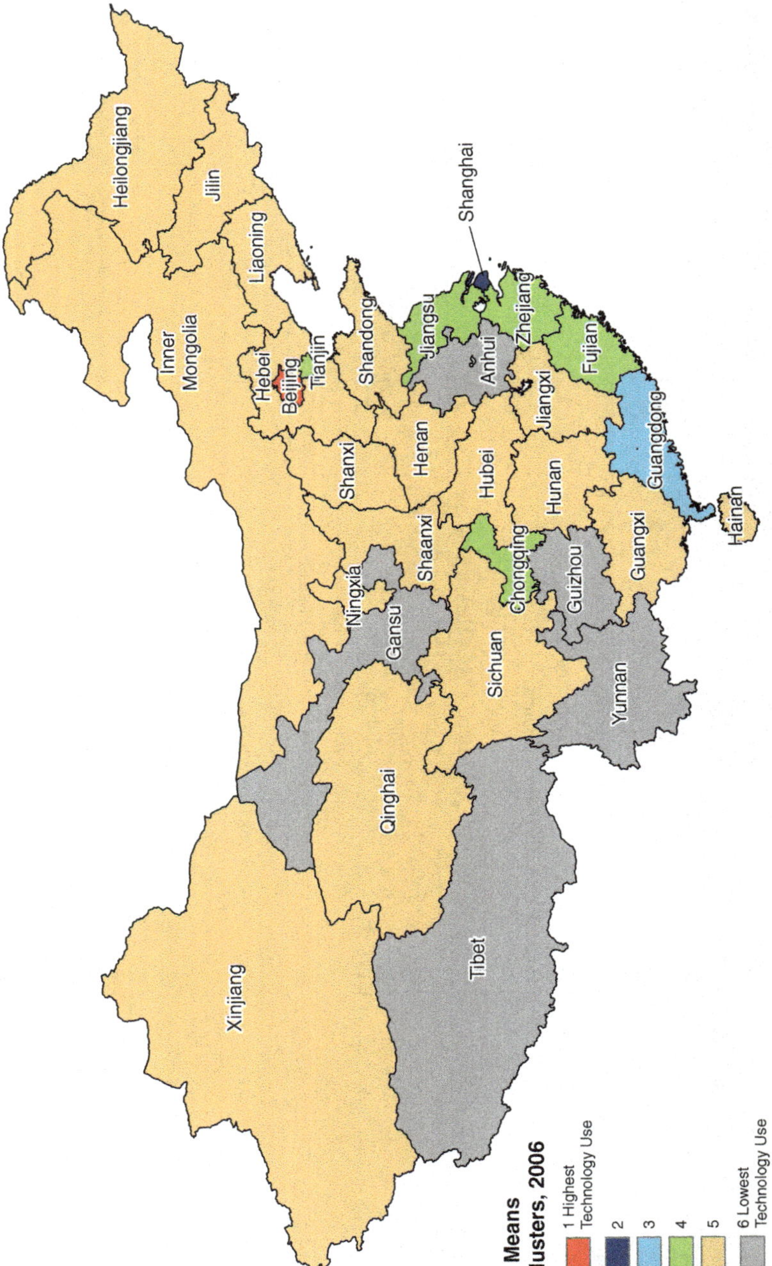

Fig. 5.4 K-means cluster analysis of use of ICTs in Chinese Provinces in 2006

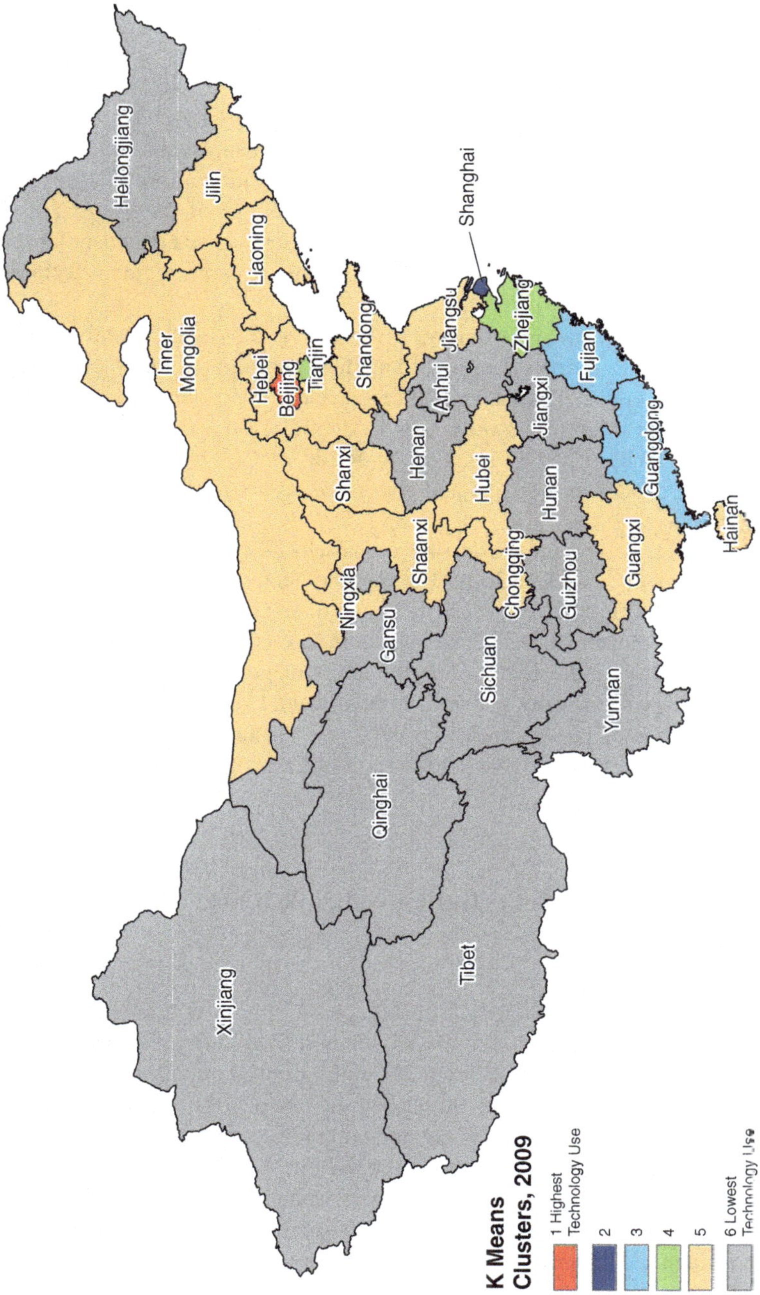

Fig. 5.5 K-means cluster analysis of use of ICTs in Chinese Provinces in 2009

not confirmatory, but is done to explore for geographically distinct technology clusters, to see how the patterns evolve over time. For six clusters in 2006 (Fig. 5.4), there were unique clusters for the provinces of Beijing, Shanghai, and Guangdong, as well as a mostly eastern coastal cluster (Fujian, Jiangsu, Tianjin, Zhejiang, and Chongqing). China's central and western regions contain the fifth and sixth clusters. The fifth consists of 18 western, central and northeastern provinces, while the sixth comprises one southwestern, two south central, one north central, and one eastern province. By far the highest technology availability and utilization are the two clusters of Beijing and Shanghai. The ratio of the average of the Beijing and Shanghai clusters to the average of the fifth and sixth clusters varied between 3.8 for mobile subscribers to 54.2 for web pages, underscoring the large regional technological differences in 2006.

Cluster findings for the Chinese provinces in 2009 (Fig. 5.5) again reveal unique, highest-technology clusters for Beijing and Shanghai; two high-technology eastern and southeastern clusters of Tianjin and Zhejian, and Fujian and Guangdong; a large low-technology cluster of 12 provinces in central east and north excluding Heilongjiang, plus Guangxi and Hainan in the south; and a large very-low-technology 13-province cluster in the west and south central areas plus Heilongjian in northern Manchuria.

The results for the two years are mostly consistent in cluster groupings, with some changes. For both 2006 and 2009, Beijing and Shanghai stand out as provinces with the highest ICT utilization. In both years, there is a set of two moderately high eastern/southern clusters and a set of two large low technology clusters in the west and north. The main difference is that the splits of provinces within each set change over time. It is noteworthy that, in the three years from 2006 to 2009, there are double the number of provinces in the very low cluster 6. Overall, the geographic groupings are consistent over time, in spite of China's rapid increase in ICT utilization.

5.6 Conceptual Model of Technology Availability and Utilization

Based on the book's general conceptual model for the digital divide (Fig. 3.8), a specific conceptual model of China's digital divide is shown in Fig. 5.6. This model has twelve independent factors that relate to eight dependent ones. The model screens for spatial autocorrelation as described in Chap. 1. For the dependent variables, exploratory cluster analysis is conducted that is spatially mapped. For the independent variables, five of the eight categories in Fig. 3.8 are represented except for demographic/ethnic, societal openness, and government support. Demographic/ethnic variables were excluded because prior research on China revealed no demographic influence while ethnic variables were not available at the provincial level. Societal openness variables are unavailable from the Chinese government or other sources due to cultural, political, and census-taking factors presently

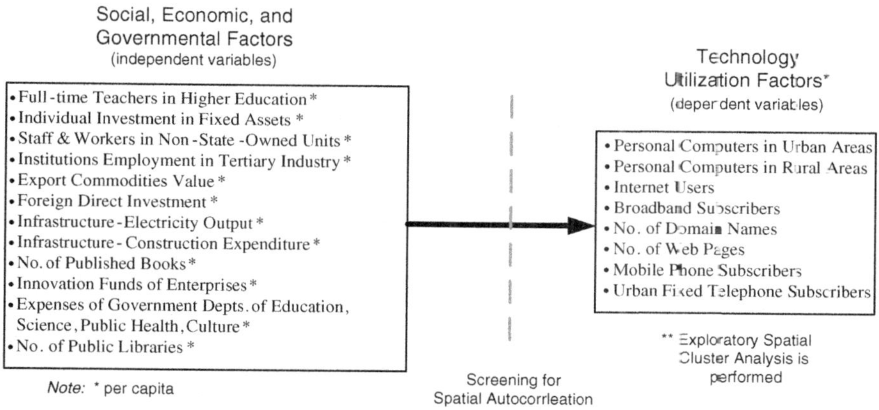

Fig. 5.6 Specific conceptual model for digital divide in China

precluding their data collection. The variable, government support and prioritization of ICT, likewise is not available, which might be due to provincial budgetary or policy quantitative indicators not being publicly provided.

The eight dependent technology utilization factors include those in the general model (Fig. 3.8) except for social media, attributes which are not presently available from the Chinese government or private sources at the provincial level. Definitions and sources of the China variables are shown in Table 5.2. Some of these variables were justified in the general model discussion in Chap. 3, including higher education, foreign direct investment, infrastructure, and innovation. For innovation, the variable of innovation funds in enterprises is available, while published books is explained below.

Eight of the independent variables—individual investment in fixed assets, workforce in non-state-owned units, tertiary industry employment, export commodities value, government expenses in education/science/public health/culture, and number of public libraries—were not justified in the sub-section on independent variables in Chap. 3, so their justification is presented here.

Individual investment in fixed assets. Cost of technology has been identified as a barrier to technology adoption in several studies of China, especially for rural areas (Fong 2009; Qiu 2010). For instance, in 2006, the average per capita expenditure for mobile telecommunications as a proportion of disposable income was 29 % in rural areas and 9 % in urban areas (Fong 2009). The fact that two thirds of the Internet connections were to homes rather than workplaces and public venues such as Internet cafes is noteworthy, as it reflects the importance of individual investments for the spread of technology (Zhu et al. 2005). In EU nations, national wealth index was a major influence on broadband use (Yates et al. 2011). Worldwide, per capita wealth is correlated with cell phone diffusion (Comer and Wikle 2008) and access to mobile networks (Ahn and Lee 1999). A study of developing nations found access cost to be the leading and inverse factor associated with reduced online public services (Vicente Cuervo and López-Menéndez 2006).

Table 5.2 Definitions of China variables

Variable	Dependent or independent	Definition	Years of data collection	Data source
PCs per 100 urban families	Dep.	Personal computers per 100 urban families	2006, 2009	CI Almanac
PCs per 100 rural families	Dep.	Personal computers per 100 rural families	2006, 2009	CI Almanac
Broadband subscribers	Dep.	Broadband subscribers per 100 population	2006, 2009	CSY
Internet users	Dep.	Internet users per 100 population	2006, 2009	CNNIC
Domain names	Dep.	Number of domain names per 100 population	2006, 2009	CNNIC
Web pages	Dep.	Number of web pages per capita	2006, 2009	CNNIC
Mobile telephone subscribers	Dep.	Number of mobile telephone subscribers per 100 pop.	2006, 2009	CSY
Urban fixed telephone subscribers	Dep.	Number of urban fixed telephone subscribers per urban population	2006, 2009	CDO
Electricity output	Indep.	Electricity output (in 100 million kilowatt hours) per capita	2006, 2009	CDO
construction enterprise output	Indep.	Construction enterprise output (in 100 million yuan) per capita	2006, 2009	CDO
Employed persons in tertiary industry	Indep.	Ratio of employed persons in tertiary industry to all employed persons	2006, 2009	CDO
Staff and workers in non-state-owned units	Indep.	Staff and workers in non-state-owned units (of 10,000 persons) per capita	2006, 2009	CDO
Individual investment in fixed assets	Indep.	Individual investment in fixed assets (of 100 milllion yuan) per capita	2006, 2009	CDO
Export commodities value	Indep.	Value of export commodities at the place of export origin (in 1000 U.S. dollars) per capita	2006, 2007	CDO
Foreign direct investment	Indep.	Foreign direct investment (in 10,000 U.S. dollars) per capita	2006, 2008	CDO
Published books	Indep.	Number of published books (in 100 million copies) per capita	2006, 2009	CDO

(continued)

Table 5.2 (continued)

Variable	Dependent or independent	Definition	Years of data collection	Data source
Innovation funds	Indep.	innovation funds of enterprises (in 100 million yuan) per capita	2006, 2009	CDO
Government expenses in education, science, public health, and culture	Indep.	Expenses of government departments of education, science, public health, and culture (in 100 million yuan) per capita	2006, 2009	CDO
Public libraries	Indep.	Number of public libraries per capita	2006, 2009	CDO
Full-time teachers in higher education	Indep.	Number of full-time teachers in higher education institutions (10,000 persons) per capita	2006, 2009	CDO

Note All variables are numerical
CI Almanac China Information Almanac (2010)
CSY China Statistical Yearbook (2010)
CDO China Data Online (2010)
CNNIC China Internet Information Center (2007, 2009)

Workforce in non-state-owned units. As China opened up to a market economy over the past two decades, the market driven (i.e. non-state-owned) sector expanded, while the state-owned sector diminished. A sizeable workforce segment of "old labor" from the state-owned sector was laid off, while the market driven economy recruited more skilled labor (e.g. knowledge workers, technologists) with information skills, and added workforce from less skilled labor migrating from rural to urban areas (Zhu and Wang 2005; Qiu 2010). Much of the expanded technology use occurred in the market driven sector (Guo and Chen 2005; Qiu 2010). We reason that higher per capita employment in the market-driven sector of a province is associated with greater technology availability and utilization. Correspondingly, we posit that the market-driven sector in China has greater compatibility with technology than the state-owned sector. Adoption-Diffusion Theory implies that non-state-owned employment, that we reason has greater trialability and observability, is positive for the spread of technology. Accordingly, the construct of *employment in non-state-owned units* is included in the conceptual model.

Tertiary industry employment has been important in a variety of studies. Scientists and engineers per capita were two dominant determinants of technology capacity in 49 nations (James and Romijn 1997). In an investigation of socioeconomic influences on technology-sector receipts and payrolls in U.S. counties, the most important determinant was professional/scientific/technical services workforce, followed by other services workforce (Azari and Pick 2005). For China, technology adoption has been shown to be mostly led by knowledge workers, that is, professionals, scientists, and highly skilled business people (Qiu 2010).

Export commodities value. China is following, at a lag, an economic model that was successful in Singapore, Taiwan, and South Korea, which emphasizes exports, and in turn stimulates ICT industries (Meng and Li 2002). This approach has resulted in major geographical differences in China's eastern, central, and western regions in per capita GDP and in ICT penetration. Even for the early year of 1999 the Internet penetration rate was 0.56 in the eastern region, 0.14 in the central region, and 0.18 in the west region, while the GDP per capita in the East was 2.39 times higher compared to the West (Meng and Li 2002). Although trade and export have infrequently been examined as correlates of technology level, an investigation of technology capability in 49 countries found that an inward trade policy reduced technology capability, although an outward trade policy did not increase it (James and Romijn 1997). Provinces with high levels of export commodities have relative advantage over other provinces for technology use, since their enterprises are more likely to be at a world supply chain level and their production processes are more likely to be complex, implying greater information flow. Adoption-Diffusion Theory's emphasis on relative advantage implies higher technology availability and utilization. The Chinese government defines export commodities value as the value of export commodities at the place of origin of export.

Number of published books. Since content is essential for Internet and Web-related dependent variables, and Adoption-Diffusion Theory posits a positive effect from observability on technology usage (Rogers 2003), number of published books originating in a province is included in the conceptual model as an innovation-related factor. It is also innovative since increasingly book knowledge content is being made available in e-published form, without printing. This e-knowledge production is an emerging innovative area in China. The number of published articles has been identified as an important R&D correlate of ICT use (Pick and Azari 2008).

Government expenses in education/science/public health/culture. Prior research has confirmed the crucial importance of government support for education, science, and innovation to technology availability and utilization in China. Furthermore, from the standpoint of Adoption-Diffusion Theory, higher government expenditure in education, science, public health and culture would provide a relative advantage and also be compatible with greater technology use. This variable is a unique one for China, and it overlaps several categories in the general model, which include education, government support, and innovation.

Number of public libraries represents a social capital variable. China's 2880 public libraries, with over 400 million volumes, had 330 million visits in 2010 (National Bureau of Statistics of China 2012; English.news.cn 2011). In 2011, the public libraries had 80,000 digital reading rooms, with 140,000 computers (English.news.cn 2011). The larger and more urban of the public libraries, led by the Beijing and Shanghai Libraries, have started to open up to free use and to provision of digital resources (Watson 2012), and the government has started to view the public libraries as "a public space where [patrons] can not only read, but also meet people and discuss ideas and issues" (Watson 2012, citing George Lossius).

This variable is regarded in the category of social capital, because as discussed in Chap. 3, social capital represents individuals' commitment to and investment in social relations and social networking including informal networking, with expectation of returns (Putnam 2001; Chen 2013). Chinese library patrons receive the benefits of social capital in physical public libraries particularly ones beginning to have free and open access policies and also through the information they access that stimulates social relations and networking.

The model's dependent variables are posited in the general conceptual model to represent technology availability and utilization. Dependent variables in the China model are similar to those in the general model, with the exception of addition of web-based variables. The pertinence of these dependent variables for China is indicated by their inclusion in studies on the country's technology growth (Zhu and Wang 2005; Davison et al. 2005; Guo and Chen 2005; Qiu 2010). Two web-based dependent variables are included. Number of domain names was utilized to demonstrate the growth of China's digital community in the early 2000s (Loo 2003) and in a study that distinguished Chinese provinces by level of Internet activities (Song 2008).

The other web variable is domain names. It was analyzed in Italy for gender, age, and provincial prevalence differences (Serrechia et al. 2005) and in a descriptive study of web pages and domain names for nations in Europe and the Middle East (Noruzi 2011). Although domain names is somewhat similar to number of Internet hosts or servers, which have been studied for their socio-economic correlates (Hargittai 1999; Baliamoune-Lutz 2003; Pick and Azari 2011), that similarity is weakened by the growing virtualization of hosts and servers. Number of web pages reflects an even broader concept of web content availability, i.e. the full amount of content available in total pages (Serrechia et al. 2005). Although number of web pages has rarely appeared in digital divide studies (Noruzi 2011), it is included to measure the amount of available web content.

5.7 Determinants of Technology Differences

Regression analysis was conducted for years 2006 and 2009, based on the sample of 31 mainland provinces. For 2006, the most important determinants of ICT use are export commodities value, significant in seven regressions, and published books, significant for six (see Table 5.3). Published books dominates for web pages and domain names. For urban PCs, the two dominant correlates are employees in non-state-owned units and tertiary sector employees. Another secondary correlate is innovation funding of enterprises.

For the 2009 regressions (see Table 5.4), the most important correlate is export commodities value (see Fig. 5.7), significant for all dependent variables, followed by published books (see Fig. 5.8), which dominates for web pages and for domain

Table 5.3 Standardized regression results for dependent variables, China, 2006

Category	Independent variable	Dependent variable							
		PCs per 100 urban families	PCs per 100 rural families	Broadband subscribers per 100 pop.	Internet users per 100 pop.	Number of domain names per 100 pop.	Number of web pages per capita	Mobile telephone subscribers per 100 pop.	Urban fixed telephone subscribers per capita
Education	Full-time teachers in high education institutions (10,000 persons) per capita								
Economic	Individual investment in fixed assets (100 million yuan) per capita								
Economic	Staff and workers by non-state-owned units (10,000 persons) per capita	0.657^{***}							
Economic	Employed persons (10,000) in tertiary industry per capita	0.403^{***}							0.444^{***}
Economic	Export commodities value (1000 US dollars) per capita		0.530^{***}	0.558^{***}	0.401^{*}	0.458^{***}	0.127^{***}	0.717^{***}	0.607^{***}
Economic	Foreign direct investment (10,000 US dollars) per capita				0.427^{**}				
Infrastructure	Electricity output (100 million kwh) per capita								
Infrastructure	Construction enterprise expenditure (100 million yuan) per capita								
Innovation	No. of published books (100 million copies) per capita		0.469^{***}	0.431^{***}	0.281^{***}	0.730^{***}	0.811^{***}	0.423^{***}	
Innovation	Innovation funds (100 million yuan) of enterprises per capita		0.227^{***}	0.274^{***}			0.283^{***}		

(continued)

Table 5.3 (continued)

Category	Independent variable	Dependent variable							
		PCs per 100 urban families	PCs per 100 rural families	Broadband subscribers per 100 pop.	Internet users per 100 pop.	Number of domain names per 100 pop.	Number of web pages per capita	Mobile telephone subscribers per 100 pop.	Urban fixed telephone subscribers per capita
Science, technology, education	Gov. depts. of culture, educ., science, and public health expenses (100 million yuan) of per capita								
Social capital, education	No. of public libraries per capita								
Regression adjusted R square and significance level		0.786***	0.953***	0.942***	0.886***	0.939***	0.993***	0.865***	0.906***
Sample size (N)		31	31	31	31	31	31	31	31
OLS regression tests									
Joint Wald Statistic		203.3***	28,299.2***	9765***	2339.7***	9204.2***	97,599.2***	96.9***	869.9***
Koenker (BP) statistic		0.729	9.169*	7.320	2.218	1.650	3.018	4.868	0.511
Jarque-Bera statistic		1.477	56.535***	16.995***	1.112	165.276***	4.234	1.108	1.139
Spatial autocorrelation of residuals									
Moran's index		0.083	0.119	−0.014	−0.111	−0.053	0.023	0.04	0.125

*Signif. at 0.05
**Signif. at 0.01
***Signif. at 0.001

Table 5.4 Standardized regression results for dependent variables, China, 2009

Dependent variable									
Category	Independent variable	PCs per 100 urban families	PCs per 100 rural families	Broadband subscribers per 100 pop.	Internet users per 100 pop.	Number of domain names per 100 pop.	Number of web pages per capita	Mobile telephone subscribers per 100 pop.	Urban fixed telephone subscribers per capita
Infrastructure	Electricity output (100 million kwb) per capita				0.484***				
Infrastructure	Construction enterprise expenditure (100 million yuan) per capita							0.401***	0.371***
Employment	Employed persons (10, 000) in tertiary industry per capita								
Employment	Staff and workers by non-state-owned units (10,000 persons) per capita	0.433***							
Individual investment	Individual investment in fixed assets (100 million yuan) per capita								
Trade	Export commodities value (1000 US dollars) per capita	0.596***	0.718***	0.765***	0.483***	0.239**	0.210***	0.588***	0.498***
Trade	Foreign direct investment (1000 US dollars) per capita						0.073*		0.211*
Innovation	No. of published books (100 million copies) per capita		0.497***	0.411***		0.824***	0.878**		
Innovation	Innovation funds (100 million yuan) of enterprises per capita					0.124			

(continued)

Table 5.4 (continued)

Dependent variable									
Category	Independent variable	PCs per 100 urban families	PCs per 100 rural families	Broadband subscribers per 100 pop.	Internet users per 100 pop.	Number of domain names per 100 pop.	Number of web pages per capita	Mobile telephone subscribers per 100 pop.	Urban fixed telephone subscribers per capita
Science, technology, education	Gov. depts. of culture, educ., science, and public health expenses (100 million yuan) of per capita								
Education	No. of public libraries per capita								
Education	Full time teachers in high education institutions (10,000 persons) per capita								
Regression adjusted R square and significance level		0.874***	0.894***	0.933***	0.810***	0.961***	0.987***	0.794***	0.913***
Sample size (N)		31	31	31	31	31	31	31	31
OLS regression tests									
Joint Wald statistic		525.9***	4991.1***	117.5***	77.7***	22,153.9***	35,101.3***	191.4***	2317.8***
Koenker (BP) statistic		0.431	7.910*	2.306	11.756**	4.296	4.817	3.588	1.838
Jarque-Bera statistic		7.571*	5.583	153.9***	1940	62.306***	7.993*	0.796	0.346
Spatial autocorrelation of residuals									
Moran's index		−0.176	−0.072	0.122	0.187*	−0.010	0.002	0.135	0.194*

*Signif. at 0.05
**Signif. at 0.01
***Signif. at 0.001

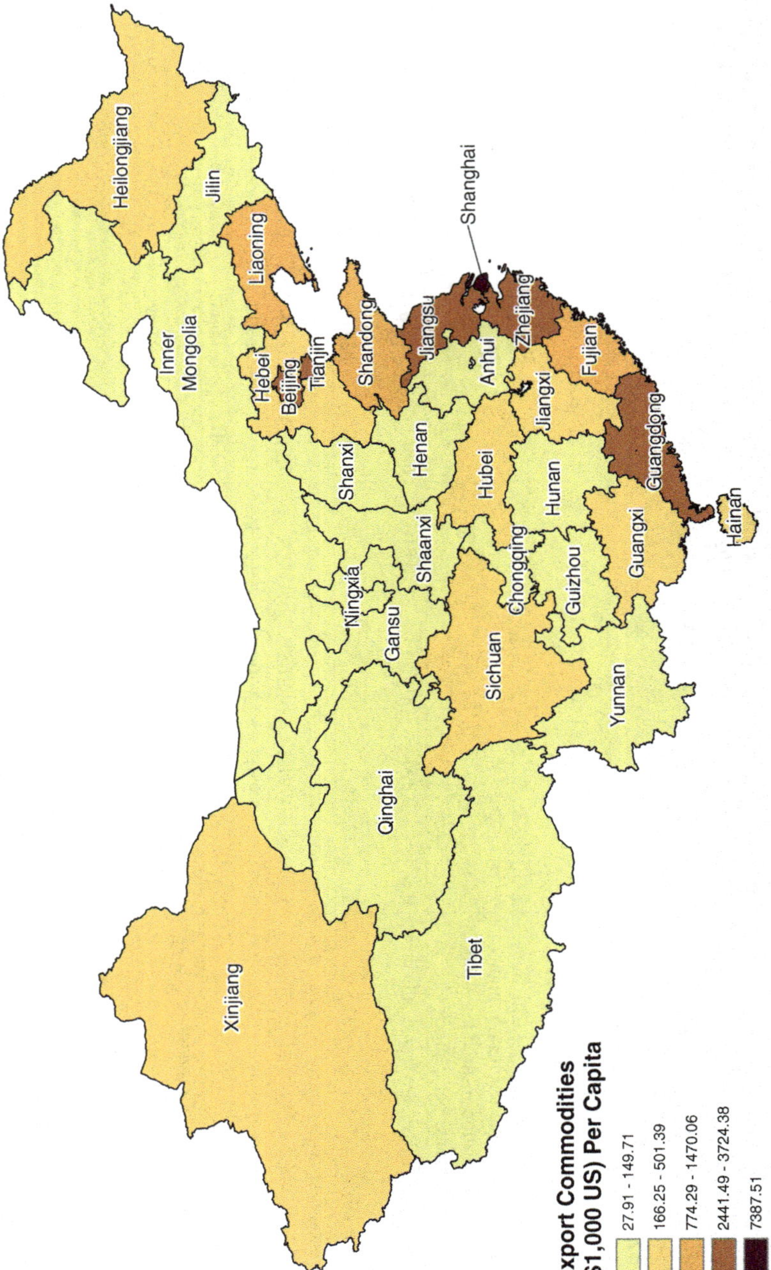

Fig. 5.7 Export commodities value (in $1000 U.S. dollars) per capita in Chinese Provinces, 2009

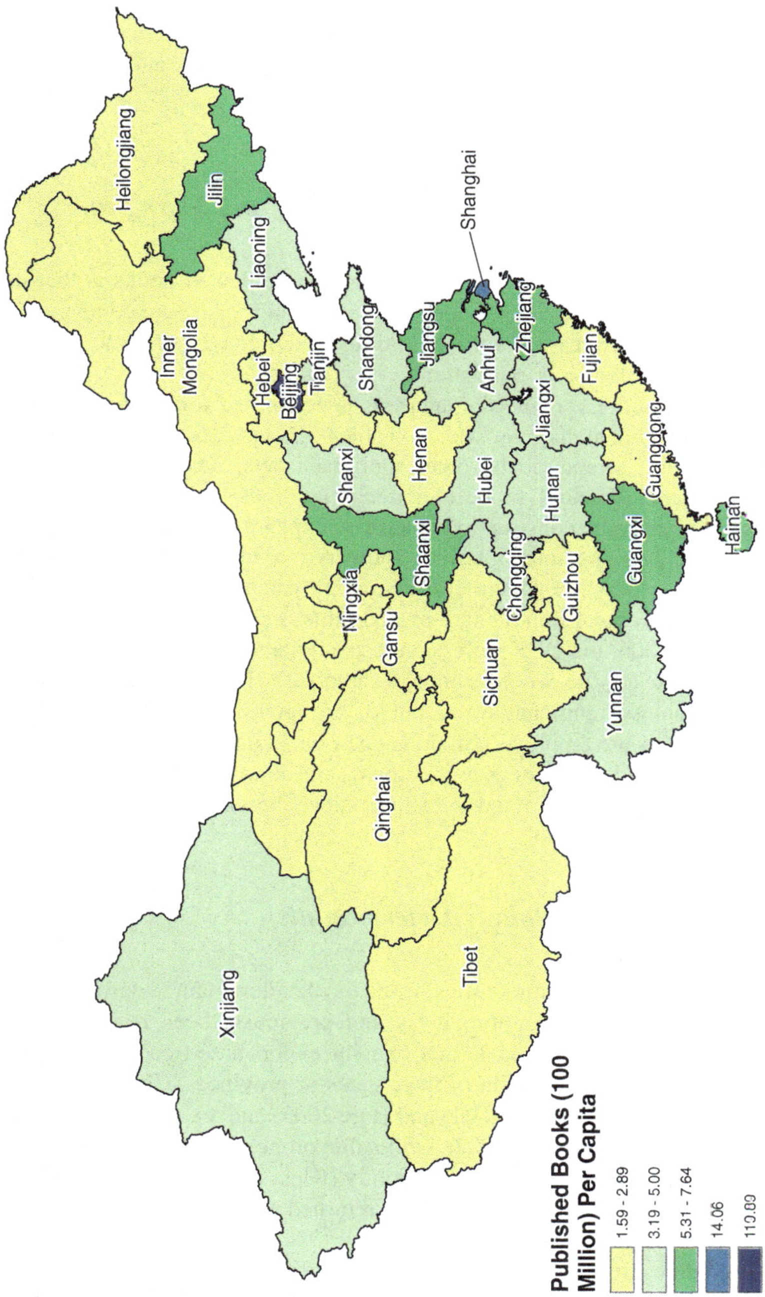

Fig. 5.8 Number of published books per 1000 persons, Chinese Provinces, 2009

names. Of lesser significance are tertiary employment (for mobile telephone subscribers and urban fixed telephone subscribers) and foreign direct investment (for web pages and urban fixed telephone subscribers). In addition, construction enterprise expenditure is a significant determinant for Internet use, and employees in non-state-owned units for PCs per urban families.

5.7.1 The Roles of Book Publishing in Beijing and Shanghai

A data analysis problem with the independent variable published books is that its values in 2006 and 2009 are remarkably high in Beijing and Shanghai, compared to other provinces. The likely reason is the long-time predominance of these cultural and intellectual capitals for book publishing.

To control for the outliers, regression analysis is performed for 2006 and 2009, excluding Beijing and Shanghai. As seen in Tables 5.5 and 5.6, published books drops out entirely and is replaced by other significant determinants. In 2006 the dominant determinant continues to be export commodities value. However, the secondary correlates become construction enterprise expenditure (for rural PCs and web pages) and employed persons in tertiary industry (for internet users and urban fixed telephone subscribers). For internet users, employment in tertiary industry is significant. Staff and workers in non-state-owned units is significant for urban PCs and broadband. For 2006, the OLS regression tests are largely valid and results are quite similar to those for the whole sample (Table 5.5). For 2009, export commodities value remains significant for 6 out of 8 regressions (Table 5.6). The secondary determinants are employees in non-state-owned units (for 4 regressions), employees in tertiary industry (for 3 regressions), and innovation funds of enterprises (for web pages and urban fixed telephone subscribers).

5.7.2 Discussion of Technology Determinants

For the regressions, export commodities value is the dominant determinant of technology availability and utilization. It has not previously been recognized in digital divide literature. One reason is that heavily export-based economies have been little studied from the standpoint of their states or provinces. China increased its exports vastly during 1990–2003 (Hale and Hale 2003), and exports continued at a high level from 2003 through 2007, while the proportion of domestic consumption in the nation's economy declined slightly (Klein and Cukier 2009). At the end of the 2007–2009 financial crisis, China returned to its prior high level of exports (Klein and Cukier 2009).

Export commodities' regression importance reflects the inter-related factors behind China's economic boom that include increasing worldwide exports, participation in the global supply chains, economic liberalization, central-government-initiated investment

Table 5.5 Standardized regression results for dependent variables, excluding Beijing and Shanghai, China, 2006

Category	Independent variable	Dependent variable							
		PCs per 100 urban families	PCs per 100 rural families	Broadband subscribers per 100 pop.	Internet users per 100 pop.	Number of domain names per 100 pop.	Number of web pages per capita	Mobile telephone subscribers per 100 pop.	Urban fixed telephone subscribers per capita
Infrastructure	Electricity output (100 million kwh) per capita							0.293**	
Infrastructure	Construction enterprise expenditure (100 million yuan) per capita								
Employment	Employed persons (10,000) in tertiary industry per capita			0.264*				0.251*	0.327***
Employment	Staff and workers by non-state-owned units (10,000 persons) per capita	0.385***	0.814***		0.273*	0.718***			
Individual investment	Individual investment in fixed assets (100 million yuan) per capita		0.254*						
Trade	Export commodities value (1000 US dollars) per capita	0.571***		0.761***	0.555***		0.682***	0.684***	0.559***
Trade	Foreign direct investment (10,000 US dollars) per capita								
Innovation	No. of published books (100 million copies) per capita								
Innovation	Innovation funds (100 million yuan) of enterprises per capita						0.346***		0.225*

(continued)

Table 5.5 (continued)

Category	Independent variable	Dependent variable							
		PCs per 100 urban families	PCs per 100 rural families	Broadband subscribers per 100 pop.	Internet users per 100 pop.	Number of domain names per 100 pop.	Number of web pages per capita	Mobile telephone subscribers per 100 pop.	Urban fixed telephone subscribers per capita
Science, technology, education	Gov. depts. of culture, educ., science, and public health expenses (100 million yuan) of per capita		0.287***						
Education	No. of public libraries per capita								
Education	Full-time teachers in high education institutions (10,000 persons) per capita								
Regression adjusted R square and significance level		0.783	0.677***	0.790***	0.811***	0.498***	0.783***	0.783***	0.787***
Sample size (N)		29	29	29	29	29	29	29	29
OLS regression tests									
Joint Wald statistic		195.5***	26.1***	128.5***	226.7***	11.9***	67.8***	180.2***	314.5***
Koenker (BP) statistic		0.582	8.211*	0.989	0.666	2.911	9.103	0.317	5.536
Jarque-Bera statistic		8.340	7.832*	0.274	0.529	8.522*	0.195	0.969	0.594
Test of spatial autocorrelation of residuals									
Moran's index		−0.163	−0.134	0.057	0.152	−0.028	0.075	0.089	0.136

*Signif. at 0.05
**Signif. at 0.01
***Signif. at 0.001

Table 5.6 Standardized regression results for dependent variables, excluding Beijing and Shanghai, China, 2009

Category	Independent variable	Dependent variable							
		PCs per 100 urban families	PCs per 100 rural families	Broadband subscribers per 100 pop.	Internet users per 100 pop.	Number of domain names per 100 pop.	Number of web pages per capita	Mobile telephone subscribers per 100 pop.	Urban fixed telephone subscribers per capita
Education	Full-time teachers in high education institutions (10,000 persons) per capita								
Economic	Individual investment in fixed assets (100 million yuan) per capita			0.254*					
Economic	Employed persons (10,000) in tertiary industry per capita				0.264**			0.251*	0.327***
Economic	Staff and workers by non-state-owned units (10,000 persons) per capita	0.385***	0.814***		0.273*	0.718***			
Economic	Export commodities value (1000 US dollars) per capita	0.571***		0.761***	0.555***		0.682***	0.684***	0.559***
Economic	Foreign direct investment (10,000 US dollars) per capita								
Infrastructure	Electricity ouput (100 million kwh) per capita							0.293**	
Infrastructure	Construction enterprise expenditure (100 million								
Innovation	No. of published books (100 million copies) per capita						0.346***		0.225*
Innovation	Innovation funds (100 million yuan) of enterprises per capita								

(continued)

Table 5.6 (continued)

Category	Independent variable	Dependent variable							
		PCs per 100 urban families	PCs per 100 rural families	Broadband subscribers per 100 pop.	Internet users per 100 pop.	Number of domain names per 100 pop.	Number of web pages per capita	Mobile telephone subscribers per 100 pop.	Urban fixed telephone subscribers per capita
Science, technology, education	Gov. depts. of culture, educ., science, and public health expenses (100 million yuan) of per capita		0.287***						
Social capital, education	No. of public libraries per capita								
Regression adjusted R square and significance level		0.783	0.677***	0.790***	0.811***	0.498***	0.783***	0.783***	0.787***
Sample size (N)		29	29	29	29	29	29	29	29
OLS regression tests									
Joint Wald statistic		195.5***	26.1***	128.5***	226.7***	11.9***	67.8***	180.2***	314.5***
Koenker (BP) statistic		0.582	8.211*	0.989	0.666	2.911	9.103	0.317	5.536
Jarque-Bera statistic		8.340	7.832*	0.274	0.529	8.522*	0.195	0.969	0.594
Test of spatial autocorrelation of residuals									
Moran's index		−0.163	−0.134	0.057	0.152	−0.028	0.075	0.089	0.136

*Signif. at 0.05
**Signif. at 0.01
***Signif. at 0.001

in infrastructure, and policymakers' emphasis on high tech (Hale and Hale 2003; Klein and Cukier 2009).

The export commodities value finding corresponds to findings of income per capita as a determinant of ICT access and use in several prior country-level studies (Hargittai 1999; Baliamoune-Lutz 2003; Guillén and Suarez 2005; Chinn and Fairlie 2007; Yates et al. 2011). The correspondence is demonstrated here by the simple correlations between export commodities per capita and urban income per capita and rural income per capita of respectively 0.884 and 0.881. Income was excluded from the present chapter due to multi-collinearity, in particular its high correlation with education.

China's exports have been stimulated by receipt of growing foreign direct investment inflow, which in 2008 was $175 billion, second largest in the world and in 2012 was $253 billion, in first place worldwide (OECD 2013). Although some problems loom for this export-driven economy including wage inflation, devaluation of the Chinese renminbi (RMB), greater domestic consumption, and some non-productive infrastructure (Economist 2011), it is likely that the export economy will continue to boom for some time. Consequently, in China's provinces the strong association of export commodities value and technology utilization is due to export-related foreign and domestic investments which have focused on technology (Hale and Hale 2003; Gao and Yu 2010), and which have been geographically heavily weighted towards China's eastern region. Although the China West policy has led to moderate reduction in the East-West range of the technology availability and utilization, in 2009, there remained large differences between East and West.

An important secondary finding is that the volume of published books is prevalent as a significant correlate. We interpret it as due to the outlier status of Beijing and Shanghai. They are the intellectual and innovation capitals of mainland China and the traditional location of book production, as well as more recently, the place for the production of e-books, web content, and and innovative technologies. Prior studies of other nations have not identified published books as important, perhaps due to those nations' provincial and state data being more keyed to consumption of knowledge products and services than to their production.

Two other secondary factors of tertiary employment and employment in non-state-owned enterprises were present mainly for the samples that exclude Beijing and Shanghai. The employment factors were significant for three regressions in 2006 and seven regressions in 2009. Tertiary employment has been rarely included in digital divide studies. An exception is that the professional/scientific/technical services workforce, which is tertiary, was a significant determinant of IS-data processing payroll per capita and IS-data processing receipts per capita for U.S counties (Azari and Pick 2005). In China's case, the export-oriented factories need engineers, ICT specialists, and business specialists to manage and improve the quality of manufacturing. Likewise, the secondary determinant of workforce in non-state-owned units can be justified as providing higher skill levels necessary for those units to be competitive in worldwide markets, and concomitantly to be more likely to utilize ICT (Guo and Chen 2005; Gao and Yu 2010).

A secondary factor in 2006 is construction enterprise expenditure (see Table 5.5). Some of this investment was for ICT infrastructure that China implemented in the late 1990s and 2000s as she opened up her economy. ICT is an important part of the domestic and foreign investment that has been essential to its economic growth (Hale and Hale 2003). By 2009, some ICT infrastructure, especially in the East, had reached a mature level, so construction enterprise expenditure was less crucial. However, in the West, construction continues to be crucial to technology availability and utilization (Davison et al. 2005), so infrastructure factors should be considered for future research that emphasizes the West. Lastly, innovation funds of enterprises is a significant determinant for web pages and fixed telephone subscriptions in 2009. This relates to China's government policy and support in the 2000s to emphasize more domestic innovation (Gao and Yu 2010; Economist 2011). It is not certain why innovation funds of enterprises is important only for these technologies. Due to the central government's long-term goal to expand domestic innovation, it is likely is that the impact of innovation on ICT use will broaden and become more significant in the future.

Two cases are now presented. The Alibaba case has the purposes of examining a well-known technology company in terms of the chapter's conceptual model for China and of comparing Alibaba's story to the model's empirical findings for China. The Google case is intended to illustrate China's reduced openness of information exchange, an important factor that could not be included in the conceptual model because provincial data are not available on it.

5.8 Case Study: Alibaba

Alibaba is a Chinese group of e-commerce and financial companies that provides businesses and consumers with web portals, e-commerce sites, and financial sites to conduct a variety of e-business transactions. In 2012, two of Alibaba's portals accounted for $170 billion in sales, a figure exceeding the combined e-Bay and Amazon (Economist 2013a, b). It dominates in China's e-business and has among the highest value of transaction volume of any company globally. The firm has been privately owned and was made public as an initial public offering (IPO) in late fall of 2014.

The company was founded in 1999 in Hangzhou, China, by Jack Ma, then of age 34. The original Alibaba.com was a business-to-business e-commerce portal between Chinese small business manufacturers and foreign buyers (Economist 2013b). The company grew with the help of an early Goldman Sachs investment and later in 2005 by a deal in which Alibaba exchanged 40 % ownership with Yahoo in order to take over Yahoo China. In its huge growth, Alibaba has divided into a number of affiliated companies that include the following. *Alibaba.com.* The original enterprise in English language now has the world's largest business-to-business (B-to-B) portal, available in nearly all countries, as well as offering a smaller Chinese domestic B-to-B portal, 1688.com.

Taobao. A consumer-to-consumer (C-to-C) portal that offers over one billion products. It resembles e-Bay but also has search functions so is a one-stop location for any type of purchase, and is one of China's most visited sites.

Tmall.com. This is a business-to-consumer shopping site opened in 2008 that somewhat resembles Amazon and includes global and domestic brands.

Alipay. Offers an online payment platform, with no transaction fees. Customers pay up-front into a type of escrow, but the funds go to the seller, only when the buyer indicates satisfaction. (Economist 2013a). It is expanding rapidly and connects with Visa, MasterCard and other Western payment systems. It also provides lending to small businesses, and is seeking a broader service market of making micro-loans to consumers.

Alibaba Cloud Computing. Started in 2009, this service offers cloud services, including simple space rental or advanced features such as data mining and analysis.

The Alibaba Group has the largest private consumer data-base in China and one of the largest globally. This means the Group is information-driven, although it needs to be careful with the government about having such vast data on the Chinese population in a nation known for government control.

The company's colorful chairman Jack Ma (see Fig. 5.9), has been entrepreneurial in seizing vast markets of Chinese consumers and businesses, and well as international Business-to-Business (B-to-B) markets. Mr. Ma came from a family of traditional Chinese theatre performers, and today is theatrical himself with the media spotlight and revered by Alibaba Group's over 40,000 employees (Economist 2013b).

This case study not only illustrates how far China has progressed in information technology, it also reinforces our China empirical findings. This includes the importance of export to ICT utilization. At the core of its B-to-B platform, Alibaba has facilitated import-export relationships, as well as a Consumer-to-Consumer (C-to-C) platform that includes import and export items. It also reinforces the importance for ICT use of non-state-owned companies, of which Alibaba is an extreme instance.

Fig. 5.9 Jack Ma, Chairman of Alibaba Minnesota. *Source* Asia Society

Further, it illustrates the significant effect of innovation funds of enterprises on number of web pages, since Alibaba consistently invested in, and demonstrated benefits from innovation in e-business and mobile business software and services which, among other things has led to a much larger number of web pages in use, consistent with our study.

Box: Google China (Hong Kong)

The conceptual model of the digital divide emphasizes the importance of societal openness (see Fig. 3.9), which refers to democracy, strong legal system, and open communications. However for China, the variables on societal openness are not available for the provinces and municipalities, and consequently are excluded from this chapter's model. Nonetheless, the question of societal openness in information and communications is a prominent one influencing the nation's software, ICT services, and user environment and experiences. This box is intended to fill in this gap by presenting the example of Google confronting societal openness in information search.

Google China was founded in 2005 by Kai-Fu Lee, a former manager of Microsoft. In 2006, a version of the Google search engine in Chinese appeared and initially complied with Chinese government censoring of particular websites. The Chinese government since 1996 has censored websites considered highly critical of the government or in other ways offensive to the government or society. After its start, Google competed directly with Baidu, the leading Chinese search engine leader, which had half of the Chinese search market. Google initially accepted censoring but was criticized, including by some legislators in the U.S. Congress who "asked how a company with the slogan 'Don't Be Evil' could conspire with China's censors" (Thompson 2006). Even though Google came late to the China market, it had the advantage of leveraging from its world leadership.

To attempt to compete with Baidu, Google broke with its standard worldwide approach and followed Baidu by hiring agents to sell keywords, often in China's prevalent internet cafes (So and Westland 2010). Google also adapted to the local market in China by such features as enabling Pinyin method of inputting i.e. being able to use Latin symbols to capture the shape of Chinese characters.

Nonetheless, Google continued to trail Baidu and also faced a smaller competitor Soso.cn, part of the giant company Tencent, which partly imitated Google. In Google's favor were the benefits of building off a worldwide platform, including advertising. At the same time, Google China was often pulled in two directions, by Global Google to conform with its standards, and on the other hand to meet local conditions and customer needs in China (So and Westland 2010). Also, Google China faced intermittent hassling by the Chinese government related to content.

Things came to a head in 2010. In January Google decided to stop self-censoring its site so it no longer conformed to Chinese government regulations.

On March 23, 2010, China placed Google under strict censorship, and Google.cn responded by referring its users to Google.hk in Hong Kong, which allows free speech. Google retained a presence in China with some other services including Google Maps, Gmail, and Google Docs. Problems for Chinese Google search users still occur, as China intermittently has used its Great Firewall to fully or partially disable mainland access to Google.hk. Overall, by 2013, Baidu dominated with about 2/3 of the search market in mainland China, with Soso and other Tencent search engines in second place and Google is relegated in China to a mere 2–3 % of the mainland market.

This example illustrates issues with democratic, free expression of information in China, an aspect that is excluded from our formal analysis due to lack of provincial data, but is nevertheless an important feature for ICT users in mainland China.

5.9 Policy Implications for China

The findings for the years 2006 and 2009 of the ICT utilization model and cluster analysis inform ICT policy for provincial and central governments of China. Ambitious private enterprises such as Lenovo and Alibaba have leapt ahead of government policies, partly using examples from Silicon Valley and other illustrious centers to expand niches of technology. However, for most of China geographically, much more ICT utilization is possible and the central and provincial governments can help that potential to be realized through longer-term planning initiatives. Four recommended policies are as follows.

1. *Expansion in web and Internet content.* Utilization for expanding web content has been associated with the publishing centers of Beijing and Shanghai. Centers for development of web content might be supported in other nearby provinces with much lower educational levels such as Heibei, Shandong, Anhui or even Western and Middle Region provinces such as Chongqing, Human, and Sichuan, which have good foundations in higher education and are more likely to be receptive to government ICT policies than in remote parts of the country.
2. *Export industries and commodities.* Since technology availability and utilization is so closely related to export commodities value, the government could stimulate technology availability and utilization in provinces outside the export-oriented east and southeast by developing export of industrial products and commodities. Currently, the Chinese government encourages "low-end industry transfer," which refers to transfer of industries from more developed to less developed areas, provinces, or regions. Several export industries will be transferred from developed provinces (e.g., Beijing and Shanghai) to the Middle and Western Regions in the period of "twelfth 5-year plan" (2011–2015). It is

expected that other remote provinces will be stimulated in this way in technology availability and utilization within 5 years.

3. *Innovation.* The 2009 finding that innovation funds of enterprises are associated with technology availability and utilization for web pages and urban fixed telephone subscribers suggests that future government policies should encourage even more innovation and R&D in enterprises in China. As the Chinese government's "twelfth 5-year plan" is implemented, innovation funds will be enhanced, and more enterprises in China will be encouraged to establish engineering research centers, innovation centers, and other R&D centers. However, there is a tendency that developed provinces have advantages in establishing R&D centers (e.g., lifestyle, environment, higher education resources), and they may enhance their advantage by providing high salaries and innovation funds to attract talented workers (e.g. Ph.D.s). Thus, to reduce the country's digital divide, Middle and Western provinces should be provided with more innovation funds or other incentives supported by the central government to retain their own talented scientists and other specialists and attract new talent to develop R&D in their enterprises. This Westward movement can be "seeded" if leading brand-name Chinese technology companies, such as Alibaba, Baidu, Tencent, and Lenovo, can be incentivized to relocate innovation centers from East to West.
4. The Western region which has by far the lowest technology availability and utilization can be stimulated by policies, which include policy support for export oriented enterprises, higher proportion of private enterprises, innovation and R&D in enterprises, and web content production. This applies also to a lesser extent to provinces in the Middle Region of China. In the context of "low-end industry transfer," more of the export-oriented and non-state-owned enterprises can be encouraged to transfer to the Western and Middle Regions (Ma and Summers 2009). In those provinces, provincial and local policies such as tax cuts and cheaper land for industry can be implemented to encourage and incentivize enterprises to migrate from the East and Southeast. With such policies, it is expected that the West's and Middle Region's levels of technology availability, utilization, and innovation will increase within the next 5–10 years.

5.9.1 Cluster Analysis and Its Policy Implications

Maps of the spatial distribution of K-Means clusters in 2006 and 2009 reveal geographic agglomerations of the technology factors, as discussed earlier in the chapter. Four regional ICT policy models, based on prior academic literature and on China's government policies, are proposed. As seen in Table 5.1, Policy Area 1 (Clusters 1 and 2) consists of Beijing and Shanghai. Policy Area 2 (Cluster 4) of Tianjin, Jiangsu, Zhejiang, Fujian, and Chongqing in 2006 drops to Tianjin and Zhejiang in 2009. Policy Area 3 (Cluster 3) comprises only Guangdong in 2006, but adds adjoining Fujian in 2009. Policy Area 4 (Clusters 5 and 6) includes the remainder of the administrative units (23 in 2006 and 25 in 2009) located in the west, north, and central regions. In both years, the digital divide gap can be observed in the set of ratios of high to low clusters in technology availability and utilization (Column 8 in Table 5.1), which varies between 2.1 times for mobile subscribers in 2009–95.3 times for web pages in 2006. These four policy areas and their transitions 2006–2009 can be interpreted chronologically by policy models, as follows.

1. "*Central-policy-oriented model.*" Beijing and Shanghai in Policy Area 1 form a high technology area. In mainland China, Beijing and Shanghai are the dominant municipalities. They also dominate in the technology factors (see Table 5.1). Compared to Shanghai, Beijing is technologically stronger. In particular it is more than two times as intensive in domain names and web pages. In these municipalities, the ICT availability and utilization development model is "policy-oriented," that is, the central government in Beijing makes the policy on state-owned industries development (e.g. ICT, publishing) in nearby provinces, known as the "Jing-Jin-Tang Old Industrial Base. " The policy encourages state-owned companies to invest in hardware, in particular Internet-related.
   Moreover, Beijing and Shanghai also have special policies from central government which may enhance the ICT availability and utilization. The first policy advantage is in the higher education field. These two municipalities have the most state-owned research institutes and top universities in China. For example, Beijing has 58 universities including Peking University, Tsinghua University, Renmin University, while Shanghai has 31 universities that include Fudan University, Shanghai Jiaotong University, and Tongji University. According to the literature and our findings, policy favoring higher education may facilitate elevated level of information technology usage (Ono and Zavodny 2007). A second policy advantage is in the area of publishing. As the political, economic, educational, and cultural centers of China, Beijing and Shanghai have the most publishing institutions (e.g., People's Publishing House). Since the number of publications is highly related to number of web pages, domain names, and broadband/Internet usage, this policy advantage enhances these two cities' overall ICT availability and utilization level. The publishing industry can be considered as a knowledge producer that provides the content basis for increased deployment of the web, broadband, and 3G/4G mobile devices.
2. "*Low-end industry transfer model.*" Policy Area 2 is larger and more important in 2006, but diminishes in 2009. In 2006 the provinces of Jiangsu, Shanghai, Zhejiang, and Fujian form the core of Policy Area 2, which emphasizes the free economic market. Shanghai is in the center of these provinces, which are mostly in the "Yangtze River Delta Economic Zone." In these provinces, the ICT availability and utilization development model is the "low-end industry transfer model." With development of Shanghai, it may transfer some low-end industries to nearby provinces, such as Anhui. These industries are mostly private light industries, which may not have yet invested in web technologies. By 2009, this distinctive cluster was no longer in place, except for Zhejiang. Its reduction in size may represent some success with the transfer of technology from the 2006

cluster to neighboring western provinces. Besides the core, in both years, the municipality of Tianjin receives low-end industry transfer from Beijing. Chongqing's inclusion in this model in 2006 is unexplained.

3. "*Overseas Model.*" Policy Area 3 has expanded in size and importance over the three-year period under study. In 2006 this model applied only to Guangdong, which represents the Pearl River Economic Zone, locus of several original SEZs that had their start in 1980s. In 2009 the model applies to Guangdong and Fujian to the north. These provinces have similar economic development models, that is, they assemble components transferred from outside mainland China and export finished products. In the case of Policy Area 3, instead of having its own core, in many respects, the core that animates its industrial development is outside of the Chinese mainland, e.g., Taiwan, Hong Kong, South Korea.
In accordance with the successful experience of Singapore and South Korea, this "overseas" economic model may enhance these provinces' ICT availability and utilization level, since they can invest in advanced communications infrastructure, and develop local ICT industries (Meng and Li 2002). However, the speed of ICT availability and utilization may decrease if these provinces do not upgrade their own local suppliers, rather than having to depend on supply from the core of a foreign country. Also, opportunities may be lost if low-end supplier industry is transferred to other nations, rather than maintained within China.
In the Suzhou metropolitan region in the state of Jiangsu, the policy dilemmas mentioned above are exemplified in two studies of multinational electronics firms which have located production there, but remain headquartered in Taiwan (Yang and Hsia 2007; Wei et al. 2011). In order to avoid the multinationals being supplied mostly from Taiwan, local suppliers need to upgrade their supply chain to be compatible with a global network (Yang and Hsia 2007). If not, the multinationals will interact only among themselves and with their home countries' suppliers. Suzhou is moving to become a globalized locus of production, with a cluster of global multinationals (Wei et al. 2011). Although the city government policies have attempted to globalize the city as a whole, the multinationals have agglomerated and threaten to ignore the local, domestic firms which are largely "small, low-tech, and locally oriented" (Wei et al. 2011). Hence, if the "Overseas Model" is to enhance the domestic companies and economy, it should not be left to the multinationals to plan for, but can be guided by a balanced approach of city, provincial, and central governments working towards integration and sharing of domestic and foreign ICT benefits.

4. "*Western, Central, and Northern model.*" The broad expanse of Policy Area 4 encompasses about 95 % of China's territory and about 78 % of its population, but has much lower technology availability and utilization than the other policy areas, as detailed in Table 5.1. Within Policy Area 4, the West/South Cluster 6, with the lowest ICT usage, somewhat distinctive in 2006, became a large and very distinctive cluster in 2009, including the addition of Heilongiang in the far northeast. The increase in size of the West/South cluster over the three years, which runs contrary to China's "Develop the West" policy of 2000, is a result of several factors.

First of all, from 2006 to 2009, the brain drain away from the western regions continued, and that may have led to lower technology availability and utilization. Considering the higher average income, richer education resources, greater prevalence of high-level industry, and better social environment in eastern regions, it is not surprising that the most talented workers with higher education would still prefer to work in eastern metropolitan areas such as Beijing, Shanghai, Guangzhou, and Shenzhen. The attraction of the East and Southeast is also seen in the large migration of lower-skilled workers from West to East over the past decade.

Second, governments in the western provinces have had lower awareness of the importance on ICT applications. While most governments in eastern regions implemented "e-government," "digital government," and "smart city" planning to promote ICT availability and utilization, western provincial governments have not recognized the importance of ICT availability and utilization and applications in industrial development.

Third, there were different responses to the global financial crisis of 2007–2010 in the West and East. In 2009 China proposed a 4 trillion yuan (approximately 590 billion U.S. dollars) economic stimulus plan and invested in hardware (e.g., highways, communication infrastructure) and social development (e.g., support for small-medium enterprises, social service, and education). In the ICT field, governments in western regions invested heavily in advanced infrastructure, while eastern governments spent more money on ICT related education, attraction of talented workers, and support of small and medium-sized enterprises. According to a worldwide investigation (Pick and Azari 2011), government support and social openness may strongly improve the socio-economic level of regions and contribute to ICT availability and utilization, while investment in technology and advanced infrastructure, ignoring educational and social advances, is less likely to relate to ICT availability and utilization. Furthermore, while ICT infrastructure is important, it is still held back in the West by its challenging physical topography, vast distances, and low transport accessibility of provinces. From 2006 to 2009, recessionary factors held back investment in ICT infrastructure and education in China's West.

Accordingly, the recommended approach in Policy Model 4 is to stimulate technology investments, attract enterprises to the area, and encourage migration of technology skilled workers. These efforts resonate with the central government initiatives to invest in social development, especially education, in western regions.

5.10 Conclusion

China has the world's largest ICT user base and has undergone an unprecedented ICT transformation. Nonetheless, many challenges remain to narrow the digital divide including fostering domestic technology innovation, extending ICT utilization advances to the vast West, increasing openness of users access to Internet content, and balancing the ICT needs of high-end export-oriented firms with the development of domestic ICT capabilities in local and regional supplier and service firms.

Spatial analysis confirms that the nation is highly uneven geographically in its information technology utilization, much more so than leading economies such as the U.S. and Japan, which are covered in Chaps. 7 and 8.

The leading determinants of provincial ICT utilization in 2006 and 2009 are shown to be published books, which is interpreted as a proxy for web content production and innovation, export commodities value, and number of non-state owned enterprises, while innovation funds of enterprises is partly influential in 2009. Published books is seen to be linked to the outlier provinces of Beijing and Shanghai. These findings for China are in accord with prior research on other nations. As China may shift to less export and more internal consumption, export-related factors should become less important in determining ICT usage. The case study of Alibaba, at the high end of ICT utilization success in China, is consistent with the findings of this study.

Policy implications for China suggest government should emphasize four different policy models depending on the region of the nation, which are Central-policy-oriented model, Low-end industry transfer model, Overseas model and Western, Central, and Northern model. Key challenges for China in future development of ICT are how to expand ICT utilization more evenly throughout the nation, how to make information more open to all its citizens, and how to stimulate greater domestic ICT innovation.

References

Azari, R., & Pick, J. B. (2005). Technology and society: Socioeconomic influences on technological sectors for United States counties. *International Journal of Information Management, 25*(1), 25–37.

Ahn, H., & Lee, J.-H. (1999). An econometric analysis of the demand for access to mobile telephone networks. *Information Economics and Policy, 11*(3), 297–305.

Baliamoune-Lutz, M. (2003). An analysis of the determinants and effects of ICT diffusion in developing countries. *Information Technology for Development, 10*, 151–169.

Bruno, G., Esposito, E., Genovese, A., & Gwebu, K. (2011). A critical analysis of current indexes for digital divide measurement. *The Information Society, 27*, 16–28.

Chen, W. (2013). The implications of social capital for the digital divides in America. *The Information Society, 29*, 13–25.

China Data Online. (2010). China Data Center. University of Michigan, Ann Arbor, MI. Available at www.chinadatacenter.org. Accessed on March 1, 2012.

China Information Almanac. (2010). *China information almanac*. Beijing: Government of China.

China Internet Network Information Center. (2007, 2009). Statistical report on Internet development in China. Various issues. Beijing: China Internet Network Information Center. Available at www.cnnic.net.cn. Accessed March 1, 2012.

China Statistical Yearbook. (2010). China statistical yearbook. Beijing: China Statistics Press. Available August 28, 2014 at www.idigitaltimes.com.

Chinn, M. D., and Fairlie, R. W. (2007). The determinants of the global digital divide: cross-country analysis of computer and internet. *Oxford Economic Papers, 59*, 16–49.

Comer, J. C., & Wikle, T. A. (2008). Worldwide diffusion of the cellular telephone. *The Professional Geographer, 60*(2), 252–269.

Davison, R. M., Vogel, D. R., & Harris, R. W. (2005). The e-transformation of Western China. *Communications of the ACM, 48*(4), 62–66.

Economist. (2011). Rising power, anxious state. Special Report on China. *Economist, 399*(8739), 1–16.

Economist. (2013a, March 23). The Alibaba phenomenon. *The Economist*.

Economist. (2013b, March 23). The world's greatest bazaar; Alibaba. *The Economist*, p. 27.

English.news.cn. (2011, October 27). Chinese public libraries see record numbers of patrons. *English.news.cn*. Beijing: Xinhuanet Corporation LTD. Available on September 5, 2014 at http://news.xinhuanet.com/english2010/china/2011-10/27/c_131216279.htm.

Fong, M. W. L. (2009). Digital divide between urban and rural regions in China. *Electronic Journal of Information Systems in Developing Countries, 36*(6), 1–12.

Gao, P., & Yu, J. (2010). Emerging markets: Has China caught up in IT? *Communications of the ACM, 53*(8), 30–32.

Guillén, M. F., & Suárez, S. I. (2005). Explaining the global digital divide: Economic, political and sociological drivers of cross-national Internet use. *Social Forces, 84*(2), 681–708.

Guo, X., & Chen, G. (2005). Internet diffusion in Chinese companies. *Communications of the ACM, 48*(4), 54–58.

Hale, D., & Hale, L. H. (2003). China takes off. *Foreign Affairs, 82*(6), 36–53.

Hargittai, E. (1999). Weaving the western web: Explaining differences in Internet connectivity among OECD countries. *Telecommunications Policy, 23*, 70–118.

ITU. (2012). *Measuring the information society*. Geneva, Switzerland: International Telecommunication Union.

ITU. (2013). *Measuring the information society*. Geneva, Switzerland: International Telecommunication Union.

James, J., & Romijn, H. (1997). The determinants of technological capability: a cross-country analysis. *Oxford Development Studies, 25*(2), 189-207.

Klein, B. P., & Cukier, K. N. (2009). Tamed tigers, distressed dragon. *Foreign Affairs, 88*(4), 8–16.

Loo, Becky P. Y. (2003). The rise of a digital community in the People's Republic of China. *Journal of Urban Technology, 10*(1), 1–21.

Ma, D., & Summers, T. (2009). *Is China's growth moving inland? A decade of 'develop the west (Chatham House Asia Programme Paper, ASP PP 2009/02)*. London: Chatham House.

McFarlan, E. F., Jia, N., & Wong, J. (2012). China's growing IT services and software industry: Challenges and implications. *MIS Quarterly Executive, 11*(1), 1–9.

Meng, Q., & Li, M. (2002). New economy and ICT development in China. *Information Economics and Policy, 14*(2), 275–295.

Moxley, M. (2009). *China: Ten years on, develop-the-west policy has mixed results*. Beijing: Inter Press Service. Available at www.globalissues.org. Accessed on June 6, 2011.

National Bureau of Statistics of China. (2012). *Statistical communiqué on the 2011 national economic and social development*. Beijing: National Bureau of Statistics of China (February 22).

Noruzi, A. (2011). Web presence of European and middle-eastern countries: A digital divide. In St. K. Amant, & B. Olaniran (Eds.), *Globalization and the digital divide*. Amherst: Cambria Press (pp. 146–160).

OECD. (2013). *FDI in figures* (Report, April). Paris: Organisation for Economic Co-operation and Development.

Ono, H., & Zavodny, M. (2007). Digital inequality: A five country comparison using microdata. *Social Science Research, 36*, 1135–1155.

Pick, J. B., & Azari, R. (2008). Global digital divide: influence of socioeconomic, governmental, and accessibility factors on information technology. *Information Technology for Development, 14*(2), 91–115.

Pick, J. B., & Azari, R. (2011). A global model of utilization of technology based on governmental, social, economic, and business investment factors. *Journal of Management Information Systems, 28*(1), 51–85.
Population Reference Bureau. (2012). *World population data sheet*. Washington, DC: Population Reference Bureau.
Putnam, R. (2001). Social capital: Measurement and consequences. *Canadian Journal of Policy Research, 2*, 41–51.
Qiu, J. L. (2010). Mobile phones, the bottom of the pyramid and working-class information society in China. *Electronic Journal of Information Systems in Developing Countries, 44*(2), 1–14.
Quibria, M. G., Ahmed, S. N., Tschang, T., & Reyes-Macasaquit, M.-L. (2003). Digital divide: determinants and policies with special reference to Asia. *Journal of Asian Economics, 13*(6), 811–825.
Raven, V. P., Huang, X., & Ki, B. B. (2007). E-Business in developing countries: A comparison of China and India. *International Journal of E-Business Research, 3*(1), 91–108.
Robison, K. K., & Crenshaw, E. M. (2002). Post-industrial transformations and cyber-space: A cross-national analysis of Internet development. *Social Science Research, 31*, 334–363.
Rogers, E. (2003). *Diffusion of innovations*. San Francisco: Free Press.
Serrechia, I., Serrechia, M., & Martinelli, M. (2005). Digital divide in Italy among individuals by using endogenous metrics. In *Proceedings of IADIS International Conference 2005* (pp. 20–25). Lisbon: International Association for Development of the Information Society, IADIS Press, pp. 20–25.
So, S., & Westland, J. C. (2010). *Red wired: China's internet revolution*. London: Martin Cavendish Business.
Song, W. (2008). Development of the internet and digital divide in China: A spatial analysis. *Intercultural Communication Studies, 17*(3), 20–43.
Thompson, C. (2006, April 3). Google's China problem (and China's Google problem). *New York Times*.
Vicente Cuervo, M. R., & López Menéndez, A. J. (2006). A multivariate framework for the analysis of the digital divide: Evidence for the European Union-15. *Information & Management, 43*, 756–766.
Watson, E. (2012). *Widening access to public libraries in China. Publishing Trends*. New York: Market Partners International. Available on September 7, 2014 at http://www.publishingtrends.com/2012/04/widening-access-to-public-libraries-in-china.
Wei, Y. H. D., Liefner, I., & Miao, C.-H. (2011). Network configurations and R&D activities of them ICT industry in Suzhou municipality, China. *Geoforum, 42*, 484–495.
Wong, P.-K. (2003). Global and national factors affecting e-commerce diffusion in Singapore. *The Information Society, 19*, 19–32.
World Bank. (2013). *World development indicators*. Washington, DC: World Bank.
World Economic Forum. (2013). *Global information technology report 2013*. Davos: World Economic Forum.
Yang, Y.-R., & Hsia, C.-J. (2007). Spatial clustering and organizational dynamics of transborder production networks: A case study of Taiwanese information technology companies in the Grater Suzhou area, China. *Environment and Planning A, 39*, 1346–1363.
Yang, C., & Liao, H. (2008). Backward linkages of cross-border production networks of Taiwanese PC investment in the Pearl River Delta, China. *Tijdschrift voor Economische en Sociale Geografie, 101*(2), 190–217.
Yates, D. J., Gulati, G. J., & Weiss, J. W. (2011). Different paths to broadband access: The impact of governance and policy on broadband diffusion in the developed and developing worlds. In *Proceedings of the 44th Hawaiian International Conference on System Sciences* (pp. 1–10). Washington, DC: IEEE Computer Society.
Zhu, J. J. H., & Wang, E. (2005). Diffusion, use, and effect of the Internet in China. *Communications of the ACM, 48*(4), 49–53.

Chapter 6
India's Digital Divide

6.1 Introduction

India, with an estimated population of 1.24 billion, is the second most populous country in the world. In recent years, it has been recognized for its information technology, IT enabled services, and knowledge services based economy. In fact, between 2009 and 2013, value added services consistently provided at least 55 % of India's GDP (World Bank 2014). Worldwide, the presence of a robust services sector has often positively influenced the adoption and utilization of information and communication technologies (ICTs). The role of ICTs in catalyzing and sustaining growth in the services sector in India and in general driving economic growth is therefore crucial. Use of ICTs in India has also expanded rapidly in recent years with an estimated 894 million mobile cellular telephone subscribers in 2013 (The Central Intelligence Agency 2014), second only to China, 195 million internet users also in 2013, behind China and the United States, and 6.75 million internet hosts in 2012. Despite a large base of ICT users, India's per capita use of ICT is low. This is reflected by a low 12.6 % internet penetration in 2012, which lags substantially behind India's BRICS peers and many Asian nations including China, South Korea, Philippines, Thailand, Australia, New Zealand, Fiji, Sri Lanka, and others (ITU 2013). This is further exemplified by India's middling rank of sixty-eighth among 144 nations worldwide in terms of the Networked Readiness Index (World Economic Forum 2013) and 121st among 157 nations in terms of the International Telecommunications Union's ICT Development Index (ITU 2013). In fact the ITU has classified India as one of 39 least connected countries along with populous nations such as Bangladesh in Asia and smaller nations such as Cameroon, Rwanda and Niger in Africa.

India's moderate ICT status can partially be attributed to its significant urban-rural divide with almost three-quarters of the population living in villages where access to ICTs remains very low. There are low levels of literacy in general and

J.B. Pick and A. Sarkar, *The Global Digital Divides*, Progress in IS,
DOI 10.1007/978-3-662-46602-5_6

computer literacy in particular, low levels of education, poverty, vast diversity in non-English languages spoken, as well as infrastructural issues such as unstable power supply, lack of access to personal computers (4.6 % PC penetration in 2013, compared to 14 % globally (Statista 2014; Maps of the World 2014)) and low national teledensity (wireline telephones plus wireless telephones per 100 population) with a significant difference between urban and rural locations, all of which act as major bottlenecks (Cecchini and Scott 2003; Rao 2005). The difference in ICT adoption between the states of India can sometimes be stark; for example, while states such as Karnataka, whose capital Bangalore is considered to the Silicon Valley of India, and Maharashtra, an economic powerhouse with the highest net state domestic product among Indian states and whose capital Mumbai (formerly Bombay) is considered to be the financial capital of India, have approximately 61 mobile telephone subscribers and 2.5 internet users per 100 population, the corresponding numbers in the traditionally agrarian states of Bihar and Uttar Pradesh in the Hindi speaking heartland of India are around 30 and 0.25 respectively (Department of Telecommunications, India 2010). Incidentally, both Uttar Pradesh and Bihar are both considered poor states; their per capita incomes were among the bottom 3 out of 31 states in 2009–10 (Reserve Bank of India 2011). The city state of Delhi, the national capital of India and home of the national government is an ICT outlier (discussed later in this chapter) with 156 mobile telephone subscribers and 10.64 internet users per 100 people in 2010 and 2009 respectively. It is therefore essential to examine the geography and determinants of ICT utilization in India at the state level to understand the extent of digital divide in India. This is the focus of this chapter.

Next, the chapter seeks to understand the evolution of ICT adoption in India over the years, as reported in the literature. Using descriptive mapping, we explore patterns of ICT utilization in India at the state level, for both individual technologies and also spatial distribution of all technologies aggregated together. Based upon this review of the literature, we tease out factors that may potentially impact ICT adoption and build our conceptual model of India's ICT utilization, which is based on the general model discussed earlier in Chap. 3. In building an India specific model, we pay attention to independent factors that are unique to the Indian digital divide context. This includes factors such as registered newspapers and periodicals and cooperative societies that represent societal openness/innovation and social capital respectively. The major determinants of ICT utilization and the digital divide in India are analyzed and discussed next. Later in the chapter, we present case studies of ICT adoption in India in diverse sectors that highlight the contribution of ICTs towards socio-economic development, poverty alleviation, and other societal benefits. The case studies serve to reinforce results from our confirmatory regression analysis and provide the foundation for ICT policy development and its implications in the Indian context. Concluding remarks and a summary of India's digital divide are in the last section of this chapter.

6.2 Historical Background

The Indian telecommunications industry was one of the first industries that was opened up for private investment following the initiation of economic liberalization in the early 1990s. Since then, telecommunications in India has come a long way with consumers of ICTs possessing multiple choices for mobile and landline-based telephone as well as internet and broadband services. The public sector companies that operated as monopolies for decades now face steep competition. Telecom reforms that started in 1992 resulted in the National Telecom Policy in 1994. Its objectives included to make telecommunications services available within the reach of all, achieve universal service covering all villages as quickly as possible and provide access to basic telecommunications services at affordable and reasonable prices. Specific targets that resulted from this policy were to make telephones available on demand, provide one public call office (a kiosk with one/multiple landline connections) per 500 people in urban areas, and cover all villages by 1997 (TRAI 2014a).

At the time, the Indian Department of Telecommunications (DoT) played the role of operator and regulator. This was deemed unfair by private sector telecom operators and subsequently the government of India created different entities, the Ministry for Communications, Telecom Commission, and the Telecom Regulatory Authority of India (TRAI). The Ministry was the key policymaker, whereas the Commission formulated policy, licensing, wireless spectrum management, R&D, equipment standards, and administered public sector telecom companies. TRAI was set up as the key regulator and ensured conformity to broader international norms and standards.

Due to far reaching changes in telecommunications, IT, consumer electronics and media industries worldwide and unfulfilled targets set in the 1994 National Telecom Policy, a new national telecommunications policy came into existence in 1999. Increasing teledensity pan-India was a key focus of this policy. Specific targets included achieving a teledensity of 7 per 100 people by 2005 and 15 by 2010. Other targets included increasing rural teledensity from 0.4 in 1999 to 4 by 2010, providing internet access to all district headquarters across India by 2000, voice and low speed data services to almost 300,000 villages by 2002, and high speed data and multimedia capability to all towns with a population of at least 200,000 by 2002 (Tangutari and Harmantzis 2007; TRAI 2014b). Teledensity targets that were set as part of this policy were largely met; rural teledensity also improved from 0.4 in 1999 to 1.7 per 100 people in March 2004 (Tangutari and Harmantzis 2007). Since 2004, spurred by the explosive growth of mobile wireless subscribers, teledensity in India increased to 80 subscribers per 100 people by February of 2012. However the urban-rural divide still persisted, with urban teledensity at 170 compared to almost 39 in rural areas (estimates include wireline and wireless teledensity).

India's first national broadband policy came into existence in 2004 as the country was struggling with low levels of broadband, internet and personal computer (PC) penetration (0.02, 0.4 and 0.8 % respectively at the end of December, 2003). Recognizing that distinctive standards for broadband existed in various nations, growth targets for internet and broadband subscription were estimated to be from 6 to 40 million for internet between 2005 and 2010, and 3–20 million for broadband during the same period. Policies for infrastructural advancement in technology options for broadband services such as optical fiber technologies and DSL on copper loop that would facilitate such growth in internet and broadband utilization also resulted from the 2004 National Broadband Policy. At the end of 2011, internet subscribers numbered 22.39 million; around the same time, the corresponding number for broadband was 13.54 million. It is thus fair to say that internet and broadband connectivity benchmarks set by the 2004 National Broadband Policy were yet to be achieved.

This set the stage for India's latest National Telecom Policy in 2012, which focused on increasing teledensity like its precursors, but also on providing high speed and high quality, reliable, affordable broadband services in the country. Specific targets include (a) increasing teledensity from the current level of approximately 39 in 2012 to 70 by 2017, and 100 by 2020, (b) providing affordable and reliable broadband by the year 2015 and achieving 175 million broadband connections by the year 2017 and 600 million by the year 2020 at a minimum download speed of 2 Mbps and making available higher speeds of at least 100 Mbps on demand, and also (c) providing high speed and high quality broadband access to all village panchayats (councils) by 2014 and progressively to all villages by 2020 (TRAI 2014c).

Overall, India is one of the world's leading nations in terms of mobile cellular users; however teledensity levels remain deficient with a significant divide between urban and rural areas. Internet and broadband usage per capita also lag behind other developing and emerging economies. With one of the lowest mobile cellular prepaid tariffs worldwide, ICT utilization in the form of mobile phones is affordable to a vast majority of India's population; however the equipment cost of a mobile phone itself continues to be high. This compounds India's digital divide which extends across the nation and across other socio-economic characteristics such as gender. In the next section, we examine state-wise patterns of ICT utilization.

6.3 State-Wide Patterns of ICT Utilization

India is a union of 29 states and 7 union territories. At the time of collecting data for this chapter, ICT utilization data were not available for some of the newer Indian states such as Chhattisgarh in central India and Jharkhand in eastern India. Moreover ICT use data for six North-Eastern states of Arunachal Pradesh, Manipur, Meghalaya, Mizoram, Nagaland, and Tripura are aggregated into one region, referred to as the North-East in the remainder of this chapter. Refer to Fig. 6.1 for a map of teledensity in Indian states.

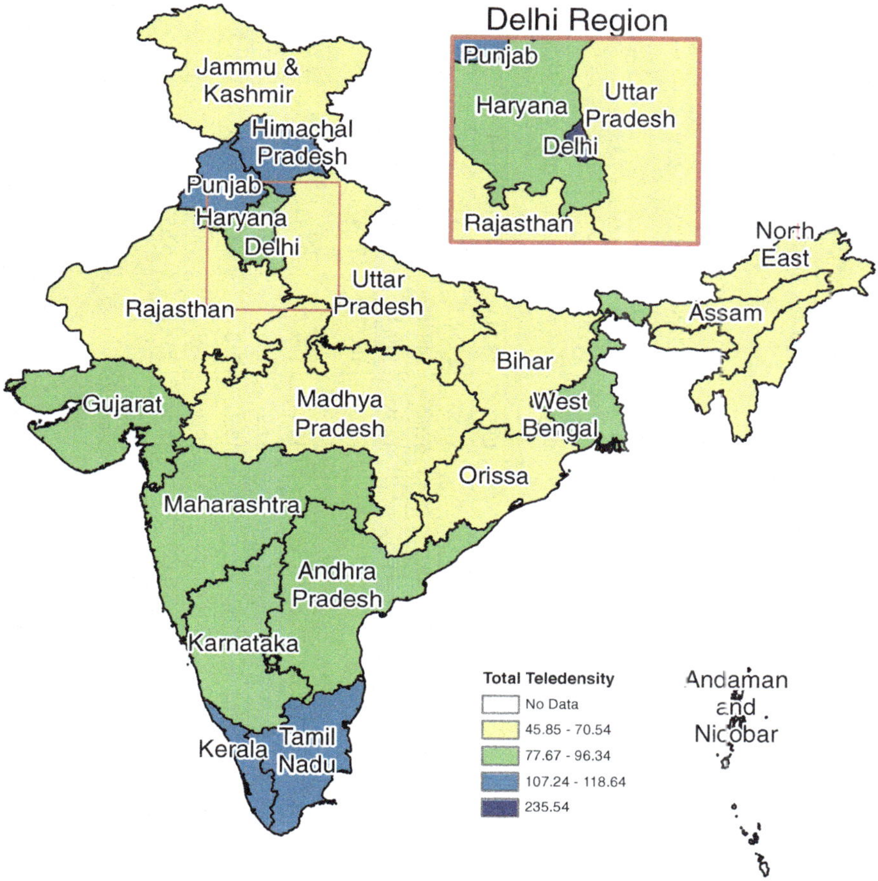

Fig. 6.1 Teledensity in Indian states, December 2011

Teledensity in Indian states

Increasing teledensity (Fig. 6.1) across the nation, both in urban and rural areas has been a consistent objective of successive national telecom policies of the Indian government going back to 1994. Acceleration in teledensity was catalyzed by the explosive growth of wireless telephony starting in 2005 and rapid strides have since been made all across India. However there is significant variation in penetration of telecom services among the Indian states. Also there is significant disparity in teledensity between urban and rural areas. In December 2011, urban teledensity was approximately 4.5 times that of rural teledensity in India (see urban/rural teledensity ratios for Indian states and India overall in Table 6.1). Among the states, the national capital and city-state of Delhi, home of the India national government as well as the state government, leads the nation with a teledensity of 235 telephone subscriptions per 100 people. This includes both wireline and wireless telephony.

Table 6.1 Teledensity in Indian states, December 2011

	Wireline teledensity			Wireless teledensity			Total teledensity			Urban/rural teledensity ratio
	Rural	Urban	Total wireline teledensity	Rural	Urban	Total wireless teledensity	Rural	Urban	All	
Delhi	N.A.	N.A.	15.27			220.27			235.54	–
Himachal Pradesh	4.07	8.94	4.61	70.85	458.16	114.03	74.91	467.1	118.64	6.24
Tamil Nadu including *Chennai*	1.98	7.13	4.85	52.31	155.1	109.44	54.3	162.33	114.28	2.99
Punjab	2.68	8.37	5.05	60.65	173.68	107.64	63.33	182.05	112.69	2.87
Kerala	8.20	12.12	9.20	48.43	242.85	98.04	56.63	254.97	107.24	4.50
Maharashtra including *Mumbai*	1.14	9.18	4.90	48.5	140.32	91.44	49.64	149.5	96.34	3.01
Karnataka	1.34	9.82	4.52	39.16	174.29	89.78	40.5	184.11	94.30	4.55
Gujarat	1.10	6.03	3.10	49.75	135.55	84.56	50.86	141.57	87.67	2.78
Haryana	1.12	4.66	2.33	52.53	143.43	83.48	53.65	148.09	85.80	2.76
Andhra Pradesh	1.10	7.06	2.75	36.27	182.66	76.90	37.38	189.71	79.65	5.08
West Bengal including *Kolkata*	0.50	5.93	2.05	41.7	160.78	75.62	42.2	166.71	77.67	3.95
Rajasthan	0.70	5.04	1.74	41.74	154.64	68.80	42.44	159.77	70.54	3.76
North-East	0.63	5.89	1.90	37.29	139.63	62.12	37.91	145.51	64.02	3.84
Orissa	0.49	4.35	1.14	32.43	207.86	62.11	32.91	212.2	63.25	6.45
Uttar Pradesh	0.29	3.62	1.03	30.1	154.64	57.94	30.38	158.27	58.97	5.21
Madhya Pradesh	0.23	3.97	1.23	25.23	123.13	51.34	25.46	127.1	52.57	4.99
Jammu and Kashmir	0.45	5.26	1.75	28.81	110.14	50.77	29.27	115.4	52.52	3.94
Bihar	0.16	2.43	0.47	24.11	189.50	46.70	24.27	191.93	47.16	7.91
Assam	0.19	3.87	0.75	28.01	140.85	45.10	28.21	144.72	45.85	5.13
All India	0.92	6.85	2.71	36.56	161.01	74.15	37.48	167.85	76.86	4.48

Source Telecom Regulatory Authority of India (TRAI 2014d)

Delhi's teledensity is more than two times that of the second-ranked state, Himachal Pradesh. Delhi is also the home of numerous foreign consulates and has a robust presence of electronic and print media that provides up-to-the-minute coverage of politics and related content. Other states for which teledensities exceed 100 are the neighboring states of Tamil Nadu and Kerala in southern India, and Punjab in Northern India. Tamil Nadu's capital city of Chennai is recognized as one of India's metropolitan cities and is a hub of advanced manufacturing, especially in the automotive sector. Teledensity is also high in the southern Indian state of Karnataka; its capital city Bangalore is home to numerous indigenous software and knowledge services companies, as well as research and development centers of well-known multinational corporations such as GE, Intel, and Bristol-Myers Squibb.

The state of Maharashtra, on the western coast of India, was ranked sixth in terms of teledensity in December 2011; its capital Mumbai, a large metropolis, is a hotspot of financial activity and is also considered to be the entertainment capital of India. Another huge city, Kolkata is the capital of the state of West Bengal, located on the eastern coast of India. West Bengal has achieved moderate success in improving teledensity but lags behind other populous and industrialized states such as Tamil Nadu, Maharashtra, Karnataka, and Gujarat. West Bengal's neighboring states in eastern India—Bihar, Orissa, and Assam and the North-East (comprised of six states excluding Assam) are among the states with lowest teledensity. Two of India's most populous states—Uttar Pradesh in northern India, and Madhya Pradesh in central India are in the bottom five in terms of teledensity.

Figure 6.2 shows that an urban-rural divide in teledensity, in fact for ICTs in general, exists across India and the effect is more pronounced in states with low overall teledensities such as Bihar, Orissa, Uttar Pradesh, and Assam

The urban-to-rural teledensity ratio in Bihar is 7.91, the highest in the nation; in fact, Delhi's teledensity is 9.7 times that of rural Bihar. Interestingly, Himachal Pradesh, whose overall teledensity is the highest in the nation if the national capital "outlier" Delhi is excluded, has fared poorly in bridging the teledensity gap between its urban and rural population; its urban-to-rural teledensity ratio of 6.24 is comparable with that of Assam. This can possibly be attributed to the mountainous terrain in state of Himachal Pradesh that is full of isolated valleys and high Himalayan ranges (Fig. 6.3).

Urban-rural ratios in the states of Haryana, Gujarat, Punjab, Tamil Nadu, and Maharashtra are between 2.75 and 3.01, indicating that while an urban-rural divide still exists, these states have made significant progress in eradicating telephone penetration issues. It is also interesting to observe that the urban-to-rural teledensity ratios for wireless in these five states are only slightly less than the corresponding overall ratios.

Internet and Broadband Use in Indian states

The number of internet subscribers has increased noticeably since the year 2000, when the number of subscribers was less than one million. In December of 2011, there were approximately 9 million narrowband connections (download speed < 256 kbps) and in February of 2012, there were approximately 13.50 million broadband connections (>256 kbps). Mobile broadband is increasingly popular in

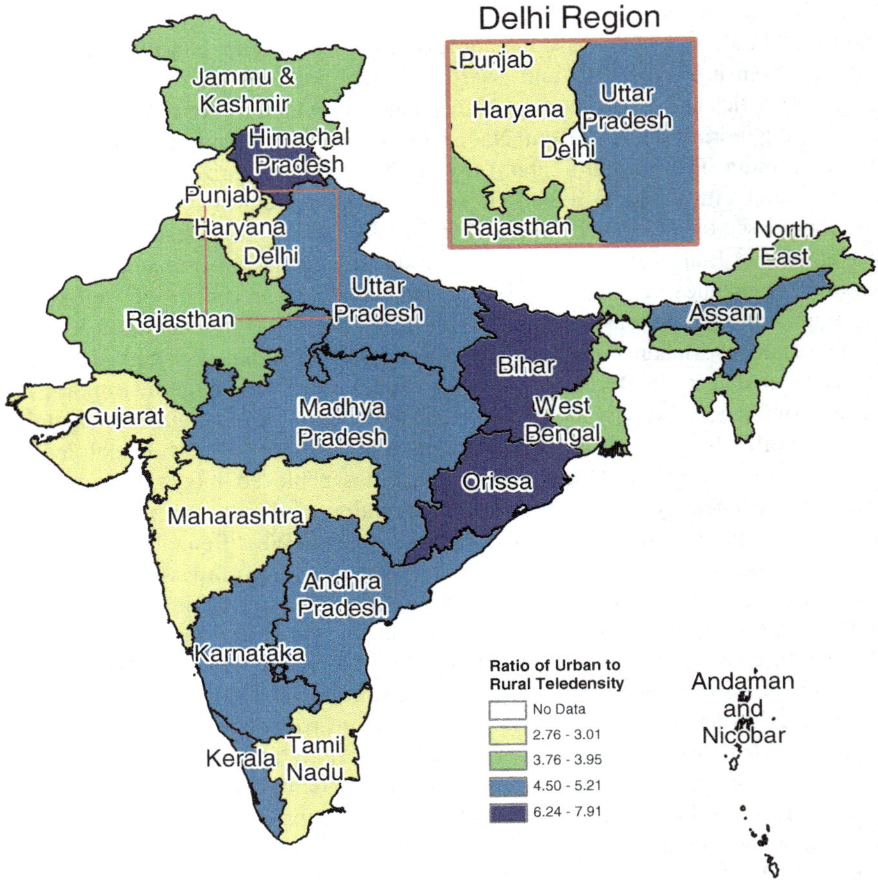

Fig. 6.2 Urban/rural teledensity divide in Indian states, December 2011

India and in December 2011, approximately 431 million wireless subscribers (almost 50 %) had subscribed for data services (TRAI 2014d).

For broadband subscribers per 100 people in 2010, as seen in Fig. 6.4, the top six states in India were Delhi (highest per capita internet subscribers), Kerala, Tamil Nadu, Karnataka, Punjab, and Maharashtra, while the bottom five were comprised of Jammu and Kashmir, states in the North East region, Uttar Pradesh, Assam, and Bihar. In March 2011, Maharashtra had India's largest share (18.6 %) of broadband subscribers, followed by Tamil Nadu, Andhra Pradesh, Delhi, and Karnataka, in that order. These five states accounted for approximately 60 % of India's broadband subscribers in 2011.

Broadband utilization in Indian states is largely consistent with the states' teledensity, with the southern Indian states of Tamil Nadu, Kerala, Andhra Pradesh, and Karnataka, as well Delhi and Punjab in the north, and Maharashtra in the west

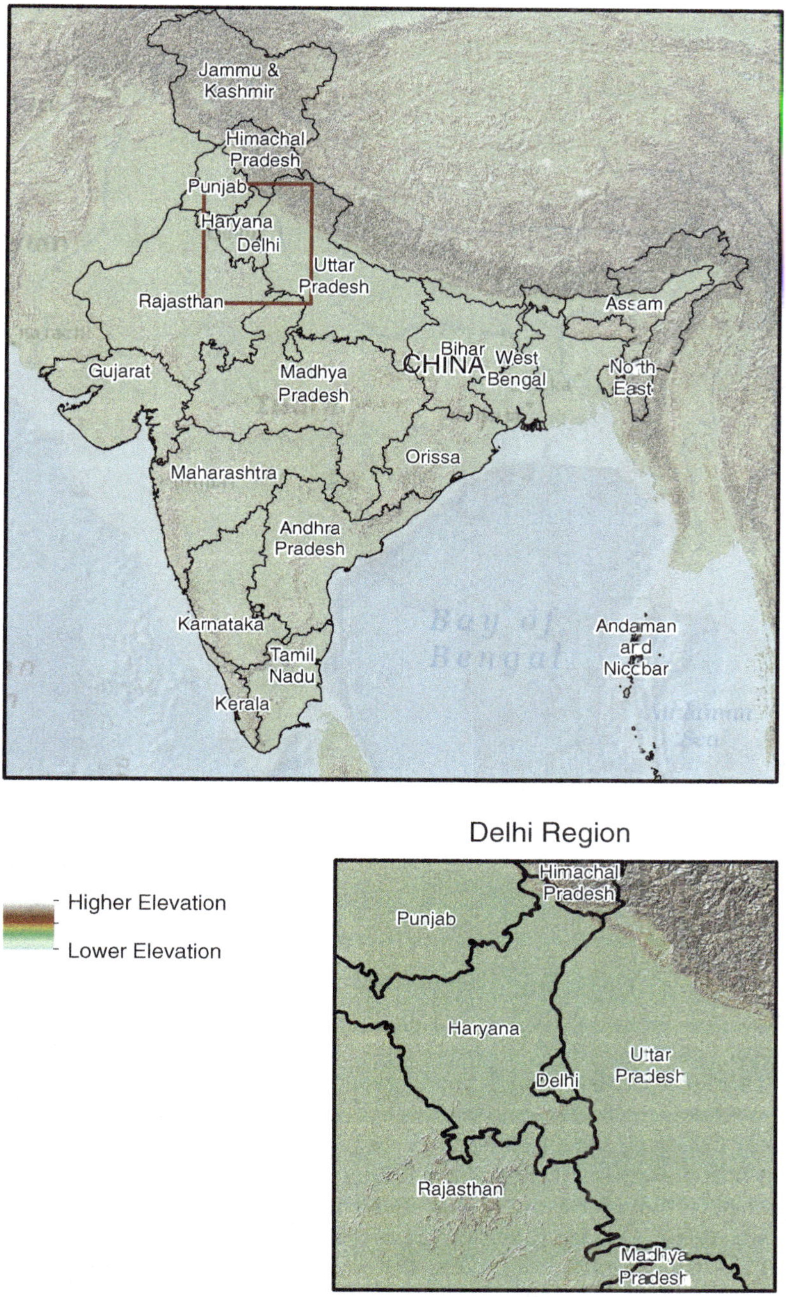

Fig. 6.3 Mountainous terrain (Himalayas) in North India

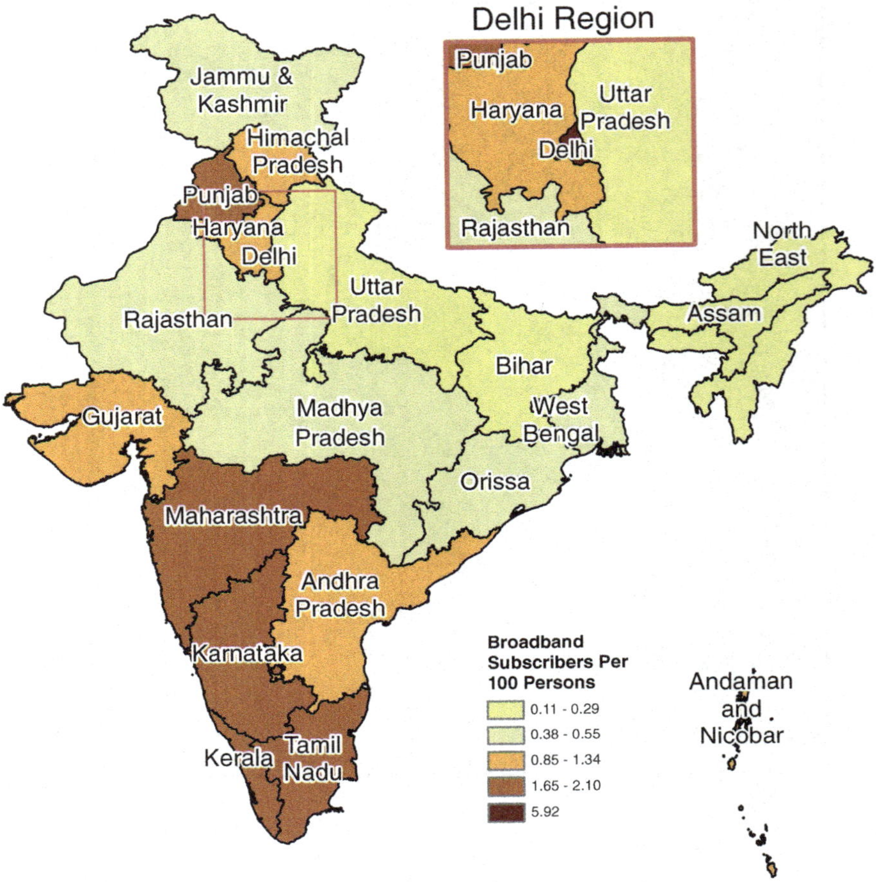

Fig. 6.4 Broadband subscribers per capita in Indian states, 2010

leading in ICT utilization. By contrast, Bihar in the east, Assam and the states in the northeast, Uttar Pradesh in the north, and Madhya Pradesh in central India are among the weakest states.

Through sustained efforts driven by the Indian government as well as governments in the states since 2009–10, broadband adoption in village panchayats in several Indian states and Union Territories has steadily increased. By March 2011, almost half (15 out of 33) Indian states and Union Territories have achieved at least 50 % broadband coverage in its panchayats (see Fig. 6.5). Pan-India coverage at the same time was around 45 %. States that have made significant progress include Kerala in south India, Haryana, Chandigarh, and Punjab in north India, and Nagaland and Tripura in the north-east. While Kerala, Punjab, and Haryana have

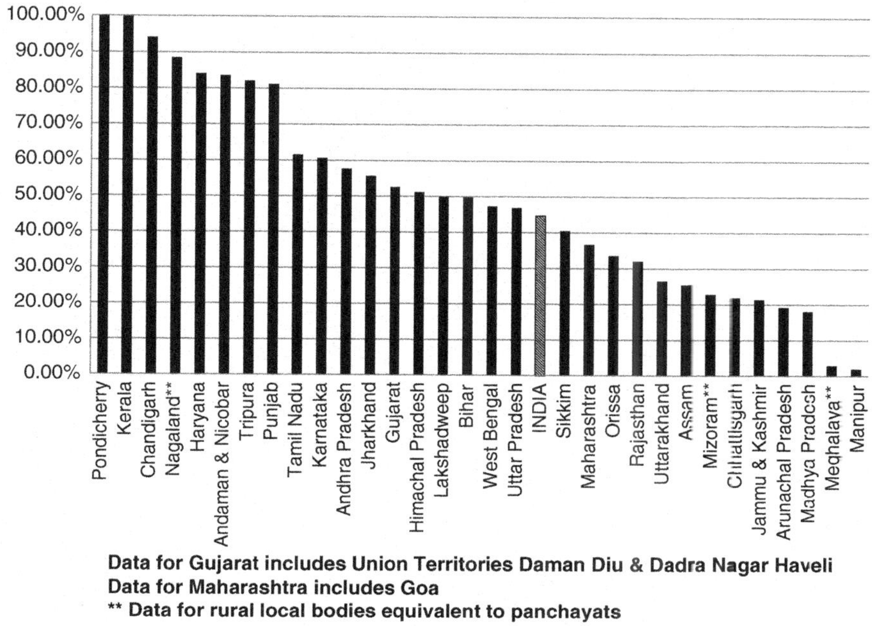

Fig. 6.5 Percent of village panchayats with broadband for 31 states and union territories, March 2011

high-moderate teledensity, the progress made by the two states in the north-east is especially encouraging given the low teledensity prevalent in the northeast. Note that Fig. 6.6 shows 31 states and union territories, whereas our chapter dataset has only 20. The discrepancy is because our ICT variables were generally only available for 20 states.

Other low teledensity states such as Bihar and Uttar Pradesh have made moderate progress, but the remaining northeastern states of Assam, Mizoram, Arunachal Pradesh, and Manipur lag far behind their peers—Nagaland and Tripura. It is pertinent to mention here that while Kerala leads the nation in terms of rural broadband coverage, only 6 % of India's broadband subscribers live in that state (TRAI 2014d).

Some of the constraints to improve rural telephony and internet/broadband adoption are (i) difficulties in acquisition of land, (ii) complications with right of way, (iii) lack of reliable power supply, (iv) low average revenue per customer which discourages service providers, (v) problems with affordability of handsets due to low income levels, and (vi) low literacy level, since most internet/broadband content is delivered in English (TRAI 2014d).

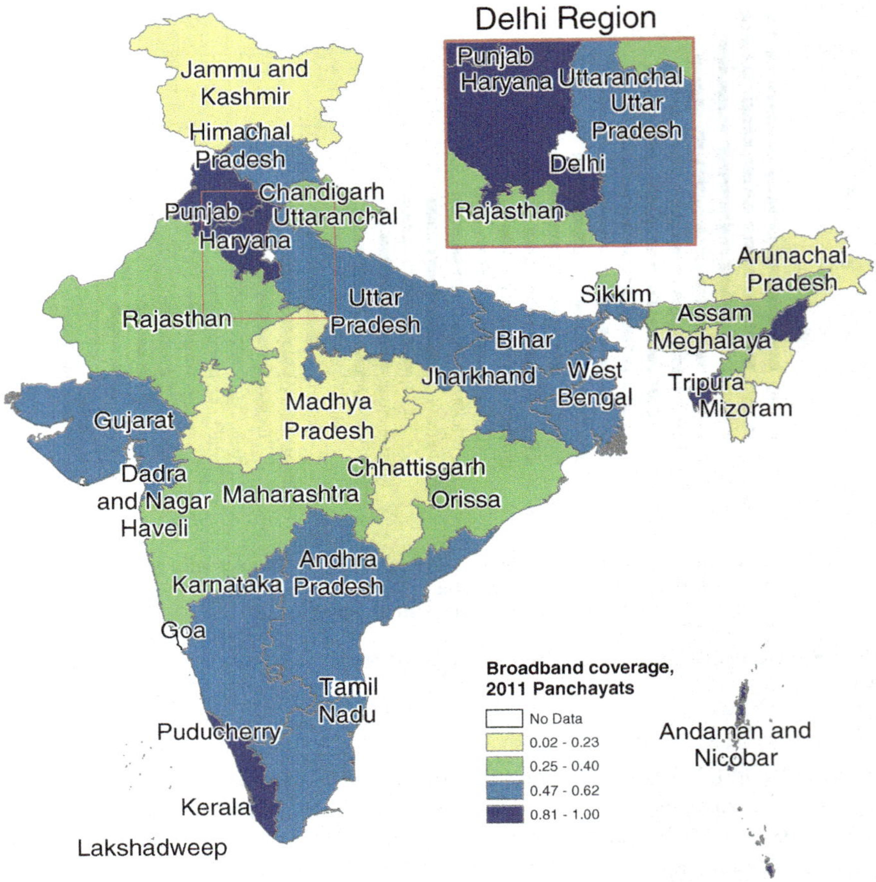

Fig. 6.6 Broadband coverage in village panchayats (percent of panchayats with broadband), 31 states and union territories, March 2011

6.3.1 ICT Clusters in India

As discussed in Chap. 1, k-means cluster analysis was conducted to classify Indian states with approximately similar ICT utilization levels based upon four ICT dependent variables: internet users, broadband subscribers, mobile telephone subscribers, and landline telephone users. K-means cluster analysis is not confirmatory; consistent with prior discussion in Chap. 1, it provides more insights about geographic patterns of ICT utilization in the states of India, based on the full set of ICT variables. Using K-means, four clusters were designated for ICT utilization in 2009–10; state membership of clusters is depicted in Fig. 6.7 and cluster characteristics appear in Table 6.2.

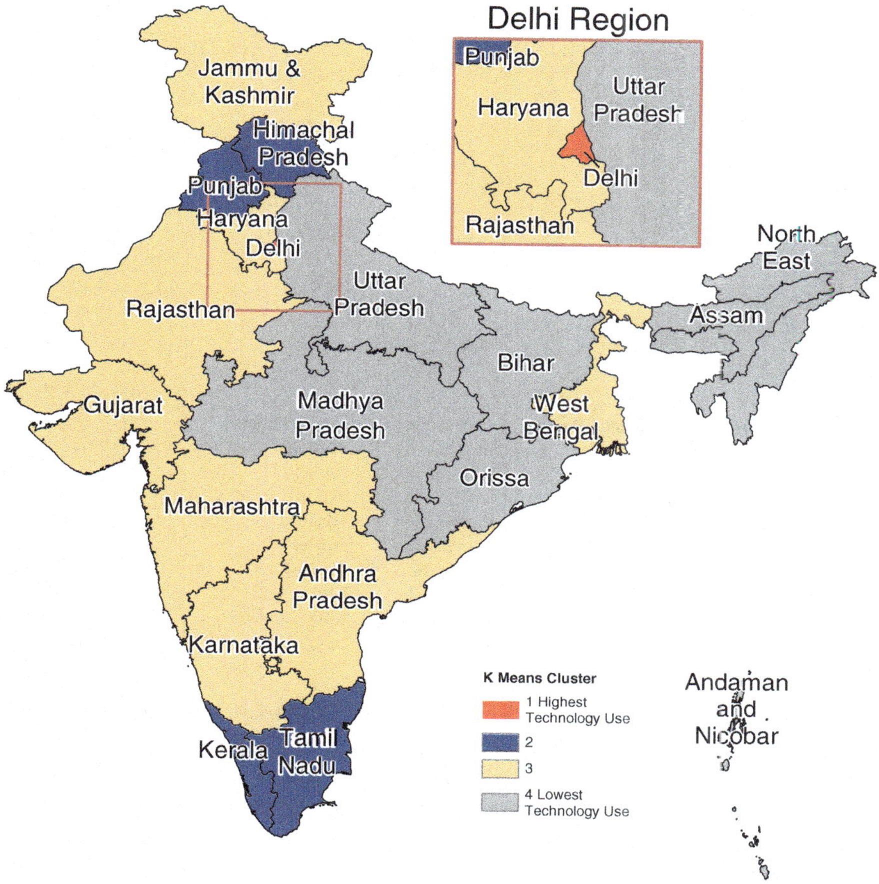

Fig. 6.7 K-means cluster of ICT utilization in Indian states, 2009–10

Cluster 1 has one member, the capital of India and the city-state of Delhi. Delhi is the world's second largest megacity with population in 2014 of 25 million (UN 2014). Its population is almost 100 % urban, and as seen in previous discussion, it is a high "outlier" in terms of teledensity, both wireline and wireless, and it has the highest internet and broadband penetration among Indian states. Cluster 2, the second highest ICT utilization cluster is comprised of four states—Himachal Pradesh and Punjab in the north, and the neighboring states of Kerala and Tamil Nadu in the south. Cluster 3 has eight members, consisting of Andhra Pradesh and Karnataka in the south, Gujarat and Maharashtra in the west, West Bengal in the east, and Haryana and Jammu and Kashmir in the north. These states account for almost 43 % of India's 2011 population and include two megacities—Mumbai in Maharashtra, the financial and entertainment capital of India, and Kolkata in West Bengal. Mumbai is the 7th largest megacity worldwide with a population of

Table 6.2 ICT clusters in India: 2009–10

ICT dependent variable	Cluster number				Ratio of high to low clusters	Ratio of high to low clusters excluding Delhi
	1	2	3	4		
Broadband subscribers	5.423	1.698	1.034	0.403	13.47	4.22
Internet users	9.840	2.290	1.440	0.600	16.40	3.82
Landline users	15.115	6.649	3.342	1.842	8.20	3.61
Mobile telephone subscribers	157.790	74.536	54.601	33.229	4.75	2.24
	Delhi	Himachal Pradesh Kerala Punjab Tamil Nadu	Andhra Pradesh Gujarat Haryana Jammu and Kashmir Karnataka Maharashtra Rajasthan West Bengal	Andaman and Nicobar Assam Bihar Madhya Pradesh North East Orissa Uttar Pradesh		

21 million and Kolkata ranked 14th at 15 million (UN 2014). Also in this cluster is the city of Bangalore, a near-megacity at 9.7 million, the information technology and knowledge services hub of India, in the state of Karnataka. Jammu and Kashmir's membership in the same cluster as Karnataka, Maharashtra, and West Bengal is somewhat surprising, given that Jammu and Kashmir (often referred to as just Kashmir) is one of the low teledensity states, and also low in broadband and internet penetration. Lastly, the states in the north-east plus Assam, often referred to as the "seven sisters", Bihar and Orissa in the east, Madhya Pradesh in central, and Uttar Pradesh in northern India, along with the union territory of Andaman and Nicobar islands in the Bay of Bengal comprise the lowest ICT utilization cluster. Close to 40 % of the Indian population resides in these states.

In comparison to the states in this lowest-ICT cluster, Delhi's telephone utilization is 4.75 times for wireless and 8.2 times for wireline telephony (see Table 6.2). The divide between Delhi and this cluster for internet and broadband is even more pronounced with Delhi's penetration in 2009–10 being 13–16 times higher. When the high outlier Delhi is excluded, the extent of the ICT digital divide between Indian states appears to be alleviated, with utilization ratios between 2.2 and 4.2 times in those states that are in cluster 2, compared to those in cluster 4.

Delhi's unique membership as a high outlier in cluster 1 is analogous to the Chinese provinces of Beijing, Shanghai, and Guangdong being unique one-province clusters (see Chap. 5) and to Tokyo prefecture in Japan (see Chap. 7). Delhi, India's national capital is home to the Indian national government and to Delhi's own state government. Numerous foreign government consulates also operate out of Delhi and it is considered to be the epicenter of Indian politics. As a natural consequence, numerous electronic and print media organizations are either headquartered in Delhi or have a significant presence in the city-state resulting in teledensity that is more than twice that of second ranked Himachal Pradesh, and broadband penetration that is almost thrice than that of second ranked Kerala.

We conclude this section by observing that our K-means findings, as evident from Fig. 6.7 largely conform to Tobler's Law of Geography which states that nearby entities are more similar than distant entities. While cluster 1 is in two parts in the north and south of India, Himachal Pradesh and Punjab, as well as Tamil Nadu and Kerala are neighboring states. Cluster 3 and 4 member states are also proximate to members in their cluster. Also our findings are largely consistent with Das (2010a) which analyzed intra-national digital divide in India and classified Indian states into three categories based upon an ICT diffusion index. States with high ICT diffusion in 2005–06 were Maharashtra, Kerala, and Tamil Nadu, states with medium ICT diffusion were Punjab, Gujarat, Haryana, and Karnataka, and states with low ICT diffusion were West Bengal, Himachal Pradesh, Andhra Pradesh, Rajasthan, Madhya Pradesh, Orissa, Uttar Pradesh, Assam, and Bihar. Delhi is its own unique cluster in our study and was not analyzed by Das (2010a). The only other difference is the classification of the northern Indian state of Himachal Pradesh as a low ICT diffusion state. This is perhaps explained by the 4–5 year lag between Das (2010a) and our study during which time Himachal Pradesh made significant progress in ICT utilization, possibly as a result of spillover effects from nearby high-ICT states of Delhi and Punjab.

6.4 Conceptual Model of ICT Utilization in Indian States

A generic conceptual model of ICT utilization has been described earlier in Chap. 3 and depicted in Fig. 3.8. This generic model, induced from extensive literature on ICT utilization and the digital divide posits that several independent socio-economic, governmental, and societal openness factors are associated with ICT utilization dependent variables. The conceptual model of ICT utilization presented in this section draws on the generic model, but is specific to India and posits that a dozen education, economic, and societal openness variables are associated with four ICT dependent variables: mobile phone, landline, internet, and broadband subscribers. Data on personal computers, social media, and internet servers at the state level were not available at the time of conducting this study. Spatial patterns of distribution of the four dependent variables and their clustering have been discussed in the previous section. In this section, we justify inclusion of independent variables in the model.

Twelve independent variables included in this chapter's conceptual model are split into four categories: education, represented by four independent variables, economic, represented by five independent variables, and remaining three variables

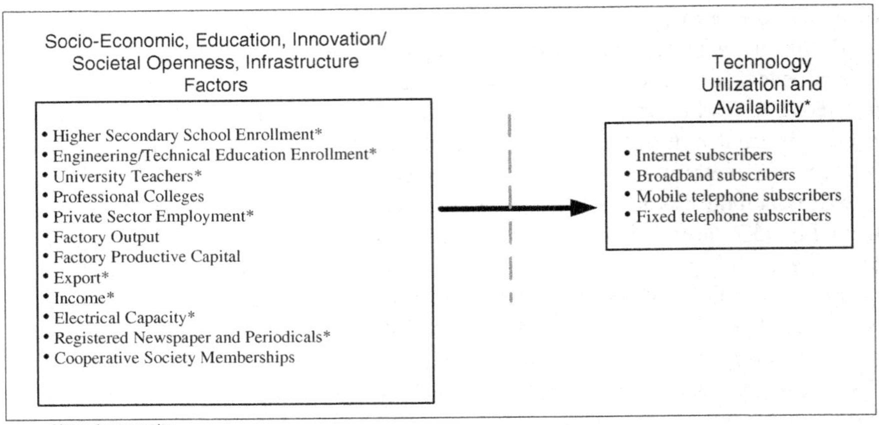

Note: * per capita

Fig. 6.8 Conceptual model of ICT Utilization in Indian States

are proxies for infrastructure, social capital, and a combination of innovation and societal openness. The conceptual model for this chapter is depicted in Fig. 6.8.

Education independent variables: As discussed in Chap. 3, education is one of the most significant correlates of ICT utilization. We include higher secondary school enrollment, engineering/technical education enrollment, university teachers, and professional colleges as independent correlates.

Economic independent variables: As discussed in Chap. 3, economic variables are often found to have strong positive association with ICT utilization. Economic variables such as income, foreign direct investment, exports and trade have commonly been included in prior studies of the digital divide. In this chapter, we posit that income, exports, factory output, factory productive capital, and private sector employment are associated with ICT use in India. We next justify inclusion of these independent variables specific to India that was not discussed in Chap. 3.

Private sector employment: In a prior study (Azari and Pick 2005) of socio-economic influences on receipts and payroll in the technological sector in US counties, two private sector workforce variables professional/scientific/technical services workforce, and other services workforce were found to have strong positive association with technology payroll and receipts. In Chap. 5, employment in non-state owned enterprises was found to be positively associated with internet use and use of PCs when high ICT use outlier provinces of Beijing and Shanghai were excluded (Pick et al. 2013). As a result, we posit private sector employment to be associated with ICT utilization in Indian states.

Factory output and capital: In charting the evolution of information technology and software development cluster in Bangalore India, van Dijk (2003) and Chaminade and Vang (2008) have commented that before IT, Bangalore was the home of high-tech manufacturing in the defense and aeronautical sector, which spawned small and medium sized enterprises (SMEs) in the manufacturing sector. We posit that the variable, factory output and capital, reflects economic activity that is engendered by use of ICTs in the manufacturing sector as well as manufacturing of ICT hardware that will facilitate the use of e-commerce and other uses of ICTs.

Now we present justification for the inclusion of three other variables—electrical capacity, registered newspapers and periodicals, and cooperative society memberships in our conceptual model.

Electrical capacity: One of the impediments identified by the Telecom Regulatory Authority of India in improving rural telephony and internet/broadband adoption is lack of reliable power supply (TRAI 2014d; Veeraraghavan et al. 2009). ICT diffusion, especially in developing or newly industrializing nations is hampered by infrastructural malaise caused by deteriorating roads, unreliable electricity, and arbitrary regulatory and judicial systems (Cooper 2002; Quibria et al. 2003). Availability of cheap electricity (and water) have been identified by van Dijk (2003) as infrastructural facilitators of Bangalore's early industrialization and subsequent development as a robust IT cluster. Hence we include electrical capacity as an independent variable and the sole proxy for infrastructure in our model.

Registered Newspapers and periodicals: It has been posited in the literature that a free, open society would be more likely to communicate, interact, exchange and share information, perhaps using ICTs. Indicators of freedom, democracy, open exchange of information, and robust legal and regulatory system are significant correlates of ICT utilization (Robison and Crenshaw 2002; Guillen and Suarez 2005; Yates et al. 2011; Pick et al. 2013). In Chap. 5, the number of published books, a proxy for societal openness, was determined to be strongly associated with broadband subscribers, domain names, webpages, and PCs per capita for Chinese provinces. Chapter 5 also reasoned that publications such as newspapers, periodicals, books, and directories also foster generation of e-content in the form of e-books, websites, softwares, games, mobile-based applications, and other web-based content, and hence are drivers of innovation. In China, book publishing was highly concentrated in the megalopolitan innovation centers of Beijing and Shanghai. We therefore include registered newspapers and periodicals as an independent variable in our conceptual model to represent both innovation and societal openness. Using the same rationale, annual publishing sales has been posited as an independent variable for the United States in Chap. 8. Figure 6.9 shows distributors packaging newspapers for newspaper vendors in the city of Kolkata, India.

Cooperative Society Memberships: In Chap. 3, the role of social capital as a facilitator of ICT usage has been discussed. Chen (2013) and Agarwal et al. (2009) have provided evidence that individual use of the internet in the United States is positively impacted by internet use peers and in communities with strong social ties. Van Dijk (2005) theorizes that "participation in society," a proxy for social capital, is a major outcome factor from access to ICT, while we posit the relationship in the opposite direction. Social capital manifests itself in different forms; in the Indian context, cooperative societies have been often been at the forefront of ICT adoption and utilization especially in rural communities (Cecchini and Scott 2003; Gupta et al. 2007; Meera et al. 2004; Veeraraghavan et al. 2009) in diverse sectors such as farming, agriculture, housing, banking, microfinance, to name a few. Use of ICTs in agricultural cooperatives across Asia has been found to provide a knowledge

Fig. 6.9 Newspapers distributors arranging newspapers by title in Kolkata, India

management system that ultimately leads to sustainable agricultural practices (Teodosio 2007).

In India, "a Co-operative is an organization which focuses on the promotion of the interests of its members in accordance with the principles of co-operation. It is a voluntary association of ten or more members residing or working in the same locality, who join together on the basis of equality for the fulfilment of their economic or business interest. The basic feature which differentiates co-operatives from other forms of business ownership is that its primary motive is service to the members rather than making profits" (Government of India 2014). From this definition, it is clear that a cooperative in India fosters linkages and forges relationships, bonds, and collaborations among people of a local population with the same interests, often economic. Given the positive association of social capital with internet usage in the US, we posit that cooperative society membership will be positively associated with ICT utilization in the Indian states.

In 2013, the top three Indian states in terms of cooperative societies were Maharashtra, Gujarat, and Karnataka, which together accounted for more than 60 % of India's total, numbering almost 400,000. However from a membership standpoint, the states of Maharashtra, Kerala, and Uttar Pradesh led the nation with a combined membership of almost 96 million members, which is approximately half

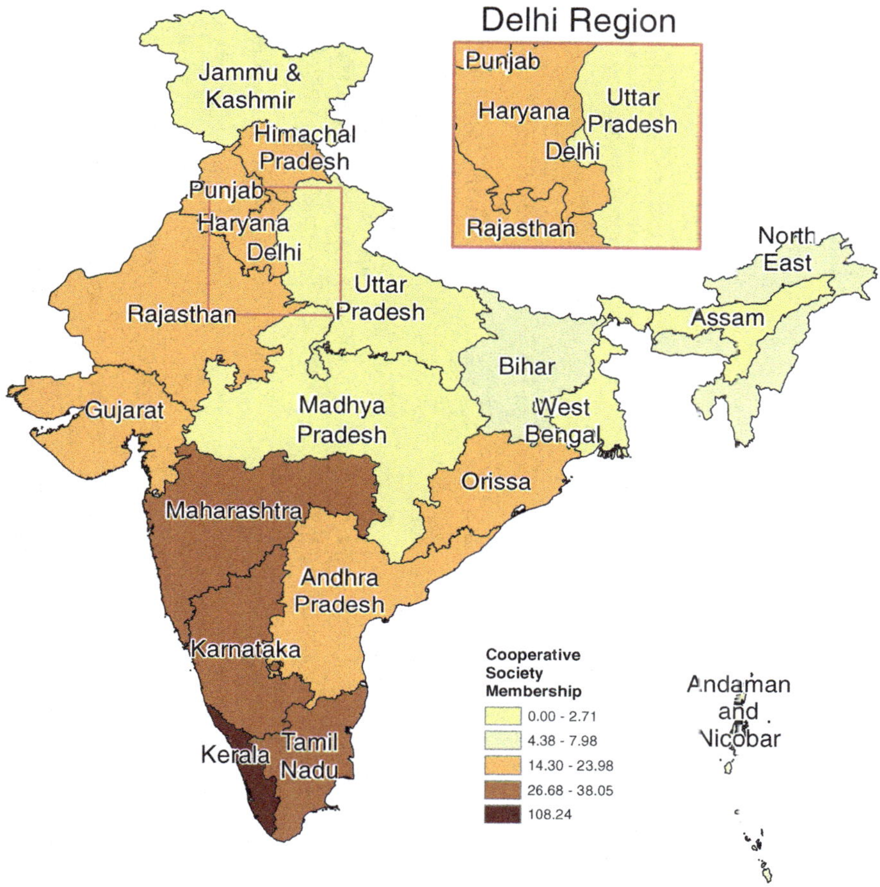

Fig. 6.10 Cooperative society membership per capita, in Indian states, 2004

of India's total cooperative society membership (Ministry of Statistics and Programme Implementation 2013). The state of Kerala has had India's highest per capita (per 100 population) cooperative society membership for the last decade. Figure 6.10 depicts per capita cooperative society membership in the Indian states in 2004.

6.5 Data Collection and Methodology

Different sources of data were used to obtain values of dependent and independent variables (see Table 6.3). Data were collected from the Indian Statistical Yearbook of 2011, the 2011 Annual Report of India's Department of Telecommunications

Table 6.3 Variable definitions and descriptive statistics, India

	Variable	Definition	Year of data collection	Data source	Average	Standard deviation
Dependent variables						
Technology utilization	Internet subscribers	Number of internet subscribers per 100 people	2009	LS	1.79 (1.32[a])	2.26 (0.91[a])
	Broadband subscribers	Number of broadband subscribers per 100 people	2010	LS	1.20 (0.95)	1.29 (0.67)
	Mobile telephone subscribers	Number of mobile telephones per 100 people	2010	T	56.8 (50.7)	31.3 (16.0)
	Fixed phone subscribers	Number of landline telephones per 100 people	2010	T	4.16 (3.51)	3.67 (2.30)
Independent variables						
Education	Higher secondary school enrollment	Enrollment in higher secondary school per 100 people	2008	SYB	3.99 (3.94)	1.10 (1.11)
	Engineering/technical education enrollment	Enrollment in engineering or technical professional education school per 100 people	2008	SYB	0.123 (0.121)	0.0959 (0.977)
	University teachers	Number of teachers in colleges and universities per 100 people	2008	SYB	0.00923 (0.00841)	0.00601 (0.00491)
	Professional colleges	Number of professional engineering colleges	2007	SYB	0.000568 (0.00058)	0.000378 (0.000385)
Economic	Percentage of private sector employment	Percentage of estimated employment in private sector to total employment	2008	CENPOP10	30.4 (30.4)	16.9 (17.3)
	Factory output	Number of total output of products in factory sector per capita	2008	SYB	24,529 (24,926)	20,163 (20,635)
	Factory productive capital	Total productive capital in factory sector per capita (rupees)	2008	SYB	10,853 (11,185)	9954 (10,113)
	Export	Export oriented units' export per capita (rupees)	2009	EPC	1389 (1423)	3729 (3828)
	Income	Per capita income (rupees)	2008	RBI	52,243 (48,359)	26,572 (20,659)

(continued)

Table 6.3 (continued)

	Variable	Definition	Year of data collection	Data source	Average	Standard deviation
Infrastructure	Electrical capacity	Electricity-installed capacity (watt per person)	2008	SYB	94.3 (96.2)	51.9 (52.6)
Innovation/societal openness	Registered newspapers and periodicals	Number of registered newspapers and periodicals per 100 people	2008	SYB	0.00836 (0.00566)	0.0126 (0.0034)
Social capital	Cooperative society membership	Membership of all types of cooperative societies	2004	SYB	17.8 (18.6)	24.0 (24.3)

SYB Statistical Yearbook 2011; *T* Department of Telecommunications, India; *LS* Lok Sabha; *RBI* Reserve Bank of India; *EPC* Export Promotion Council

[a]Average and standard deviation computed excluding the state of Delhi which is an outlier for technology utilization variables

under the Ministry of Communications and Information Technology, Census of India of years 2001 and 2011, the Reserve Bank of India, Export Promotion Council of the Department of Commerce, and Lok Sabha Proceedings from 2007 to 2010.

Lok Sabha is the House of People, an elected body similar to the US Congress. While it would have been ideal to obtain data estimates for all variables for one unique year, we often encountered a time lag of up to 3 years for certain variables. However except for cooperative society membership, data for all variables belong to the period 2007–2010. Furthermore, we were constrained by missing data for certain states. This can be attributed to the fact that certain states such as Bihar and Madhya Pradesh have been bifurcated recently. As a result, data for all variables were only available for the parent state (for example, Bihar) but data were missing for the newly created states (for example, Jharkhand). Furthermore, for the seven northeastern states, state level data were available only for Assam, while data for the remaining six states were aggregated by government agencies as North-East.

Consequently, the sample size of states is limited to 20. Descriptive statistics of all relevant variables in our study are in Table 6.3. It is pertinent to note that an econometric sub-national study (Kathuria and Kedia-Jaju 2012) of the impact of internet in India considered a sample of 19 states.

The methodology adopted in this chapter follows the generic methodology adopted in this book that is discussed in Chap. 1. To screen for spatial autocorrelation, the dependent variables were mapped using the Local Moran's I statistic (Moran 1950). This Local Indicators of Spatial Association (LISA) based method shows local patterns of spatial association. Local Moran's I was used for this chapter to account for the reduced sample size (n = 20 states) for India in comparison to sample sizes of 29, 47, and 50 states/provinces for China, Japan, and US respectively, and 50 nations for Africa. Local Moran's I statistic indicates presence of spatial outliers and clusters, which are positive or negative relative to surrounding values. The interpretation of Local Moran's I is to show positive "hot spots" or negative "low spots" of a variable relative to its neighbors (Anselin 1995). In this analysis the neighbors of an Indian state are identified as those states contiguous with a state or cluster of states. Since Andaman and Nicobar are a group of islands in the Indian Ocean without contiguous neighboring states, they were excluded from LISA analysis. An example of the application of LISA for ICT can be found in Ding et al. (2010) which applied LISA to determine agglomeration of Chinese provinces with similar cell phone diffusion speed.

For broadband subscribers, internet subscribers, and landline subscribers, the LISA analysis for India revealed one local positive outlier—Delhi (which is shown in the inset of Fig. 6.11a). The figure reflects that Delhi's broadband use is 2.7 times that of the second highest state Kerala and 5.8 times the average use for the other states. When Delhi is removed and the LISA analysis repeated, there is a high cluster of the states of Karnataka, Kerala, and Tamil Nadu in the south, and a low local value for Bihar (Fig. 6.11b). For mobile telephone subscribers the patterns are the same, except the southern high-value cluster consists only of Kerala and Tamil Nadu.

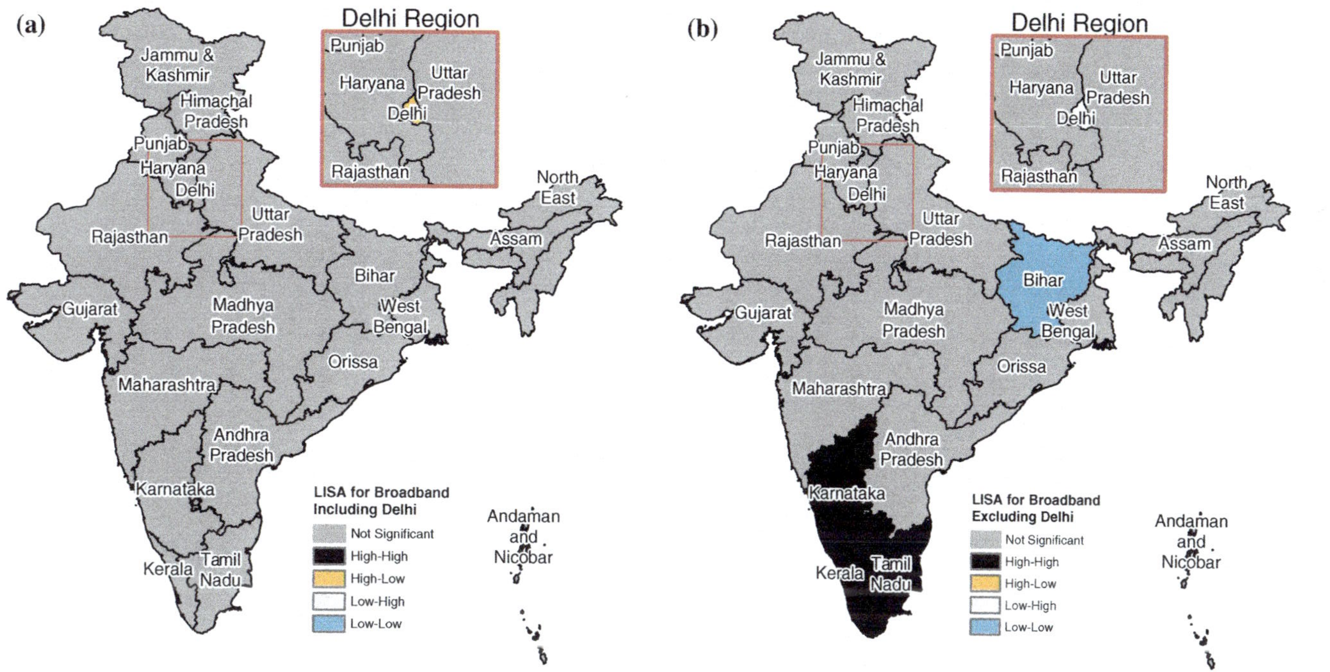

Fig. 6.11 LISA (local Moran's I) for broadband subscribers **a** including Delhi, and **b** excluding Delhi. *Note* In the legend, "High-High" indicates a state with high broadband subscription per capita surrounded by other states with similarly high values. In other words, "High-High" indicates a statistically significant cluster of states with high broadband subscription per capita. Other items of the legend can be interpreted similarly

6.6 Determinants of ICT Utilization in Indian States

The OLS regression analysis was conducted for the 20 Indian states, utilizing stepwise regression and allowing a maximum of two independent variables to enter. The number of entering variables was restricted to two, as implied by the small sample size (Halinski and Feldt 1970; Sawyer 1982). Stepwise regression was utilized because of the scarcity of studies of technological utilization at the state level and because of the constraint of a maximum of two entering variables. This approach is somewhat similar to a prior study of communications diffusion in India (Das 2010a, b). In Tables 6.4 and 6.5, all regressions are significant at the 0.05 level and 83 % are highly significant at the 0.001 level. OLS regression diagnostic tests indicate that regression assumptions are met for almost 90 % of regressions.

As seen in Table 6.4, our regression findings for the dependent variables in 2010 show that the most influential factor for all technology subscriber levels is registered newspapers/periodicals. Second in importance are the higher education variables of university teachers (for mobile telephone) and engineering/technical education enrollment (for internet and broadband). State-wise distribution of engineering/technical education enrollment is shown in Fig. 6.12. Cooperative society membership is significant for landline telephone subscribers. The newspaper/periodicals finding corresponds to a similar finding for published newspapers for the provinces of China discussed in Chap. 5. The education findings have had consistent support in the digital divide literature for nations (Robison and Crenshaw 2002; Kiiski and Pohjola 2002; Simon 2004; Chinn and Fairlie 2007; Ono and Zavodny 2007; Pick and Azari 2008, 2011; Yates et al. 2011).

Mapping and geographical analysis for the technology dependent variables indicates unique high outlier values for Delhi; in particular, Delhi's level of registered newspapers and periodicals was over seven-fold the level for the next highest state of Maharashtra. Because of its outlier status, we computed an alternative series of stepwise regressions excluding Delhi from the sample (see Table 6.5).

As expected from Delhi's exclusion, registered newspapers/periodicals is eliminated as a factor. The significant correlates shift to engineering/technical education enrollment (significant for broadband and mobile phones), cooperative society membership, and electrical capacity (significant for landlines and internet). The dominant influence of income on broadband is explained by the relatively high cost of broadband that is utilized in India by wealthier individuals and businesses in 2010 (Dharmakumar 2010). Cooperative society membership has the same explanation as findings just reported. Electrical capacity is explained since availability of electricity is an infrastructural requirement necessary to support the devices of the large number of internet and landline users (Veeraraghavan et al. 2009).

Table 6.4 Standardized regression results for dependent variables, 2010, India

Dependent variable					
Correlate	Independent variable	Internet subscribers, 2009	Broadband subscribers, 2010	Mobile telephone subscribers, 2010	Landline telephone subscribers, 2010
Education	Higher secondary school enrollment				
	Engineering/technical education enrollment	0.267***	0.385***		
	University teachers			0.449***	
	Professional colleges				
Economic	Percent of private sector employment				
	Factory output				
	Factory productive capital				
	Exports				
	Income				
Infrastructure	Electrical capacity				
Innovation/societal openness	Registered newspapers and periodicals	0.905***	0.841***	0.615***	0.893***
Social capital	Cooperative society membership				0.504***
Regression adjusted R square and significance level		0.941***	0.924***	0.840***	0.916***
Sample size (N)		20	20	20	20
OLS regression tests					
Joint Wald statistic		3130.3***	6568.2***	1595.9***	2329.9***
Koenker (BP) statistic		1.093	1.237	0.441	0.387
Jarque-Bera statistic		0.28	1.766	0.111	8.469*

Note Non-significant estimates not include in table. Refer to Chap. 1 for interpretation of statistics
*Signif. at 0.05
**Signif. at 0.01
***Signif. at 0.001

Table 6.5 Standardized regression results for dependent variables, 2010, India, excluding Delhi

Dependent variable					
	Independent variable	Internet subscribers, 2009	Broadband subscribers, 2010	Mobile telephone subscribers, 2010	Landline telephone subscribers, 2010
Correlate	Higher secondary school enrollment				
Education	Engineering/technical education enrollment		0.547**	0.534**	
	University teachers				
	Professional colleges			0.452**	
	Percent of private sector employment				
Economic	Factory output				
	Factory productive capital				
	Exports				
	Income		0.515***		
Infrastructure	Electrical capacity	0.545***			0.400***
Innovation/societal openness	Registered newspapers and periodicals				
Social capital	Cooperative society membership	0.653***			0.787***
Regression adjusted R square and significance level		0.829***	0.760***	0.675**	0.877***
Sample size (N)		19	19	19	19
OLS regression tests					
Joint Wald statistic		321.2***	93.5***	74.0***	226.1***
Koenker (BP) statistic		4.596	2.534	1.245	4.089
Jarque-Bera statistic		0.799	1.437	0.344	1.297

Note Non-significant estimates not included in table. Refer to Chap. 1 for interpretation of statistics
*Signif. at 0.05
**Signif. at 0.01
***Signif. at 0.001

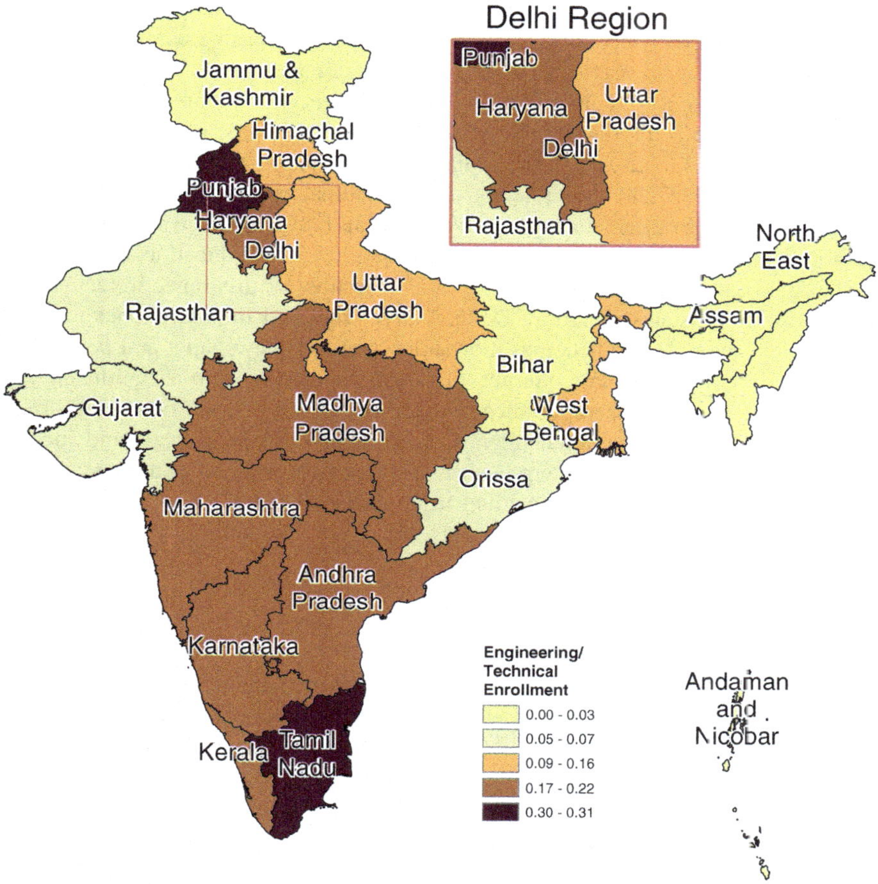

Fig. 6.12 Engineering/technical enrollment per capita, 2008

6.6.1 Findings for Changes in Technology Levels 2008–2010

To examine the influential factors on changes in technology levels from 2008 to 2010, stepwise regressions were computed using the same independent variables, but with the dependent variables calculated as follows: Difference in dependent variable = value of technology utilization variable in 2010 (or 2009, for internet subscribers) per capita − value of technology utilization variable in 2008 per capita.

In an alternative series of regressions, all of which are highly significant, the correlates of the longitudinal technology change in broadband 2008–2010 are similar to those for the cross-sectional analysis, namely income and engineering/technical education. For the three comparison technologies, newspapers/periodicals, higher education (university teachers), and cooperative society membership (for internet subscribers) are significant. Cooperative society membership is significant for internet subscribers which validates the increasing use of the internet by cooperatives for a variety of purposes such as conducting financial transactions and managing supply chains. The dominant influence of income on broadband utilization change is also noticeable in this set of regressions; increasing adoption and utilization of broadband in the Indian states is clearly a function of wealth.

As before, in an alternative series of longitudinal regressions, when Delhi is excluded due to its outlier status, regressions for all four dependent variables turned out to be highly significant. Broadband correlates (higher education and income) remain the same when compared to regression results for difference in technology utilization for all states including Delhi. For internet and mobile phones, cooperative society membership becomes more important, and newspaper and periodicals has reduced association. For internet, electrical capacity is also a significant correlate thus reinforcing the importance of electrical capacity for all other states (Van Dijk 2003). However for fixed phones, newspapers and periodicals and income reduce utilization, which is interpreted as displacement of fixed phones by mobile phones among wealthier users residing in content-rich states.

6.6.2 Discussion of ICT Determinants

The research has indicated that technology levels in India are associated significantly with newspapers and periodicals, cooperative society memberships, and university education.

Registered Newspapers and Periodicals

Newspapers and periodicals' very strong relationship with technology levels, due to the outlier status of Delhi, is corroborated in Chap. 5 where the dominance of published books is attributed to a similar outlier status for Beijing and Shanghai (see also Pick et al. 2013). In both cases, populous, educated, and relatively wealthy metropolitan areas are associated with production of knowledge content, albeit mostly in traditional hardcopy form. These content production centers understandably foster innovation related to web content production and hence positively impact the subscription base for broadband, the internet, and mobile phones. For India, the importance of newspapers and periodicals has an additional dimension; they reinforce India's status as a nation with democratic values woven into the socio-cultural fabric, values that encourage Indians, young and old to freely participate in society, express, challenge, support positions and opinions democratically, which results in a thriving, free, and open society. This has implications for bridging the digital divide.

Cooperative Societies

The association with cooperative society memberships has not been previously reported for India or other nations. A large proportion of India's cooperative societies have agricultural objectives, as seen by India's agricultural cooperative society membership of 236 million in 2007 (Teodosio 2007). Agricultural cooperatives have growing uses of technologies, including information and knowledge management, customized systems and specialized innovative applications (Teodosio 2007). An example is for the rural community of Warana in the western Indian state of Maharashtra, discussed later in the chapter in greater detail. The local agricultural cooperative adopted substantial ICT to improve efficiency and productivity of growing and harvesting sugar cane (Cecchini and Raina 2002). The Indian national government supported the cooperative in the village and surrounding areas. The case study does point out the challenge in dealing with providing technology services as the cooperative grew significantly (Cecchini and Raina 2002). Additional potential for cooperatives is to foster technologies among small businesses and rural population, since in rural India in 2007 the penetration rate for computers and laptops was quite low, even though mobile phones were much more prevalent (Gupta et al. 2007). Our study represents the only instance when the association of social capital, as manifested in the form of cooperative society memberships, with ICT usage has been reported for India. Similar to cooperatives, local clubs in neighborhoods serve as a venue for citizens to congregate during social occasions, religious festivals, or other non-organized settings to participate in group activities, discuss various topics of a neighborhood's interest, for example petitioning the local government agency to construct a speed-breaker on a neighborhood street, and thus foster and encourage development of social capital. One such club in the city of Kolkata, India is depicted in Fig. 6.13.

Higher Education

Higher education factors are associated with broadband, internet, and mobile phones for the full sample. When Delhi is excluded, higher education only influences broadband and mobile phones. This same trend is present for longitudinal

Fig. 6.13 Neighborhood club in the city of Kolkata, India

change in the technologies. Delhi has a very highly educated citizenry, which likely strengthened the overall association for the "older" forms of technology represented by fixed phones and internet. The implication is that higher education matters less for these traditional forms outside of Delhi. Otherwise the results for higher education are consistent with many digital divide studies already cited. Higher education exposes its recipients to training and experience in technologies. This has a multiplicative effect on the general population through the training and education provided by public programs of universities and by the career contributions of tech-knowledgeable graduates to stimulate and provide training in technologies.

For India, the higher education variable engineering/technical education enrollment's positive association with internet and broadband for the full sample and broadband and mobile in the sub-sample excluding Delhi are interesting. Engineering/technical education enrollees in India are almost always younger adults between the ages of 18 and 25 years old, accounting for approximately 35 % of India's population in 2010 (see Fig. 6.14). Therefore, the disaffection with fixed phones, manifested by the lack of association with this legacy form of ICT seems intuitive. This young generation, raised, trained, and educated in ICT-enabled environments are key adopters and users of ICT in India. A majority of these early adopters presently find employment in India's knowledge services workforce and it is essential from a development planning perspective to focus on the rapid development and deployment of infrastructure for ICTs to meet present and future ICT demands of this generation, both in urban and rural India (case in point is the rural state of Madhya Pradesh, which has high engineering/technical student enrollment per capita as shown in Fig. 6.12). Figure 6.15 shows uniquely designed training facility at the sprawling Mysore Development Center of India's software giant,

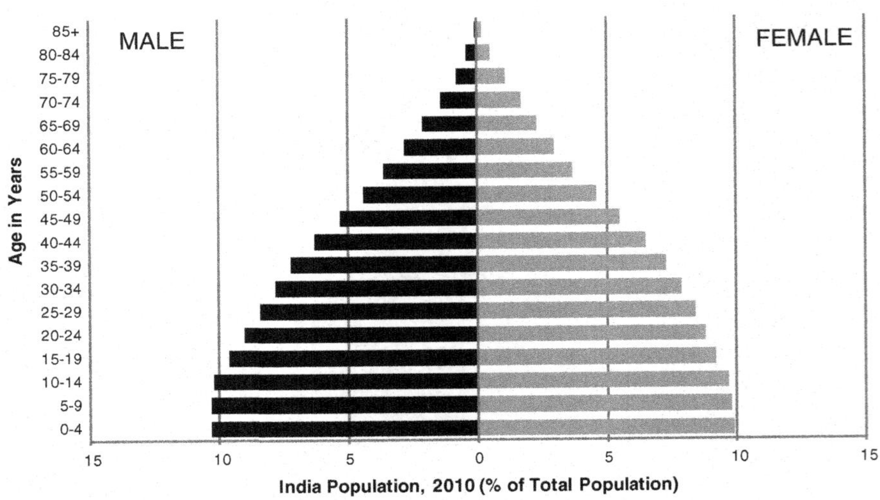

Fig. 6.14 Population pyramid, India, 2010

Fig. 6.15 Infosys development center, located at Mysore (south-west of Bangalore), India

Infosys, which provides post-baccalaureate and pre-employment training and education to hundreds of thousands of young engineers hired from engineering/technical colleges all over India, and across the world.

6.7 Case Studies of ICT Utilization in Indian States

Case studies of ICT adoption, diffusion, and utilization in India have provided examples of technology usage and projects that help bridge the digital divide in various settings such as agriculture (Veeraraghavan et al. 2009; Cecchini and Raina 2002), dairy services (Rao 2005; Chakravarty 2000), agricultural land administration (Pentland et al. 2004; Prakash and De 2007), microenterprises (Chew et al. 2010; Chew et al. 2011), telemedicine (Qiang et al. 2009), and e-governance (Warschauer 2003). Case studies have documented use by individuals (Tarafdar et al. 2012) and communities with diverse socio-economic characteristics (Warschauer 2003; Sharma et al. 2007), discussed costs, benefits, and challenges (Kumar 2004; Pentland et al. 2004; Prakash and De 2007), and debunked myths (Pentland et al. 2004) about ICT adoption. Two cases are discussed in this chapter.

6.7.1 DakNet and Bhoomi

DakNet, developed by researchers in MIT Media Lab, is an ad hoc network that uses wireless technology to provide asynchronous digital connectivity (Pentland et al. 2004). DakNet enabled creation of Bhoomi (which means land in English), a first of its kind e-governance initiative in India pioneered by the state government of

Karnataka, in which physical land records are computerized. Villagers who need their land records do not have to travel to the district headquarters, up to 40 km away. Instead they can go their own village kiosk and request a copy of their land records; the request is sent electronically to the district headquarters, where the server processes and delivers the record. The DakNet bus delivers the record to the village kiosk where the kiosk manager prints it out. Out of pocket cost to the villager is Rupees 15 (25 cents U.S. in 2014) per land record. Not only does the villager save the time and money that it would take to travel to the district headquarters, s/he is able to avoid possible harassment, unfair pricing, and corruption. Farmers were able to sell and transfer agricultural land through the automated mutation (transfer or change of title) system. This stimulated faster transactions of land parcels, especially in economically advancing areas closer to Bangalore. Land agents were able to use Bhoomi to verify land records before purchases actually took place. Furthermore, land ownership certificates that could be printed with an official seal were useful for procuring bank loans. However, no statistically significant increase in rural credit was observed (Prakash and De 2007).

Overall, DakNet-Bhoomi provides villagers access to digital messaging, digital documents, email, etc., and has been enthusiastically received by villagers along the bus routes who found it to be a better alternative than installing a community telephone. The cost of providing DakNet-Bhoomi services was estimated to be approximately US$243 for each village (Pentland et al. 2004).

The case study exemplifies an early attempt in India to improve rural digital connectivity at affordable costs to everyone—both for project sponsors as well as the individual consumer. Remote villages are able to leapfrog past traditional higher cost connectivity solutions, and DakNet was subsequently used in other settings (e.g. rural Cambodia) for solar-powered village schools, telemedicine clinics, and a providing connectivity to a governor's office. This project to restore efficiency of agricultural land administration emerged as a model of ICT use for development not just in India but other developing nations.

From an ICT policy perspective, the Bhoomi project did not find any significant influence of computerization on increase in individual capacity and freedom of choice (Prakash and De 2007). The same authors have also critiqued policy planners in India as sometimes being fixated by the notion of increasing income, especially in rural areas. Hence they tend to narrowly view ICTs as a catalyst for development that would ultimately impact income. These authors recommend that the use of ICTs for development (ITD4D) must account for contextual variations between individuals and communities; this is especially important for those who are marginalized from mainstream development programs. While not included as an independent factor in this chapter due to lack of availability of reliable data at the state level, this case example reinforces the importance of governmental support and prioritization of ICT, which is posited in Chap. 3's generic model of ICT utilization to be positively associated with ICT usage.

6.7.2 Warana Village Project

This case example deals with the use of ICTs in a large sugarcane cooperative in the sub-district of Warana, located in the western Indian state of Maharashtra (Veeraraghavan et al. 2009). The sugarcane cooperative is a large enterprise with a membership of almost 50,000 farmers from 75 villages spread over 25,000 km^2. The objective of this large-scale project funded jointly by India's central government (50 %), the state government of Maharashtra (40 %), and the cooperative (10 %) was to computerize land records, allow farmers to look up market prices in real time, and provide expert agricultural advice as well as internet access.

To "wire" the cooperative, 54 village kiosks were set up at no cost to the villagers; each kiosk has a PC, printer, modem, and backup power supply. Internet connectivity was provided by landline telephone dial-up at slow speeds no more than 10 kbps. The kiosks were manned by trained computer operators who assisted farmers seeking information. The main usage of kiosks was to transfer information between an individual farmer and the main processing center, which could be 2–50 km away depending on the location of a village relative to the processing center (Veeraraghavan et al. 2009). Information sent from the farmer to the processing center included amount of fertilizer and water used and validity of a harvesting permit, while information sent from the kiosk to the farmer included quantity of sugarcane produced and payment schedule, which were crucial to farmers. Benefits to the farmer were the time saved by not having to visit the central processing plant. This annually saved around $20, not a trivial sum for a farmer in rural India. An intangible benefit was the farmers' ability to access their produce and payment information at any time; hence they perceived the system to be more transparent. For the cooperative, the benefit was competitive advantage over other cooperatives in trying to retain and attract more farmers based upon the availability of kiosks and their related benefits.

However this "wiring" paradigm of connecting farmers from remote villages to a processing plant using kiosks encountered several challenges due to rising maintenance costs caused by obsolescence of hardware, intermittent power supply, poor internet connectivity, and minimal use of PCs. To overcome these challenges, Short Message Service (SMS) based system was piloted in 7 out of the 54 existing kiosks, in which the PCs were replaced by mobile phones (Veeraraghavan et al. 2009). This "unwiring" of Warana was a practical solution to overcome the problem of increasing maintenance costs, concerns about reliable power supply, and low use of PCs in general. Kiosk operators were still retained and trained in the use of SMS since none was familiar with text messaging a priori. Potential benefits of unwiring were a reduction of cooperative fees of $10 per farmer (Rupees 500–550 at the time of the project), which is neither a trivial sum, nor a life-changing saving to the individual farmer. However the cooperative itself could potentially save one million rupees (approximately US$25,000) which is a significant sum. These savings could be used for a variety of purposes including betterment of the processing plant, thereby impacting the individual farmer (Veeraraghavan et al. 2009). Also, within seven months of the launch of the pilot study using one cell phone at each kiosk, 61 additional cell phone numbers had been registered with the processing facility of the Warana cooperative. These were mobile phones belonging to individual farmers who requested their numbers to be registered to the database of cell phone numbers. An encouraging outcome from the "unwiring" of Warana was increased level of ICT use due to replacement of PCs by mobiles.

This case study illustrates an instance of adoption/diffusion of ICTs (see Chap. 3), facilitated by a large cooperative society. Earlier in this chapter, cooperative societies emerged as a significant correlate of the use of fixed phones and internet, especially when the national capital of Delhi, a high technology outlier, was excluded from our state sample. It also outlined a novel approach in which a mobile-phoned based system replaced a PC-based system for the transfer of a small, yet essential amount of data via SMS. Rural PC usage has the potential for socio-economic development, yet in this instance, the minimal use of PCs coupled with rising maintenance costs proved to be an impediment. Replacing PCs by inexpensive mobile phones not only reduced costs, but also democratized access to information, leading to increased diffusion of mobile phones among rural farmers.

Overall, the two ICT case studies in the Indian local context have largely succeeded in addressing digital divides that exist in rural India among poor, marginalized, and disadvantaged communities. Local contextualization of information and technology needs has been recommended for such communities (Cecchini and Scott 2003; Prakash and De 2007). Prakash and De (2007) have cautioned that the "desire for technical solutions to development problems should not take on a life of its own where we forget that development is about people and what they think and how they feel matters (p. 276)." Cecchini and Scott (2003) have advised that ICT projects should focus on a limited number of well-researched, easy-to-learn pro-poor services rather than develop and offer a whole suite of information services that lay mostly unutilized due to lack of demand; this has also been documented by Veeraraghavan et al. (2009). Cecchini and Scott (2003) and Pentland et al. (2004) have sought to debunk myths that poor, rural users are largely ignorant about their digital needs; in fact, such users are acutely aware of their daily challenges, circumstances, and aspirations, and therefore demand access to locally-contextualized information. In other words, for the digital divide in rural India, and generally worldwide, context matters!

6.8 Policy Implications

The influence of registered newspapers and periodicals on broadband was accentuated for the populous and mostly urban state of Delhi in our regression analysis findings. Policies in urban areas to encourage generation of content through publishing of newspapers, magazines and books, in traditional hardcopy or growing electronic forms, are recommended, as well as publishing of quality content in the rapidly growing social networks and blogosphere environments. Since electronic publishing and social networks will become more available for rural populations, these policies will have broader reach in the future. With the anticipated shift in India to much wider use of high-speed broadband as a consequence of the "Digital India" initiative (discussed in the next section), greater volume and diversity of content that is more multimedia and mobile-oriented will be necessary and will stimulate in turn the need for cable, fiber and 3G/4G cell phones—a pattern seen in the US and some other high-tech nations today.

Government IT policies that encourage engineering and technical higher education are also seen to have large potential benefit for broadband. These policies help in extending the benefits of education and in turn of IT usage to rural areas. The benefit of educational progress over time to IT utilization has been demonstrated in case studies of small nations such as Estonia (Dutta 2007). As India's universities can be encouraged by policies to develop interactive. video-based and collaborative virtual learning, it will enrich learning opportunities and also stimulate need for higher-capacity broadband.

Income, associated with broadband use and growth, is a prominent factor because broadband costs remain high in India (Gumaste et al. 2009; Dharmakumar 2010) and outside of the reach of the lower middle class, urban working classes and village populations. IT policies and regulations can encourage reduction in cost of broadband through market mechanisms and by effective partnering in private-public initiatives (Tanguturi and Harmantzis 2007).

The role of Indian cooperative societies and their members in fostering technology utilization is a novel finding of this work. The implications for policy makers in Indian federal and state governments are clearly immense. Our data reveal that a significant disparity exists in the number of cooperative societies and cooperative society memberships per capita between states in India. This can perhaps be explained by differing levels of social capital in Indian states, which is largely unexplored in the academic literature. In spite of the potential of cooperatives to bridge the technology divide as well as contribute to larger issues of prime national interest such as poverty reduction and alleviation of social inequity, ICT policies at the state and national level are largely devoid of any reference to cooperatives. We recommend that India's national and state governments leverage local knowledge of priorities, needs, and realities that cooperatives often possess through their membership. This can be facilitated by increased cooperation and collaborations between the Government of India's Department of Agriculture and Cooperation and Department of Telecommunications, as well as similar departments at the state level.

Since the Digital India initiative aspires to connect all Indian villages by broadband within three years, strengthening the ICT capabilities of cooperatives and thus factoring in the local context would empower cooperative members who often come from underserved rural populations. Such initiatives can be funded by India's central government, local state governments, and perhaps the cooperatives themselves, as in the case study of Warana. Whether such rural users travel to village telecenters, use mobile phones or acquire a simple low-cost computer, the cooperative societies can play a grassroots-based intermediary role (Cecchini and

Scott 2003) that fosters social capital, assists and supports users, steers them to needs-oriented applications, and leverages their local knowledge in developing the cooperative's own networks and databases.

This chapter has also revealed sharp unevenness of availability and use of broadband and the other technologies among Indian states, as seen in the LISA and cluster findings. The national capital Delhi's base of broadband and internet users is 13–16 times higher compared to states such as Bihar, Madhya Pradesh, Uttar Pradesh, and the northeast in 2009–10; the gap in mobile is lesser, with Delhi's usage being approximately four times higher than these states. It is noticeable that Delhi has attracted a majority of FDI inflows in telecom between 1991 and 2008 followed by states such as Maharashtra, Tamil Nadu, Himachal Pradesh, and Karnataka (TRAI 2014d). Unsurprisingly, these states are ICT leaders among Indian states. Also, these states formulated their ICT strategies and policies to attract private investment much earlier than others; for example, the state of Karnataka was the first state to present an IT policy as early as 1997.

In the 1990s, ICT policies and projects were within the purview of India's national government. However the role of state governments in formulating ICT policies and initiatives has become more central in recent years. Several states now have their own ICT policies that focus on private investment, development of physical infrastructure, investment in human resources including research, training, and education, and identification of needs in various sectors that can drive the development of ICT applications and content (Mitra 2000; Viswanathan and Pick 2005). Accordingly, development of a state ICT policy, perhaps as part of a broader IT framework is recommended as key starting point for states that are currently ICT laggards in India. One such state, Bihar, has published its ICT policy recently which has outlined the state government's vision to develop Bihar as an attractive destination for IT investment through good governance, rapid infrastructure development, and investment in human and intellectual capital (Government of Bihar Department of Information Technology 2011).

Needless to say that establishing an ICT policy, while necessary, is hardly sufficient to bridge the digital divide between and within Indian states. It is important for states to manage and implement ICT policies; states such as Andhra Pradesh and Tamil Nadu not only have explicit and comprehensive ICT policies, but also strong commitment and experience in implementing ICT at all levels of government (Mitra 2000). Often a state Chief Minister's (first in command in the state) personal intervention would ensure implementation in rural areas; this was evident during the implementation of the Bhoomi project in the state of Karnataka (Prakash and De 2007). Such demonstration of commitment to implementation can be beneficial while attracting investment in the IT sector.

On a related note, we recommend cross-pollination of experiences among states for the development of policy, design, funding, implementation, and all other aspects of ICT projects (Mitra 2000). Such interactions, especially in areas such as higher education, generation and dissemination of content via newspapers and periodicals, and the role of cooperatives—all significant correlates of ICT in Indian states, would be especially beneficial. Practical considerations such as geographic proximity may facilitate interactions across state boundaries. Based upon lessons learnt from the chapter case studies, we also recommend that contextualization is important for ICT policy development. India is a vast country, and demographic, socio-economic, cultural, geographic, and political differences between its states can be staggering. Equally challenging are disparities between urban and rural India. While national level ICT policies provide broad guidance from a pan-India outlook, the states, especially the laggards must share responsibility with the national government of bridging the digital divide, as they have intimate knowledge of the ground realities, physical conditions, and priorities at the local level.

6.9 Conclusions

Telecommunications and ICT use in India have come a long way since economic liberalization of the early 1990s. Today, India has the second largest base of mobile cellular subscribers and third largest base of internet subscribers worldwide. This has been made possible through a combination of technological change, regulatory initiatives, and key changes in the governance of the telecom sector in India. However challenging problems with ICT penetration persist and India's ICT penetration rates especially in internet and broadband lag significantly behind rates of China and other BRICS nations. Compounding factors are infrastructural, such as unreliable electricity, and also regulatory, such as complexities with land acquisition and right of way.

Despite these obstacles, a handful of Indian states have made significant progress in ICT adoption and utilization. The nation's capital Delhi, a state and megacity, leads in terms of teledensity, and internet and broadband subscribers per capita; in fact, our analysis reveals Delhi to constitute a singularly high ICT use cluster, followed by Himachal Pradesh and Punjab, both approximately 400 km from Delhi, and the southern Indian states of Kerala and Tamil Nadu. At the other end of the spectrum are states in central, northeast, and eastern India, Bihar, Uttar Pradesh, and Madhya Pradesh that are largely rural, and also India's northeast. Our confirmatory analysis has revealed that key correlates of ICT utilization in India are (a) registered newspapers and periodicals, represented in our conceptual model of ICT utilization under the categories of societal openness and innovation, and (b) educational attributes. This chapter's finding about societal openness is consistent with similar findings for Japan and the United States. When the high technology outlier state of Delhi is excluded, confirmatory analysis reveals the association of a key infrastructural variable, electrical capacity, with internet and landline phone subscribers. Case studies of ICT adoption in India, especially rural areas have identified the lack of reliable power supply as a key impediment for internet and broadband penetration. Another interesting finding of this chapter is the association of social capital, represented in this study by cooperative societies with internet and landline, when Delhi is excluded. This finding is consistent with the strong association of social capital with several ICTs, notably internet and broadband in the US.

Cooperative societies in India, as exemplified by the Warana case study in rural Maharashtra provide a forum to members of a community with similar interests, often economic, to interact socially, influence peers, and exchange information. From a development standpoint, the local context is important to consider in endeavors to bridge the digital divide, as exemplified in rural parts of India.

In August 2014, the Union Cabinet of India sanctioned Rupees 113,000 crores (US$18.83 billion approx.) for "Digital India", an initiative to build a nationwide fiber optic network that connects village panchayats across the country in three years (Press Information Bureau Government of India 2014; Forbes 2014). The vision for this initiative includes digital infrastructure as a utility to every citizen, governance and services on demand, and digital empowerment of citizens. While certainly ambitious, poverty, literacy, infrastructure development, and general social inclusion of marginalized communities (Warschauer 2003) represent significant challenges that merit attention in relation to India's digital divide.

References

Agarwal, R., Animesh, A., & Prasad, K. (2009). Social interactions and the 'digital divide': Explaining variations in internet use. *Information Systems Research, 20*(2), 277–294.

Anselin, L. (1995). Local indicators of spatial association-LISA. *Geographical Analysis, 27*(2), 93–115.

Azari, R., & Pick, J. B. (2005). Technology and society: Socioeconomic influences on technological sectors for United States counties. *International Journal of Information Management, 25*(1), 25–37.

Chaminade, C., & Vang, J. (2008). Globalisation of knowledge production and regional innovation policy: Supporting specialized hubs in the Bangalore software industry. *Research Policy, 37*(10), 1684–1696.

Cecchini, S., & Raina, M. (2002). Warana: The case of an Indian rural community adopting ICT. *Information Technology in Developing Countries, 12*(1). Retrieved October 1, 2014, from http://www.iimahd.ernet.in/egov/ifip/apr2002/article3.htm.

Cecchini, S., & Scott, C. (2003). Can information and communications technology applications contribute to poverty reduction? Lessons from rural India. *Information Technology for Development, 10*, 73–84.

Census of India. (2011). *Rural urban distribution of population*. Retrieved November 1, 2014, from http://censusindia.gov.in/2011-prov-results/paper2/data_files/india/Rural_Urban_2011.pdf.

Chakravarty, R. (2000). IT at milk collection centers in cooperative dairies: The national dairy development board experience. In S. Bhatnagar & R. Schware (Eds.), *Information and communication technology in rural development*. Washington, DC: World Bank Institute.

Chen, W. (2013). The implications of social capital for the digital divides in America. *The Information Society, 29*, 13–25.

Chew, H. E., Ilavarasan, P. V., & Levy, M. (2010). The economic impact of information and communication technologies on microenterprises in the context of development. *The Electronic Journal of Information Systems in Developing Countries, 44*(4), 1–19.

Chew, H., Levy, M., & Ilavarasan, V. (2011). The limited impact of ICTs on microenterprise growth: A study of businesses owned by women in urban India. *Information Technologies & International Development, 7*(4), 1–16.

Chinn, M. D., & Fairlie, R. W. (2007). The determinants of the global digital divide: A cross-country analysis of computer and Internet penetration. *Oxford Economic Papers, 59*(1), 16–44.

Cooper, F. (2002). *Africa since 1940: The past of the present*. New York: Cambridge University Press.

Das, B. (2010a). Diffusion of information communication technology across Indian states and its determinants: An empirical analysis. In *Annual Conference of IASSI on Frontier Issues in Technology, Development and Environment, Madras School of Economics, Chennai, India.*

Das, B. (2010b). Intra-national digital divide in India: An analysis on the diffusion of information and communication technologies across states. In *Proceedings of 8th Globelics International Conference, University of Malaya, Kuala Lumpur, Malaysia.*

Department of Telecommunications, India. (2010). *Department of Telecommunications Reports/Statistics*. Retrieved October 15, 2014, from http://www.dot.gov.in/content/reports-statistics.

Dharmakumar, R. (2010). The long arm of broadband. *Forbes India, February 1*. Retrieved November 1, 2014, from http://forbesindia.com.

Ding, L., Haynes, K. E., & Li, H. (2010). Modeling the spatial diffusion of mobile telecommunications in China. *The Professional Geographer, 62*(2), 248–263

Dutta, S. (2007). Estonia: A sustainable success in networked readiness. In S. Duttal & I. Mia (Eds.), *The global information technology report 2006–2007: Connecting to the networked economy* (pp. 81–90). New York: Palgrave Macmillan.

Forbes. (2014). *India wants to build massive digital infrastructure to cover 800 million rural citizens by 2019*. Retrieved October 15, 2014, from http://www.forbes.com/sites/saritharai/2014/09/18/india-wants-to-build-massive-digital-infrastructure-to-cover-800-million-rural-citizens-by-2019/.

Government of Bihar Department of Information Technology. (2011). *Information and Communication Technology Policy 2011*. Retrieved October 15, 2014, from http://gov.bih.nic.in/Documents/IT-Policy-2011-English.pdf.

Government of India. (2014). *Co-operative Organisations*. Retrieved November 1, 2014, from http://www.archive.india.gov.in/citizen/co_operatives/co_operatives.php?id=3.

Guillen, J. F., & Suarez, S. I. (2005). Explaining the global digital divide: Economic, politica and sociological drivers of cross-national internet use. *Social Forces, 84*(2), 681–708.

Gumaste, A., Gokhale, P., & Dhar, A. (2009). On the state and guiding principles of broadband in India. *IEEE Communications Magazine, 47*(8), 42–48.

Gupta, S., Rizvi, H., Dearden, A., Polzin, C., Fu, X., & Matthews, P. (2007). Information and communication technology (ICT) survey of District Central Cooperatives Banks (DCCB) and Micro Finance Institutions (MFIs) in India. Report of Rural E-Services Project, Saral Services, Andhra Pradesh, India. Retrieved 15 October 2014, from http://www.microfinancegateway.org/sites/default/files/mfg-en-paper-information-and-communication-technology-ict-survey-of-district-central-cooperatives-banks-dccb-and-micro-finance-institutions-mfis-in-india-sep-2007_0.pdf.

Halinski, R. S., & Feldt, L. S. (1970). The selection of variables in multiple regression analyses, *Journal of Educational Measurement, 7*(3), 151–158.

ITU. (2013). *Measuring the information society*. Report. Geneva, Switzerland: International Telecommunication Union.

Kathuria, R., & Kedia-Jaju, M. (2012). *India: The impact of internet* (research report). New Delhi: Internet and Mobile Association of India (IAMAI). Retrieved November 1, 2014, from http://www.iamai.in/.

Kiiski, S., & Pohjola, M. (2002). Cross-country diffusion of the Internet. *Information Economics and Policy, 14*, 297–310.

Kumar, R. (2004). eChoupals: A study on the financial sustainability of village internet centers in rural Madhya Pradesh. *Information Technologies & International Development, 2*(1), 45–74.

Maps of the World. (2014). *Top ten countries with highest number of PCs*. Retrieved October 1, 2014, from http://www.mapsofworld.com/world-top-ten/world-top-ten-personal-computers-users-map.html.

Meera, S. N., Jhamtani, A., & Rao, D. U. M. (2004). Information and communication technology in agricultural development: A comparative analysis of three projects from India, network paper 135. *Agricultural Research & Extension Network*, Retrieved November 1, 2014, from http://www.odi.org/sites/odi.org.uk/files/odi-assets/publications-opinion-files/5186.pdf.

Ministry of Statistics and Programme Implementation. (2013). *Statistical Year Book, India*. Retrieved November 1, 2014, from http://www.odi.org/sites/odi.org.uk/files/odi-assets/publications-opinion-files/5186.pdf.

Mitra, R. (2000). Emerging state-level ICT development strategies. In S. Bhatnagar & R. Schware (Eds.), *Information and communication technology in rural development*. Washington, DC: World Bank Institute.

Moran, P. A. P. (1950). Notes on continuous stochastic phenomena. *Biometrika, 37*, 17–23.

Ono, H., & Zavodny, M. (2007). Digital inequality: A five country comparison using microdata. *Social Science Research, 36*(3), 1135–1155.

Pentland, A., Fletcher, R., & Hasson, A. (2004). DakNet: Rethinking connectivity in developing nations. *Computer, 37*(1), 78–83.

Pick, J. B., & Azari, R. (2008). Global digital divide: Influence of socioeconomic, governmental, and accessibility factors on information technology. *Information Technology for Development, 14*(2), 91–115.

Pick, J. B., & Azari, R. (2011). A global model of utilization of technology based on governmental, social, economic, and business investment factors. *Journal of Management Information Systems, 28*(1), 51–85.

Pick, J. B., Nishida, T., & Zhang, X. (2013). Determinants of China's technology utilization and availability 2006-2009: A spatial analysis. *The Information Society, 29*(1), 26–48.

Prakash, A., & De, R. (2007). Importance of computerization of land records in India. *Information Technology & People, 20*(3), 262–281.

Press Information Bureau, Government of India. (2014). *Digital India—A programme to transform India into digital empowered society and knowledge economy*. Retrieved November 1, 2014, from http://pib.nic.in/newsite/PrintRelease.aspx?relid=108926.

Qiang, C., Rossotto, C., & Kimura, K. (2009). Economic impacts of broadband. In *Information and Communications for Development 2009: Extending Reach and Increasing Impact* (pp. 35–50). Washington, DC: The World Bank.

Quibria, M. G., Ahmed, S. N., Tschang, T., & Reyes-Macasaquit, M.-L. (2003). Digital divide: Determinant and policies with special reference to Asia. *Journal of Asian Economics, 13*, 811–825.

Rao, S. S. (2005). Bridging digital divide: Efforts in India. *Telematics and Informatics, 22*, 361–375.

Reserve Bank of India. (2011). *Handbook of Statistics on Indian Economy, 2010–11*. Retrieved October 1, 2014, from http://rbidocs.rbi.org.in/rdocs/Publications/PDFs/FHB160913FLS.pdf.

Robison, K. K., & Crenshaw, E. M. (2002). Post-industrial transformations and cyber-space: A cross-national analysis of Internet development. *Social Science Research, 31*, 334–363.

Sawyer, R. (1982). Sample size and the accuracy of predictions made from multiple regression equations. *Journal of Educational Statistics*, *7*(2), 91–104.

Simon, J. S. (2004). Critical success factors for electronic services: Challenges for developing countries. *Journal of Global Information Technology Management, 7*(2), 31–53.

Sharma, C., Sharma, S., & Subhedar, U. (2007). Putting ICTs in the hands of the women of Kanpur and the Chikan embroidery workers of Lucknow. *Information Technologies & International Development, 4*(2), 11–16.

Statista. (2014). *Global PC penetration per capita from 2000 to 2015*. Retrieved October 1, 2014, from http://www.statista.com/statistics/203677/global-pc-penetration-per-capita-since-2000/.

Tanguturi, V. P., & Harmantzis, F. C. (2007). Broadband in India: Strategic investment opportunities. *Technology in Society, 29*, 431–440.

Tarafdar, M., Anekal, P., & Singh, R. (2012). Market development at the bottom of the pyramid: Examining the role of the information and communication technologies. *Information Technology for Development, 18*(4), 311–331.

Teodosio, V. A. (2007). Agricultural cooperatives and information communication technology in an emerging Asia. *Living the information society: The impact of ICTs on people, work, and communities in Asia, Manila, Philippines*. Retrieved October 15, 2014, from http://unpan1.un.org/intradoc/groups/public/documents/eropa/unpan030894.pdf.

Agency, The Central Intelligence. (2014). *The world factbook*. Washington, DC: Central Intelligence Agency.

TRAI. (2014a). *National telecom policy 1994*. Retrieved November 1, 2014, from http://www.trai.gov.in/Content/telecom_policy_1994.aspx.

TRAI. (2014b). *New telecom policy 1999*. Retrieved November 1, 2014, from http://www.trai.gov.in/Content/telecom_policy_1994.aspx.

TRAI. (2014c). *National telecom policy 2012*. Retrieved November 1, 2014, from http://www.trai.gov.in/WriteReadData/userfiles/file/NTP%202012.pdf.

TRAI. (2014d). *Telecom sector in India: A decadal report*. Retrieved November 1, 2014, from http://www.trai.gov.in/WriteReadData/Publication/Document/201304121052403536675NCAER–Report08june12.pdf.

UN. (2014). *World urbanization prospects, 2014 revision*. New York, NY: Department of Economic and Social Affairs, United Nations.

van Dijk, M. P. (2003). Government policies with respect to an information technology cluster in Bangalore, India. *The European Journal of Development Research, 15*(2), 93–108.

van Dijk, J. A. G. M. (2005). *The deepening divide: Inequality in the information society*. Thousand Oaks, CA: SAGE Publications.

Veeraraghavan, R., Yasodhar, N., & Tayoma, K. (2009). Warana unwired: replacing PCs with mobile phones in a rural cooperative. *Information Technologies & International Development, 5*(1), 81–95.

Viswanathan, N. K., & Pick, J. B. (2005). Comparison of e-commerce in India and Mexico: An example of technology diffusion in developing nations. *International Journal of Information Technology, 31*(1–2), 2–19.

World Bank. (2014). *The World Bank data*. Retrieved October 1, 2014, from http://data.worldbank.org/indicator/NV.SRV.TETC.ZS?order=wbapi_data_value_2013–wbapi_data_value+wbapi_data_value-last&sort=desc.

Warschauer, M. (2003). Demystifying the digital divide. *Scientific America, 289*(2), 42–47.

World Economic Forum. (2013). *The global information technology report*. Geneva: Switzerland.

Yates, D. J., Gulati, G. J., & Weiss, J. W. (2011). Different paths to broadband access: The impact of governance and policy on broadband diffusion in the developed and developing worlds. In *Proceedings of the 44th Hawaiian International Conference on System Sciences, Hawaii, USA* (pp. 1530–160).

Chapter 7
Japan's Digital Divide

7.1 Introduction

Japan is one of the leading industrialized nations in the world, well-known for technology, innovations, and world leadership in manufacturing, especially in the automobile sector. In 2013, Japan's population of 127.3 million made it the tenth most populous nation in the world (Statistics Japan 2014); in the same year, Japan's GDP was US$4.902 trillion (World Bank 2014) and when ranked by gross national income (GNI) per capita, Japan ranked thirteenth worldwide with a GNP per capita of $46,140 (World Bank 2014). Japan is classified as a very high human development nation by the United Nations.

Japan is also recognized as a leader in information and communications technology (ICT) adoption, diffusion, and use, and was ranked twelfth in 2012 in terms of the ICT Development Index of the International Telecommunications Union (ITU 2012), ahead of most other industrialized and developed nations such as the United States, and much ahead of developing nations in Asia such as China and India. Japanese citizens started using the commercial internet in 1993, and almost two decades later, in 2013, Japanese internet subscribers crossed the 100 million threshold, resulting in a penetration rate of 82.8 per 100 population aged 6 years or older. Japanese citizens use a variety of devices to access the internet, such as PCs, mobile phones, smart phones, tablets, gaming devices, with PCs being the most common enabler of internet access. In March 2014, broadband subscribers in Japan numbered almost 90 million, behind the United States and China, with more than 50 % of users accessing broadband using their 3-G phones, and with 28 % who are fiber-to-the-home broadband subscribers. In terms of fixed broadband, Japan had 35 million subscribers in 2012, preceded by China and the US. Also, between 1995 and 2014, the landscape of telephony in Japan has shifted dramatically; landline telephone subscriptions declined by more than 50 % from almost 60 million in 1995 to 26 million in 2014, while mobile phone subscribers grew from 4 million to almost 150 million between 1995 and 2014, while IP phone (voice phone services

J.B. Pick and A. Sarkar, *The Global Digital Divides*, Progress in IS,
DOI 10.1007/978-3-662-46602-5_7

that use Internet Protocol technology across part or all of the communication network) subscribers increased from 8.3 million in 2005 to almost 34 million in 2014 (Statistics Japan 2014). Clearly, Japan has made significant strides as a technology society.

Despite the tremendous progress made, Japan's ICT utilization and the digital divide within its prefectures, i.e. administrative unit similar to states or provinces, has its share of challenges. A rapidly aging population, unfamiliarity with the English language keyboard (Nakayama 2002), a significant discrepancy in ICT adoption and use between rural and urban areas, a largely mountainous topography spread across the nation (Arai and Naganuma 2010), and occurrences of large-scale weather events such as typhoons, hurricanes, and tsunamis provide a tapestry of issues that impact the digital divide within Japan.

Like previous and subsequent chapters, we focus on ICT adoption and utilization at the sub-national level, in this case prefectural level, in Japan and examine patterns of ICT adoption and influences of various socio-economic, educational, infrastructural, innovation, and openness factors on technology usage. Next we discuss the historical evolution of ICT and the evolution of the ICT landscape in Japan, followed by dimensions of technology use. We then posit our conceptual model of ICT utilization in Japanese prefectures and analyze patterns of ICT usage including examination of high- and low-tech clusters. The major determinants of ICT utilization and the digital divide in India are examined and discussed next. We find that innovation, represented by patents registered by Japanese citizens, innovation and openness represented by newspaper circulation, education represented by students and pupils per capita and household expenditures on education, urbanization, and aged population structure are significantly associated with ICT adoption in Japanese prefectures. We then present two case examples of ICT usage in Japan that reinforce findings of our confirmatory analysis, and proceed to discuss policy implications of our findings, followed by concluding remarks.

7.2 Historical Background

The modern history of telecommunications in Japan can be traced back to 1985, when the state-owned telecom provider Nippon Telegraph and Telephone (NTT) was privatized, about a decade ahead of similar privatization of telecommunications in India. Liberalizing telecom had two fundamental goals as outlined by the Telecommunications Business Law: (a) increase competition resulting in lowered tariffs for the consumers, as well as achieve diversified and convenient content, and (b) soundly develop the telecom industry (Omura 1997). When the law went into effect, a Type 1 telecom business was required to make a massive initial investment to construct a network. Economies of scale and factors such as network externalities made it difficult for service providers to enter into the market. Another hindrance was the requirement that private operators categorized as Type 1 would be required to procure a license (Shimizu et al. 2008). As a result, despite competition, lowered tariffs worked in tandem with infrastructural investment and development in the Japanese private sector, although the same strategy did not work so well in Japan's initial efforts to privatize its telecom sector (Omura 1997).

The late 1990s witnessed a flurry of important changes in the telecommunications landscape of Japan to deal with the faltering of the initial privatization efforts. First, revisions were enacted in the Telecommunications Business Law of 1997 when a requirement for obtaining permission/license to operate as a Type 1 telecom provider was eased, ostensibly to facilitate the entry of new carriers into the market. Second, in 1998, restrictions on foreign investment in telecom were abolished. And third, in 1999, NTT, the state-owned incumbent i.e. dominant firm, was restructured under holding corporations into three separate companies—a regional company for eastern Japan, a regional company for western Japan, and a long-distance network company (Shimizu et al. 2008).

The early 2000s witnessed sequential ICT policies in Japan with the aim of creating a networked society at large. The Information Technology Basic Law came into existence in 2000, and 2001 saw the creation of the first national e-Japan strategy whose specific target was to upgrade Japan's internet network, to achieve speeds of 30–100 mbps within 5 years, and make this available for all citizens. Private-sector driven development was aimed at enforcing fair competition, e-commerce and e-government systems development, as well as human resources development. As a result of this initiative, broadband infrastructure development progressed rapidly and the number of households with DSL, cable TV, and fiber-to-the-home connectivity reached 35, 23, and 17.7 million respectively by 2003. Phase II of the e-Japan strategy starting in 2003 focused on ICT usage in specific sectors—namely, healthcare, lifestyle, small businesses, knowledge services, to name a few. The u-Japan strategy followed soon thereafter in 2004–05 with the aim to create a ubiquitous networked society in Japan by 2010 with focus on balancing infrastructure development, in particular on ubiquitous networks that can be used seamlessly across wireless and fixed platforms, with broadened reach, so ICT usage could more readily solve social issues.

As a result of these aforementioned initiatives, telecom providers increased in number, and in 2006, there were 314 registered and more than 13,000 overall telecom providers in Japan (Shimizu et al. 2008). We now briefly outline developments in two major ICT sectors—mobile and internet, specifically as they connect to Japanese people and Japanese society at large.

Mobile phones in Japan can be traced back to NTT's car mobile phone service in 1979 for the elite. Following the development of portable, handheld phones known as *keitai* in Japanese, cell phones entered into the general public domain and appealed to a much larger market (Nakayama 2002). A *keitai* phone can be seen in Fig. 7.1.

In 1995, the PHS or personal handyphone system was launched; PHS is a digital cordless phone system that uses the existing analog telephone network and ISDN (Integrated Service Digital Network) to offer services priced lower than those of the cellular phone, since PHS does not use the complex network used by cell phones. This triggered fierce competition between *keitai* and PHS, and resulted in cell

Fig. 7.1 *Keitai* phone, Japan. *Source* Flickr

phones being given away even for free. PHS went on to become a fashion accessory for Japanese youth, and general use of *keitai* and PHS both skyrocketed. A seminal event was the restructured NTT DoCoMo's launch of the *keitai* i-mode (internet mode) service in 1999, in other words, the advent of mobile internet in Japan. Needless to say, Japanese youth took to i-mode as fish takes to water. However for the older population exposed to cell-phones, traditional calls still constituted the primary use (Nakayama 2002). By 2005, approximately 70 million Japanese accessed the internet from their mobile phones, exceeding the number of users who accessed internet from PC terminals (66 million) for the first time. A 2001 survey indicated that Japanese used Internet on the cell phone to invoke search engines, transportation/routing maps, catch up with news and weather, search for information on sports, music concerts, play games, receive fortune telling, and utilize matchmaking (Ishii 2004).

Presently telecom providers offer subsidies to encourage the sale of high-end cell phones in Japan, much like the United States. Users are thus able to purchase cell phones that offer a high degree of functionality relatively cheaply. As Nakayama (2002) commented, people, mostly older usually read something physical—a book, newspaper, or magazine on a Japanese commuter train, or use the time to rest, but Japanese youth are found to be gazing at their cell phones, playing games, emailing, or generally accessing information from the web on their phones. This societal trend is depicted in Fig. 7.2. Nakayama (2002) has further pointed out that while PCs remain the preferred mode of web access for Japanese academics and professionals, the general population largely uses the cell phone to access the internet, especially the youth and those who cannot afford or are not able to access a PC. Mobile internet in Japan also does not impose any prerequisites such as those imposed by a traditional QWERTY keyboard of a PC. While cautioning that sophisticated use of the internet is hardly possible on cell phones, Akiyoshi and Ono (2008) have provided evidence that cell phones in Japan provide internet access to traditionally marginalized users—women, the less educated, and the less affluent.

Fig. 7.2 Japanese youths using cellphone in commuter train

Akiyoshi and Ono (2008) further contend that similar to cell phones leapfrogging fixed telephones in many nations, mobile based internet in Japan has the potential to leapfrog internet access using PCs by personalizing content in local languages. This has implications not just for Japan, but other non-English speaking nations as well.

7.3 Current Dimensions of ICT Use for Japan

As discussed in Sect. 7.1, Japan is one of the world's leading information societies. Per capita ICT usage in Japan has also grown steadily since 2000 as depicted in Fig. 7.3. The only exception is fixed phones, for which per capita use declined steadily between 2000 and 2008, to be followed by a spike in 2008, followed by another gradual decline until 2013. Figure 7.3 does not show mobile penetration since mobile subscription data from the ITU does not include personal handyphone systems, quasi cell-phones in Japan. However estimated calculations based upon mobile cellular subscription data from Statistics Japan (2014) indicate that mobile cellular subscriptions per 100 population for Japan were 44.80, 71.64, 91.12, and 111.08 in 2000, 2005, 2010, and 2013 respectively, indicating rapid growth between 2000 and 2013. Furthermore, IP telephony grew rapidly in Japan and appeals as an alternative to traditional fixed phones and to mobile phones, due to IP phone's high reliability and expanding features (Ida and Sakahira 2008).

While the growth of ICT usage in Japan is impressive, potential hindrances to ICT utilization in Japan are due to the nation's age structure and language as well as keyboard skills (Akiyoshi and Ono 2008). Firstly, Japan is the world's oldest nation

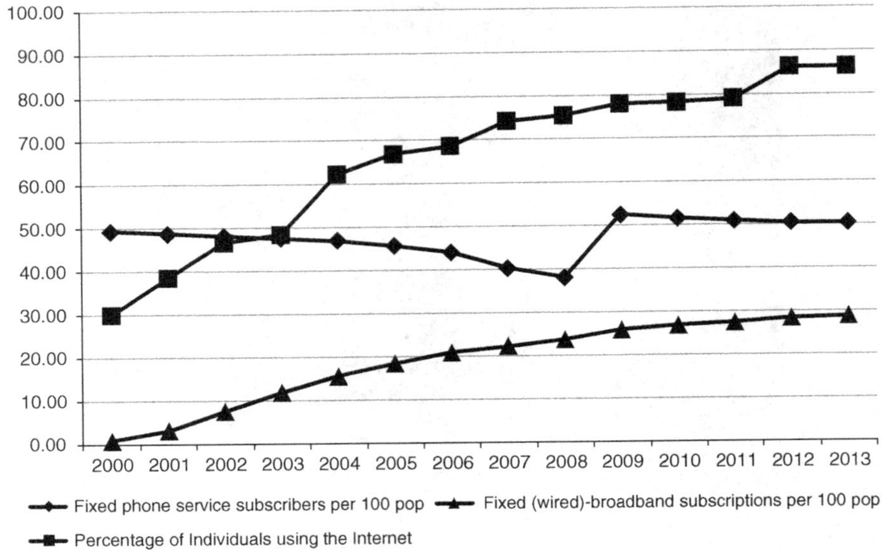

Fig. 7.3 ICT use per 100 population, Japan, 2000–2013 (*Data Source* ITU 2014)

among countries with a population of at least 100 million, with an estimated median age and elderly dependency ratio of 46.1 years and 42.1 % respectively in 2014. Such an age structure (depicted in the population pyramid in Fig. 7.4 using 2010 population) has an adverse effect on ICT usage (Akiyoshi and Ono 2008).

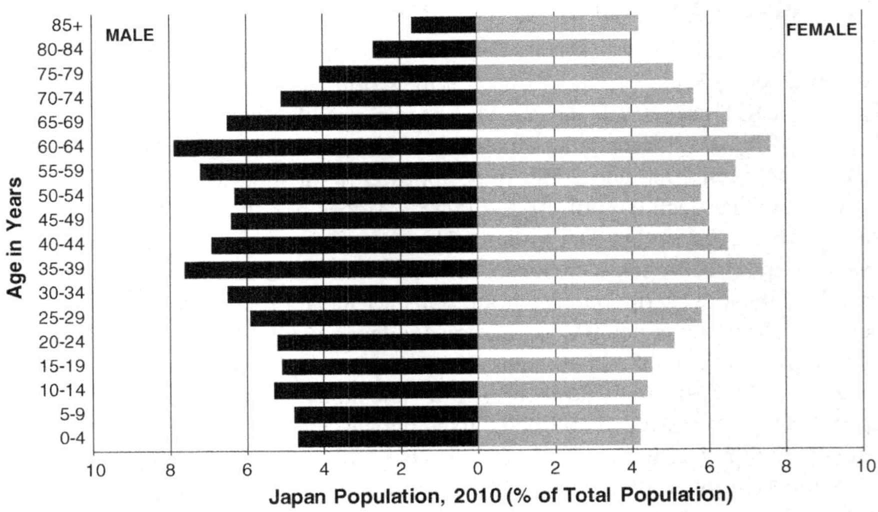

Fig. 7.4 Population pyramid for Japan, 2010

For example, individual internet usage rate exceeded 90 % among people in each age group between ages 13 and 59, although the rate drops steadily as the age increases (Statistics Japan 2014), alluding to the possibility of the existence of a technophobic demographic. Akiyoshi and Ono (2008) further contended that while older persons are just as likely to own computers as younger persons, mobile phone ownership declines steeply with age.

Another hindrance for Japan is the language barrier. The English alphabet keyboard has been an almost insurmountable problem for a majority of far East Asian nations, including Japan, since Japanese is the predominant language of the nation (Nakayama 2002). Since English is the predominant language of the Internet, Japanese users could feel discouraged thus negatively impacting PC and internet use. Due to the relative simplicity of mobile phone keyboards, Japanese users increasingly use i-mode (internet mode) on their mobile phones to access the internet (Ishii 2004; Akiyoshi and Ono 2008).

Geographically, Japan's ICT adoption and diffusion has hardly been even. Part of the problem has been the usual heavy adoption of ICTs in urban metropolitan areas such as Tokyo and its surrounding prefectures. For example, there are regional differences in broadband access by type. DSL penetration tends to be higher in East Japan, whereas cable modem use is greater in several regions in West Japan. The three major metropolitan regions, Tokyo, Nagoya, and Osaka, and the major provincial city regions lead in FTTH (Fiber to the Home) use (Arai and Naganuma 2010).

Maps of the individual dependent variables of broadband subscribers (Fig. 7.5) and Facebook subscribers (Fig. 7.6) show similar patterns, with the highest subscriber levels in the large metropolitan prefectures, including prefectures of Tokyo and surrounding ones of Saitama, Chiba, and Kanagawa; Aichi that includes the metropolis of Nagoya; Osaka; Kyoto; Hiroshima; and Fukuoka. The lowest subscriber levels are in rural prefectures in the north of the main island of Honshu and in some southeastern coastal prefectures. Japan has lagged other advanced nations in adoption of Facebook. In Tokyo prefecture, 15 % of 2011 population subscribes to Facebook, accounting for 51 % of Japan's Facebook subscribers. Note that apart from the map in Fig. 7.7, the north axis of all map figures in this chapter is tilted by about 30 degrees.

These and most other dependent variables are correlated with prefectural percent of urban population (see Fig. 7.7); this is reinforced in some inverse associations of farm population in the regression analysis in the next section. Japan's largest urban agglomerations, shown on the map by green bars, such as Tokyo and Osaka, are consistently among the highest in technology levels.

Another important aspect is Japan's largely mountainous terrain, shown in Fig. 7.8. This has the potential to impose significant infrastructure development obstacles, especially in providing connectivity for the last mile. In a case study later in the chapter, we describe how a rural, remote mountainous village in Hokkaido prefecture was able to construct a full fiber-optic information network with support from Japan's government.

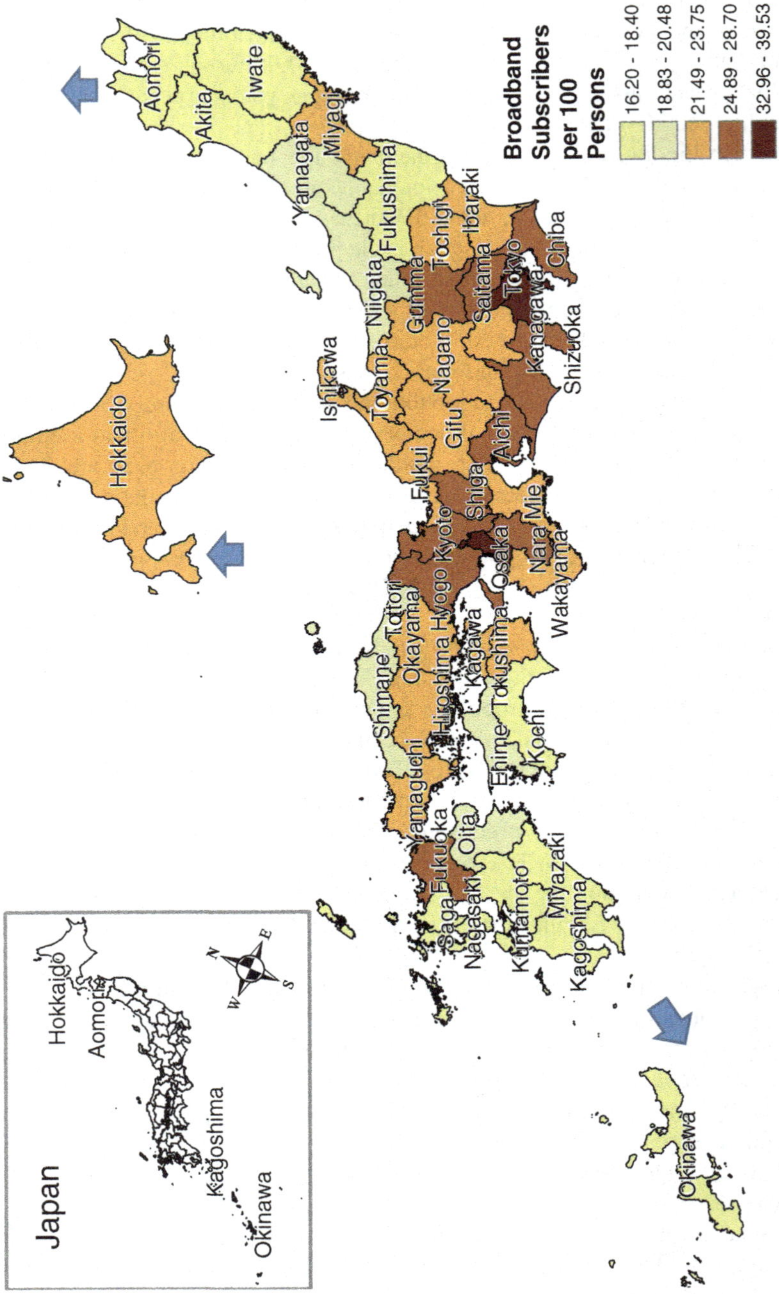

Fig. 7.5 Broadband subscribers per capita, Japan, 2010

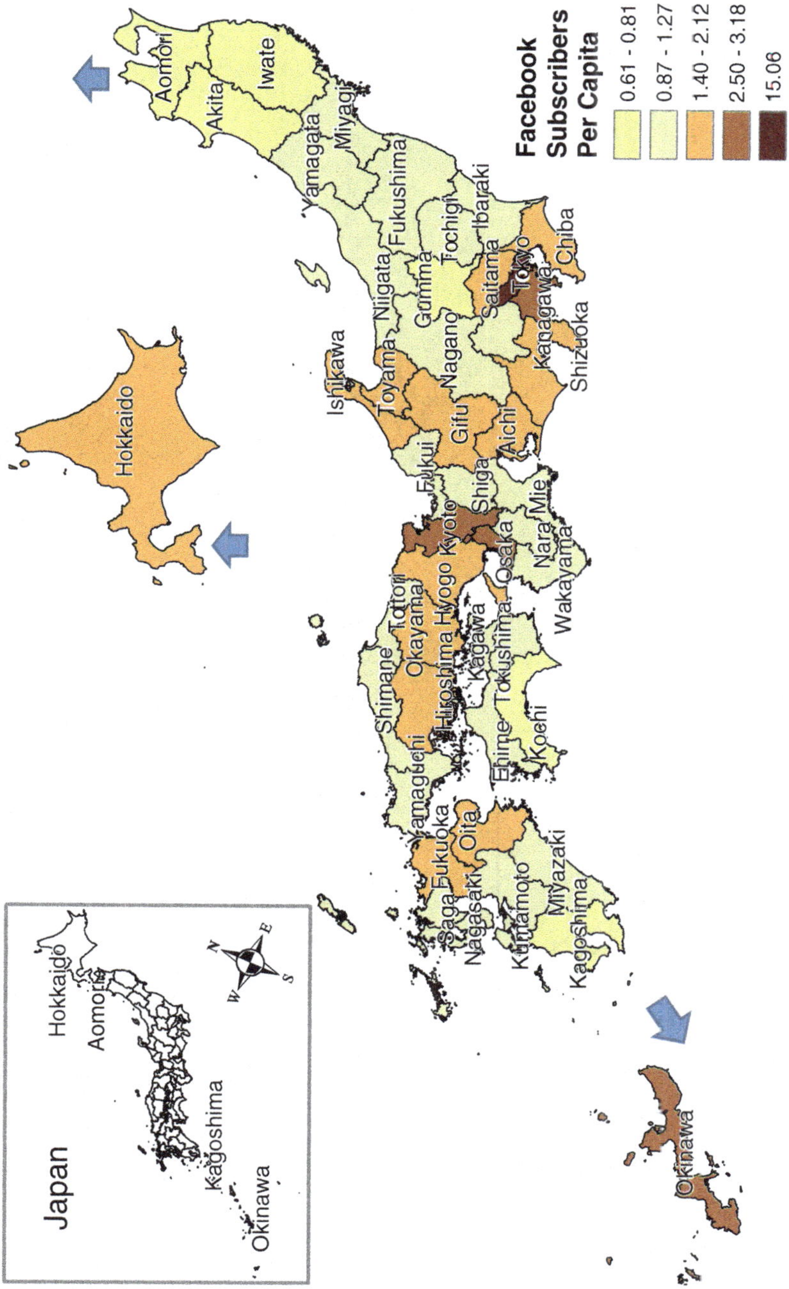

Fig. 7.6 Facebook subscribers per capita, Japan, 2011

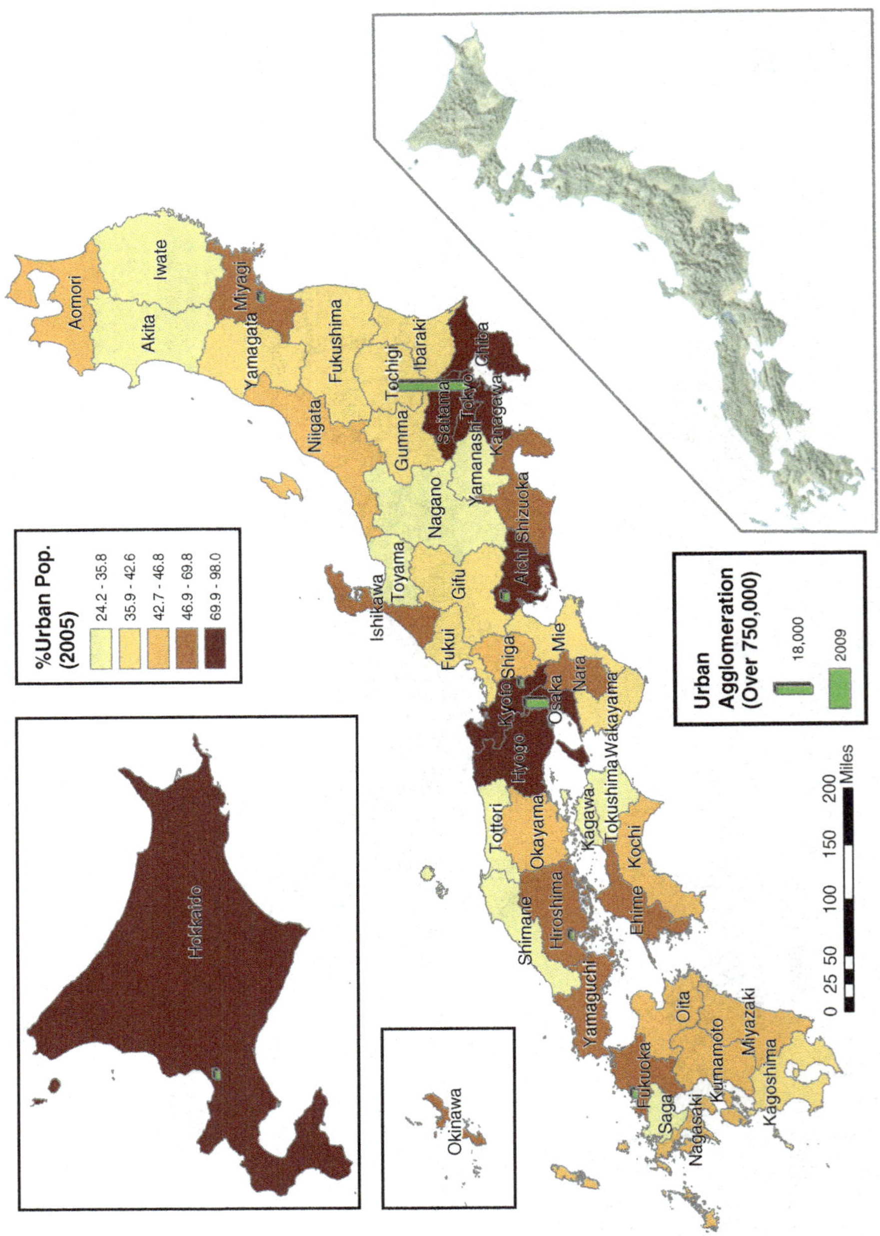

Fig. 7.7 Urban populations and urban agglomerations by prefecture, Japan, 2009

Fig. 7.8 Mountainous terrain in large parts of Japan

7.4 Conceptual Model of ICT Utilization in Japanese Prefectures

In our conceptual model of ICT utilization for Japanese prefectures (Fig. 7.9), we posit twelve independent variables that represent socio-economic, education, income, infrastructural, innovation, and societal openness factors to be associated with ten ICT dependent variables. The justification for including these variables in the model is discussed next. Since justifications of including these broad categories of independent variables have been presented earlier in Chap. 3 and in the more specific context of Chinese provinces in Chap. 5 and of Indian states in Chap. 6, we provide brief justifications in this section supplementing them with context-specific evidence from prior literature on Japan's digital divide.

Demographic influences: Demographic variables are known to impact ICT utilization, irrespective of geography. We include the variables farm household population and young dependency ratio (ratio of dependent children to working adults) since ICT utilization in Japan seemed to trail off as a person grows older (Statistics Japan 2014), we posit that both young dependency ratio will be positively associated and farm household population will be inversely associated with ICT use. We include the variable working age population ratio since Japan is one of the leading industrialized nations, and the technology workforce is mainly comprised of young individuals living in metro areas. Like other countries such as India (Chap. 6), Japan's digital divide stems partly from the urban-rural divide. Compounding this divide in Japan is Japan's largely mountainous terrain, discussed earlier, which hinders infrastructural development.

Income is a well-known correlate of technology adoption and use (Baliamoune-Lutz 2003; Dasgupta et al. 2005; Sipior et al. 2011; Lengsfeld 2011). Large survey studies of Japanese households indicated that income per capita related to PC experience and mobile phone ownership (Ono and Zavodny 2007), while income

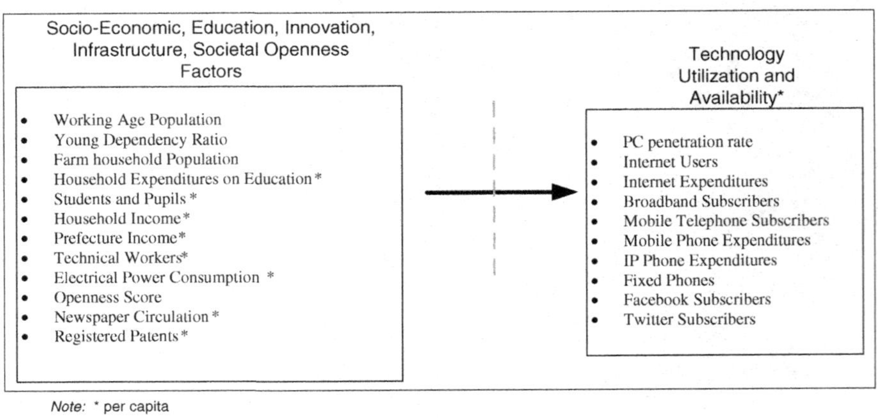

Fig. 7.9 Conceptual model of ICT utilization for Japanese prefectures

level was among the most important determinants of fiber-to-the-home broadband internet subscription (Ida and Sakahira 2008). Kawaguchi (2006) identified that a PC at home facilitates the possibility of full-time employment among women and is positively correlated with salaries of both men and women in Japan. We reason that individuals and households with higher income are better able to afford the subscription and purchase costs of technologies, while higher prefectural income stimulates more capital investment in ICT by organizations. Hence, *household income and prefecture income are posited to increase technology utilization and expenditures.*

Infrastructure: Although Japan is a leader worldwide in its overall use of technology and is the locus of leading global technology companies, its ICT infrastructure is challenged in many parts of the nation due to mountainous terrain and high investment costs. The national government's strategies such as the "e-Japan Plan" contributed to improved internet and broadband throughout the country, but geographic inequalities in infrastructure remain (Arai and Naganuma 2010). Constraints of broadband infrastructure in rural areas were keyed to reduced broadband use (Yuguchi 2008). For PCs per capita for countries, electric power consumption per capita was found to be a significant determinant when capped at 6,000 kW h per capita (Chinn and Fairlie 2007). The electrical power system is fundamental for access, use, and reliability across various technologies. *Accordingly, electrical power consumption per capita is posited to be associated with technology utilization and expenditures.*

Education: Education, apart from income, has been found to be associated with ICT utilization in many prior digital divide studies (see Chap. 3). For a Japanese national sample of individuals, college education was among the most important determinants of mobile phone and PC use (Akiyoshi and Ono 2008). We reason that greater emphasis on education both in the household and by the prefecture fosters digital knowledge/skills and the confidence to make use of them, which stimulates the readiness of people and organizations in the prefecture to adopt and use technologies. *Consequently, two determinants are posited to increase technology utilization and expenditures: household expenditures on education and students and pupils per capita.*

Innovation: An investigation of influences on technology usage, expenditure, and infrastructure worldwide found the most important determinant to be scientific and technical journal articles (Pick and Azari 2011). Likewise, as shown in Chap. 4, regression analysis of technology utilization for nations revealed capacity for innovation to be a dominant correlate worldwide and in developed nations. In Japan, there is government impetus to support innovation at the national and prefectural levels. Shimizu et al. (2008) has contended that the Japanese ICT industry has supported the growth of the national total factor productivity which is a leading element in technological innovations. In this environment of a high-tech nation emphasizing national and regional innovation, R&D activity is known to consume, use, and stimulate technology. *Accordingly, the innovation factor of patents registered by Japanese citizens is posited to increase technology utilization and expenditures.*

Table 7.1 Variable definitions, data sources, and descriptive stats

Category	Variable	Source[a]	Year	Definition	Average[b]	Standard deviation[b]
Dependent variables						
Technology use	PC penetration rate for families	NCS	2009	Personal computer penetration rate for 100 families which have at least a PC	65.34	6.63
	Internet users/100 pop.	MIC	2010	Internet users less than 6 years old per 100 population	75.26	4.66
	Internet expenditures/family	NCS	2009	Internet expenditures monthly per family (yen)	1757	316
	Broadband subscribers/capita	MIC	2010	Broadband subscribers per 100 population	22.7	4.72
	Mobile telephone subscribers/capita	TCA	2009	Mobile telephone subscribers per 100 population	80.06	10.6
	Mobile phone expenditures/family	NCS	2009	Mobile telephone expenditures monthly per family (yen)	7656	578
	IP phone expenditures per family	NCS	2009	IP telephone expenditure monthly per family (yen)	15.97	4.53
	Fixed phone subscribers/capita	TCA	2009	Fixed phone subscribers per 100 population	29.77	2.74
	LN (Facebook subscribers/capita)	Facebook	2011	Natural log (Facebook subscribers per 10 population)	2.543	0.538
	Twitter subscribers/capita	ULI	2011	Twitter subscribers per 100 population	0.6847	0.351
Independent variables						
Demographic	Working age population (Pop. 20–64)/total population	SYB	2010	Ratio of population age 20–64 to total population	0.5677	0.0211
	Young dependency ratio (Pop. 0–19/Pop. 20–64)	SYB	2010	Ratio of population age 0–19 to population age 20–64	0.3182	0.0247
	Farmers per 100 population	SYB	2010	Number of farmers per 100 population	7.904	4.33

(continued)

Table 7.1 (continued)

Category	Variable	Source[a]	Year	Definition	Average[b]	Standard deviation[b]
Education	Household expenditures on education per family	NCS	2009	Household expenditure on education per family (1000 yen)	9271	2657
	Students and pupils per capita	SYB	2005	Students and pupils per 100 population	5.38	0.454
Economic	Household income per family	NCS	2009	Household income per family (1000 yen)	5446	638
	Prefecture income per capita	SYB	2008	Prefecture income per capita (1000 yen)	2657	383
	Technical workers per capita	SYB	2005	Technical workers per 100 population	3.176	0.354
Infrastructure	Electrical power consumption per capita	SYB	2009	Electric power consumption (in millions of kilowatt hours) per capita	2285	171
Societal openness	Openness score for prefectures	ACO	2009	Multi-dimensional index of the openness of the prefectural government	63.7	7.36
	Newspaper circulation (mornings) per capita	SYB	2010	Total circulation of morning newspapers per 100 population	37.3	4.1
Innovation	Patents registered by Japanese citizens per capita	JPO	2009	Number of patents registered by Japanese citizens per 100 population	0.0529	0.106

[a] *ACO* Japan Association of Citizen Ombudsman; *JPO* Japan Patent Office; *MIC* Japan Ministry of Internal Affairs and Communications; *NCS* Japan National Consumer Survey (in Japanese); *SYB* Japan Statistical Yearbook; *TCA* Japan's Telecommunications Carriers Association; *ULI* User Local Inc.

[b] $n = 47$

Openness: Societal openness was emphasized in studies that have included democracy and freedom (see Chap. 3). We reason that a democratic, free, and lawful society fosters more communication of ideas and collaborative activity (Robison and Crenshaw 2002; Baliamoune-Lutz 2003; Simon 2004; Guillén and Suárez 2005; Yates et al. 2011) and hence posit that openness will be positively associated with ICT use in Japanese prefectures.

Standard ICT dependent variables were included, consistent with prior empirical work on examining influences on ICTs. Note that similar to Chap. 8 (US digital divide), we have also included social media platforms of Facebook and Twitter users per capita in our conceptual model.

7.4.1 Data Used for This Study

A variety of data sources were accessed, as shown in Table 7.1. Four of the sources are official publications of the Japanese government, specifically Japan Patent Office, Japan Ministry of Internal Affairs and Communications, Japan National Consumer Survey, and Japan Statistical Yearbook. In addition, information came from the Japan Citizens' Ombudsman Association (JCOA) and from User Local Inc. Data on prefectural openness were obtained from the JCOA website and translated into English. The User Local Inc., (http://social.userlocal.jp) is a well-known and reliable website in Japan that provided prefectural data on social networking. These data sources at the prefecture level have not been utilized in digital divide research previously and hence provide a unique study aspect.

As seen in Table 2, all variables were for the years 2009–2011, except one attribute for 2008 (prefecture income per capita) and two attributes for 2005 (technical workers, and students and pupils). We were constrained and had to use a three-year window because the data collection system for many factors is not yearly. We feel the three-year window for 19 out of 22 variables provides sufficient time simultaneity. Since the variables are per capita, we do not feel there is large error by treating these variables as a cross section.

7.5 Spatial Arrangements of ICT Use in Japanese Prefectures

7.5.1 Spatial Autocorrelation of ICTs

Findings on spatial autocorrelation for the dependent variables indicate that they mostly have high, positive spatial autocorrelations (Table 7.2). For Japan, autocorrelated ICT variables are PC, internet, broadband, and mobile subscribers, as well as internet expenditures. Largely similar findings have been reported for China

Table 7.2 Moran's I values for Japanese prefectures, 2009, African Nations, 2010, U.S. States, 2010, and Chinese Provinces, 2009 (Pick et al. 2015, 2014a, 2013)

Moran's I values for Japan, N = 47, 2009										
PC penetration rate for families	Internet users per 100 pop.	Internet expenditures per family	Broadband subscribers per capita	Mobile telephone subscribers per capita	Mobile phone expenditures per family		IP phone expenditures per family	Fixed phone subscribers per capita	LN Facebook subscribers per capita	Twitter Subscribers per capita
0.519***	0.392***	0.553***	0.506***	0.220***	0.146		0.111	0.085	0.216*	0.266***
Moran's I values for African Nations, N = 51, 2010										
	Internet users per 100 pop.		Broadband subscribers per capita	Mobile telephone subscribers per capita				Fixed phone subscribers per capita	Use of virtual social networks	
	0.059		0.288***	0.277**				0.324***	0.001	
Moran's I values for US, N = 50, 2010										
PC penetration rate for families	Log of internet access at home		Broadband adoption in household	Persons in cellphone only household		Mobile wireless high-speed devices		Persons in fixedphone only household	Facebook subscribers per capita	Twitter subscribers per capita
0.547***	0.471***		0.457***	0.621***		0.230***		0.648***	−0.004	0.069
Moran's I values for China ᵃN = 31, 2009										
PCs per 100 urban families	Internet users per 100 pop.		Broadband subscribers per capita	Mobile telephone subscribers per capita				Fixed phone subscribers per capita		
0.316***	0.264**		0.272**	0.205*				0.252**		

*signif. at 0.05; **signif. at 0.01; ***signif. at 0.001

ᵃOther dependent variables are also considered for China (see Chap. 5)

in Chap. 5, United States in Chap. 8, and Africa in Chap. 9. The only notable exception is internet expenditures, which was randomly diffused among African nations.

Three dependent variables for Japan, mobile phone expenditures/family, IP phone expenditures, and fixed phone subscribers, however, are not spatially autocorrelated. For the least agglomerated, fixed phone subscribers, it can be reasoned that the historical adoption/diffusion process led to fairly random penetration across the country; since this technology is becoming displaced today, former "hot spots" have diminished leading to an even more random pattern. The lack of agglomeration for the former two variables is surprising, given that they are current technologies still being rapidly improved. Overall, the large extent of spatial clustering of the dependent variables highlights the need in the regression analysis to control for spatial bias.

Local Indicators of Spatial Association (LISA) analysis of mobile subscription (Fig. 7.10) shows Tokyo, a high mobile use prefecture neighbors other high mobile use prefectures such as Chiba, while Akita prefecture in the north, a low mobile use province is spatially proximate to Aomori prefecture, also low in mobile use.

7.5.2 Clustering of ICTs

The results of the k-means cluster analysis for 5 clusters, based on the 7 dependent variables are shown in Fig. 7.11. This map excludes three dependent variables that measure expenditures, since those variables are somewhat dependent on regional differences in tariffs. For example, while Tokyo has the highest per capita mobile subscribers, monthly per family mobile expenses are the lowest possibly reflecting lower tariffs charged by service providers to a much larger critical mass of users.

Findings indicate a unique cluster 1 for Tokyo prefecture, Cluster 2 that is comprised of the other metropolitan areas of the main island of Honshu including Tokyo's neighboring prefectures, as well as Aichi, Kyoto, and Osaka; Cluster 3 with 22 prefectures mostly in the center of the main island of Honshu and the northernmost island of Hokkaido, and clusters 4 and 5 (18 prefectures total) which are found to encompass most of the rural prefectures in the north, southeast including the large southern islands of Kyushu (except Fukuoka which is part of cluster 3) and Shikoku (except Kagawa and Fukushima prefectures). The southernmost island of Okinawa and Tokushima prefecture in the small island of Shikoku comprise cluster 5.

The five clusters can be characterized by averaging their technology subscription indicators (see Table 7.3). In terms of digital divide, Tokyo prefecture is highest in technology levels, followed by Cluster 2 comprising other major metropolitan prefectures. Four out of seven national universities—University of Tokyo (shown in Fig. 7.12), Osaka University, Kyoto University, and Nagoya University (in Aichi)—are located in clusters 1 and 2.

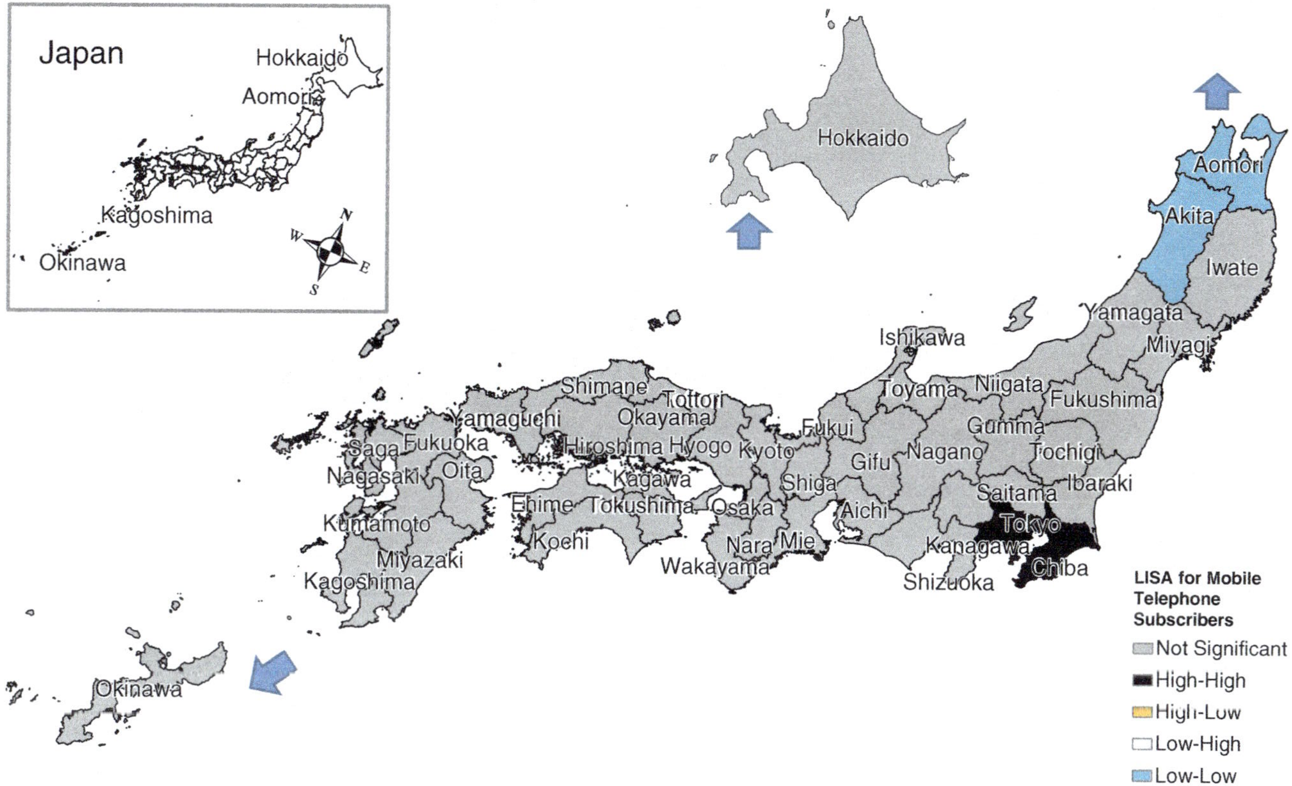

Fig. 7.10 LISA (Local Moran's I) for mobile telephone subscribers. *Note* In the legend, "High-High" indicates a prefecture with high mobile phone subscription per capita surrounded by other prefectures with similarly high values. In other words, "High-High" indicates a statistically significant cluster of prefectures with high mobile phone subscription per capita. Other items of the legend can be interpreted similarly

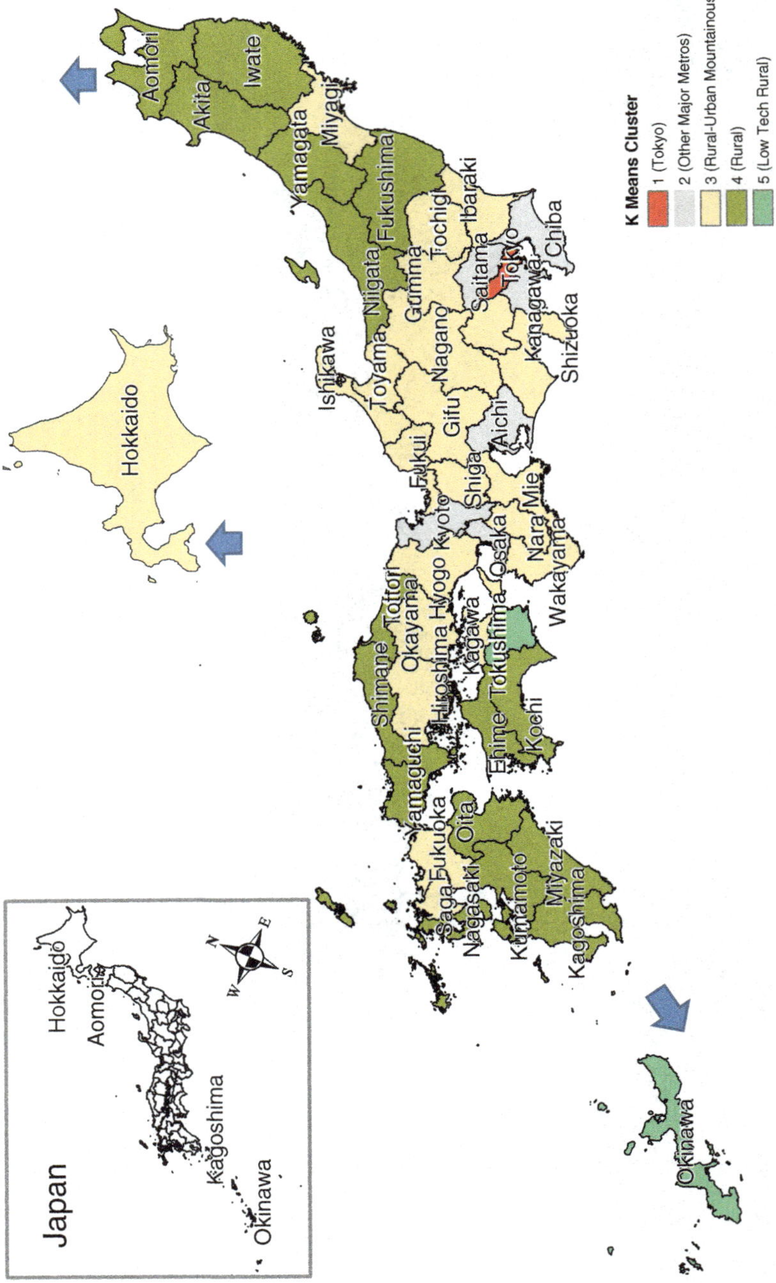

Fig. 7.11 K-means clusters of ICT utilization in Japan, 2009–2011

Table 7.3 K-means cluster characteristics for Japan 2009–2011

	Clusters 2009–2011						
ICT dependent variable	Cluster 1 mean values	Cluster 2 mean values	Cluster 3 mean values	Cluster 4 mean values	Cluster 5 mean values	Ratio of high to low clusters	Ratio of high to low clusters excluding Tokyo
Personal computer penetration rate for families (in percent)	66.3	70.15	68.83	60.43	51.3	1.37	1.37
Internet users per 100 population	83.6	81.63	76.58	70.21	77.9	1.19	1.16
Broadband subscribers per 100 families	39.53	29.68	23.14	18.82	19.44	2.1	1.58
Mobile phone subscribers per 100 population	141.8	87.14	80.4	73.37	77.8	1.93	1.19
Fixed phone subscribers per 100 population	34.39	28.8	28.76	31.39	28.55	1.2	1.1
LN (Facebook subscribers per 10 population)	5.01	3.12	2.51	2.2	2.7	2.28	1.42
Twitter subscribers per 100 population	2.48	1.08	0.64	0.49	0.63	5.02	2.19
Number of prefectures	1	6	22	16	2		

(continued)

Table 7.3 (continued)

	Clusters 2009–2011								
ICT dependent variable	Cluster 1 mean values	Cluster 2 mean values	Cluster 3 mean values		Cluster 4 mean values		Cluster 5 mean values	Ratio of high to low clusters	Ratio of high to low clusters excluding Tokyo
Member prefectures	Tokyo	Kyoto	Mie	Hiroshima	Oita	Kumamoto	Tokushima		
		Chiba	Saga	Tochigi	Miyazaki	Fukushima	Okinawa		
		Saitama	Hyogo	Shiga	Yamaguchi	Akita			
		Osaka	Hokkaido	Ishikawa	Yamagata	Nagasaki			
		Aichi	Wakayama	Fukui	Iwate	Aomori			
		Kanagawa Nara	Nara	Fukuoka	Shimane	Kochi			
			Miyagi	Gumma	Ehime	Tottori			
			Toyama	Ibaraki	Niigata	Kagoshima			
			Yamanashi	Nagano					
			Gifu	Shizuoka					
			Okayama	Kagawa					

Fig. 7.12 Tokyo University campus. *Source* Kazunori Mizuno

A partial explanation is that these eminent research universities are "hotspots" of ICT adoption and utilization for cutting edge research, education, and innovation, and also spawn a critical mass of highly educated and technologically proficient users. In third place is the central mountainous Cluster 3 which is a mix of urban and rural areas, and the island of Hokkaido, while the largely rural Clusters 4 and 5 are lowest. Cluster 4's position partly reflects its largely mountainous terrain, which may limit deployment of supporting infrastructure, as seen by its low broadband and mobile subscriber levels that are approximately half of the corresponding levels for Tokyo.

These four geographic areas (clusters 1–4) can be interpreted with respect to both the ICT policies of the national government of Japan (Sect. 7.2) and our policy recommendations based on the present study (Sect. 7.8).

1. "*Tokyo*"—This prefecture includes the central city (city proper) and very densely populated area to the east inside the Tokyo megalopolis and had population of 37.8 million as the world's largest megalopolis in 2014 (UN 2014). It is the center for national government and many major companies resulting in high demand for ICT. It is the highest cluster of all ICT factors except PC penetration rate and especially high on broadband, mobile phone subscribers, Facebook, and Twitter, for which its ratio to the lowest cluster is between 1.93:1 and 5.02:1 (Table 7.3). Tokyo's predominance in social media usage, especially Twitter is hardly surprising since Twitter users are concentrated in large metropolitan areas worldwide. Moreover, Tokyo ranks third among twenty-five metro areas in terms of

Twitter usage (Takhteyev et al. 2012). Overall, this cluster is analogous to U.S. "broadband core" regions—spatial clusters of individual zip codes displaying high levels of broadband availability and competition (Grubesic 2006).

2. "*Other metropolitan areas*"—Cluster 2 includes the Tokyo's metropolitan periphery surrounding the Tokyo prefecture (Chiba, Kanagawa, and Saitama prefectures) and the largely metropolitan prefectures of Aichi, Kyoto, and Osaka parts of which are connected by the Shinkansen bullet train route. This prosperous area has technology levels comparable to the Tokyo cluster, except for reduced levels for broadband, mobile phones, Facebook and Twitter but still much higher than clusters 3, 4, and 5. Also the fixed phone level is lower, possibly pointing to an area less metropolitan and infrastructure-concentrated than central Tokyo.
3. "*Rural-urban mountainous region*"—The 22 prefectures for Cluster 3 are distributed between largely metropolitan prefectures of Fukuoka, Hokkaido, and Yamanashi and other rural and less prosperous parts of Japan, which have older population. While personal computer adoption exceeds that for Tokyo and compares favorably with Cluster 2 levels, this cluster lags behind Clusters 1 and 2 in all other ICT forms including ubiquitous ones such as broadband and mobile phones, emerging social media technologies, and legacy technologies such as fixed phones. While some parts of this cluster such as the small plains near Sapporo, Sendai, and Hiroshima possess favorable "lay of the land," the lag in broadband and mobile phone adoption can be attributed to the largely mountainous landscape. This increases the cost of infrastructure development, for example for constructing base stations for wireless mobile, or laying fiber-optic cabling to provide FTTH service to access broadband.
4. "*Rural Areas*"—This area comprising Clusters 4 and 5 includes most of the rural prefectures in the north, and southeast including the large southern islands of Kyushu (except Fukuoka which is part of cluster 3) and Shikoku (except Kagawa and Fukushima prefectures), as well as the southernmost island of Okinawa and Tokushima prefecture in the island of Shikoku, which comprise cluster 5. This cluster is analogous to the "broadband periphery" in the U.S., identified in Grubesic (2006) as clusters of mostly rural zip codes that display low levels of broadband availability and competition. In Japan, these areas have lost young migrants to Japan's cities, lowering the areas' demand for ICT.

7.6 Determinants of ICT Use in Japan

OLS regression analysis was conducted based on the sample of Japan's 47 prefectures to test associations of twelve independent correlates with ten ICT dependent variables. The findings (see Table 7.4) indicate high overall significance for all the regressions. OLS regression diagnostic tests of Joint Wald, Koenker (BP) and Jarque-Bera reveal no problems, with the single exception of positive Jarque-Bera

Table 7.4 Standardized regression results for dependent variables, Japan

		Dependent variable									
Category	Independent variable	Personal computer penetration rate for families	Internet users/100 pop.	Internet expenditures/family	Broadband subscribers/capita	Mobile telephone subscribers/capita	Mobile phone expenditures/family	IP phone expenditures/family	Fixed phone subscribers/capita	LN Facebook subscribers/capita	Twitter subscribers/capita
Demographic	Working age population (pop. 20–64)/total population		0.585***			0.313***			−0.481**		
	Young dependency ratio (pop. 0–19/pop. 20–64)		−0.239**		−0.183**		0.459***		−0.529***		−0.213***
	Farm household population/total population	−0.168*	−0.352**		−0.445***		0.244*		−0.585***	−0.383***	−0.169**
Education	Household expenditures on education per capita			0.648***				0.552***	−0.651***		
	Students and pupils per capita									0.146*	0.307***
Economic	Household income per capita	0.845***					0.554***				
	Prefecture income per capita				0.305***						
	Technical workers per capita			0.329***							

(continued)

Table 7.4 (continued)

		Dependent variable									
Category	Independent variable	Personal computer penetration rate for families	Internet users/ 100 pop.	Internet expenditures/ family	Broadband subscribers/ capita	Mobile telephone subscribers/ capita	Mobile phone expenditures/ family	IP phone expenditures/ family	Fixed phone subscribers/ capita	LN Facebook subscribers/ capita	Twitter subscribers/ capita
Infrastructure	Electrical power consumption per capita			0.270**		0.098*					
Societal openness	Openness index for prefectures										
	Newspaper circulation (mornings) per capita		0.250**		0.176**			0.256*			
Innovation	Patents registered by Japanese citizens per capita				0.200*	0.753***				0.620***	0.686***
Regression adjusted R square and significance level		0.744***	0.757***	0.700***	0.883***	0.906***	0.395***	0.440***	0.608***	0.819***	0.910***
Sample size (N)		47	47	47	47	47	47	47	47	47	47
OLS regression tests											
Joint Wald statistic		138.5***	209.5***	154.3***	423.5***	214.4***	45.7***	37.3***	136.3***	423.0***	0.910***
Koenker (BP) statistic		2.421	3.939	5.685	5.906	10.873*	4.562	2.477	1.037	7.637	4.613
Jarque-Bera statistic		0.296	1.045	7.825*	0.874	1.463	2.212	2.945	1.421	0.828	1.123
Test of spatial autocorrelation of residuals											
Moran's index		−0.159	−0.113	0.097	0.011	0.073	0.022	−0.212	−0.034	0.092	−0.173

*signif. at 0.05; **signif. at 0.01; ***signif. at 0.001

Note Refer Chap. 1 for interpretation of statistics

test ($p = 0.05$) for internet expenditures per family, indicating lack of correspondence of residuals to the normal distribution. Hence, that regression will be regarded cautiously.

For the demographic factors, the ratio of farm population is inversely related to levels of technology utilization and expenditures, consistent with spatial findings that more urban locations have increased technology levels. The economic factors of income have limited effect. Household income relates to PC use and mobile phone expenditures, while higher prefecture income correlates with increased broadband subscribers. The infrastructure factor of electrical power consumption relates to internet expenditures per family and mobile telephone subscribers. The two education variables are positively associated with internet, IP phones, Facebook, and Twitter, but inversely associated with fixed phone expenditures. The latter reflects more educated households' preference for contemporary phones.

Innovation, as expressed by patents of Japanese citizens, is associated with the more modern attributes of broadband, mobile phones, Facebook subscribers and Twitter subscribers. The innovation proxy factor of newspaper circulation is associated with internet users, broadband subscribers, and IP phone expenditures, while the Openness Index is not significant. Overall, across the ten dependent variables, the most important factors are demographic i.e. age structure and rural-urban status; education; and innovation.

7.6.1 Discussion

While some findings support outcomes known in the digital divide literature but newly reported for Japan, others are entirely novel. Among the novel findings at the sub-national level are correlates for the utilization of technology platforms of IP phones, Facebook, and Twitter, and an openness index, innovation measured by patents, and a group of demographic indicators. Further, new spatial analysis findings provide insights into the geographical distribution of information technology utilization and expenditures throughout Japan.

Working Age Population and Young Dependency Ratio

The demographic result that working age population is associated with internet users and mobile phone subscribers corresponds to the generational emphasis on the internet for Japanese of working age and on mobile phones for younger working-age adults (Nakayama 2002). Young dependency ratio is associated with mobile phones that are mostly 3-G and 4-G (Nakayama 2002), so its association with mobile phone expenditures and internet users is expected.

Perhaps surprisingly, young dependency ratio is inversely related to broadband and Twitter. The explanation is that elevated young dependency ratio is more likely to occur in rural areas, which have reduced access to broadband (Yuguchi 2008), and where Twitter has not caught on as much. The strong inverse relationship of farm population (Japan's spatial distribution of farm household population depicted in Fig. 7.13) to technology utilization for 6 of 10 dependent factors points to rural Japan lagging in technology. Similar to India, as described in Chap. 6, the rural population has strikingly lower access and use of technologies (Pick et al. 2014b).

Income

Household and prefectural incomes are positively related to PCs, broadband subscribers, and mobile phone expenditures, findings well supported in the literature (Baliamoune-Lutz 2003; Dasgupta et al. 2005; Agarwal et al. 2009; Sipior et al. 2011; Lengsfeld 2011). Since these technologies tend to be higher priced relative to the others in Japan, income would be more important.

Education

Education is influential for the newer technology platforms of IP phones, Facebook, and Twitter, and for internet expenditures. Since they tend to be lower-cost technologies, the effect of education may be relatively higher because income's influence is reduced. Education's positive association with ICTs is widely known (Robison and Crenshaw 2002; Kvasny and Keil 2006; Ono and Zavodny 2007; van Dijk 2009; Pick and Azari 2008; Pick et al. 2015, 2014a; Sipior et al. 2011; Vicente and Lopez 2011; Yates et al. 2011) and corresponds to results in Chaps. 4, 6, 8 and 9. However, the inverse effect for fixed phones is explained by the increasing substitution of fixed phones by mobile and IP phones in Japan.

Infrastructure and Innovation

Significant relationships of electrical power consumption with internet users and mobile phone subscribers are justified by the large proportion of Japan's population using these two technologies, leading to elevated prefectural volumes of electrical consumption. Innovation, as measured by patents (Fig. 7.14 depicts its spatial distribution in Japan), is significantly associated to broadband, mobile phones, Facebook, and Twitter, forward-looking forms of technologies, which are driven by R&D. Although the innovation relationship is known for U.S. counties and other nations (Azari and Pick 2005; Pick and Azari 2008), this is the first confirmation of R&D's importance for Japanese prefectures. It is a factor that the central government and prefectures can stimulate by supporting R&D and encouraging investment in scientific education. As in some prior studies (Pick and Azari 2008; Pick and Azari 2011), these factors represent generic innovation (not IT-specific innovation). The explanation might be that R&D leads to geographic agglomeration of technical skills and creation of prefectural knowledge communities that spill over and stimulate expanded use of IT (Moretti 2012).

Newspaper Circulation

The importance of newspaper circulation corresponds to findings in Chap. 6 for India, in which high technology usage level in Delhi state is strongly related to registered newspapers and periodicals. We explain these effects as due to traditional forms of publishing being generators of printed content and increasingly e-content (Shiraishi 2004), which fosters innovation and provides openness of content

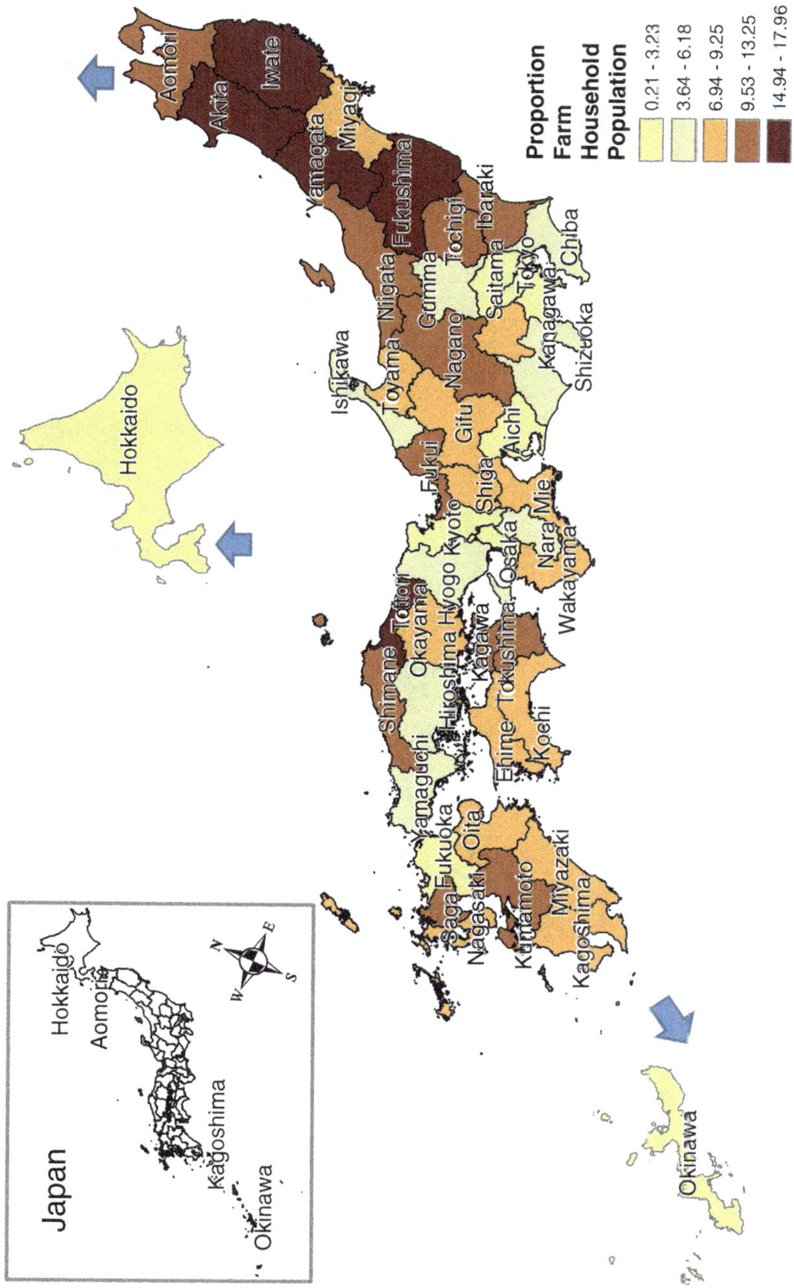

Fig. 7.13 Proportion of farm household population, Japan, 2010

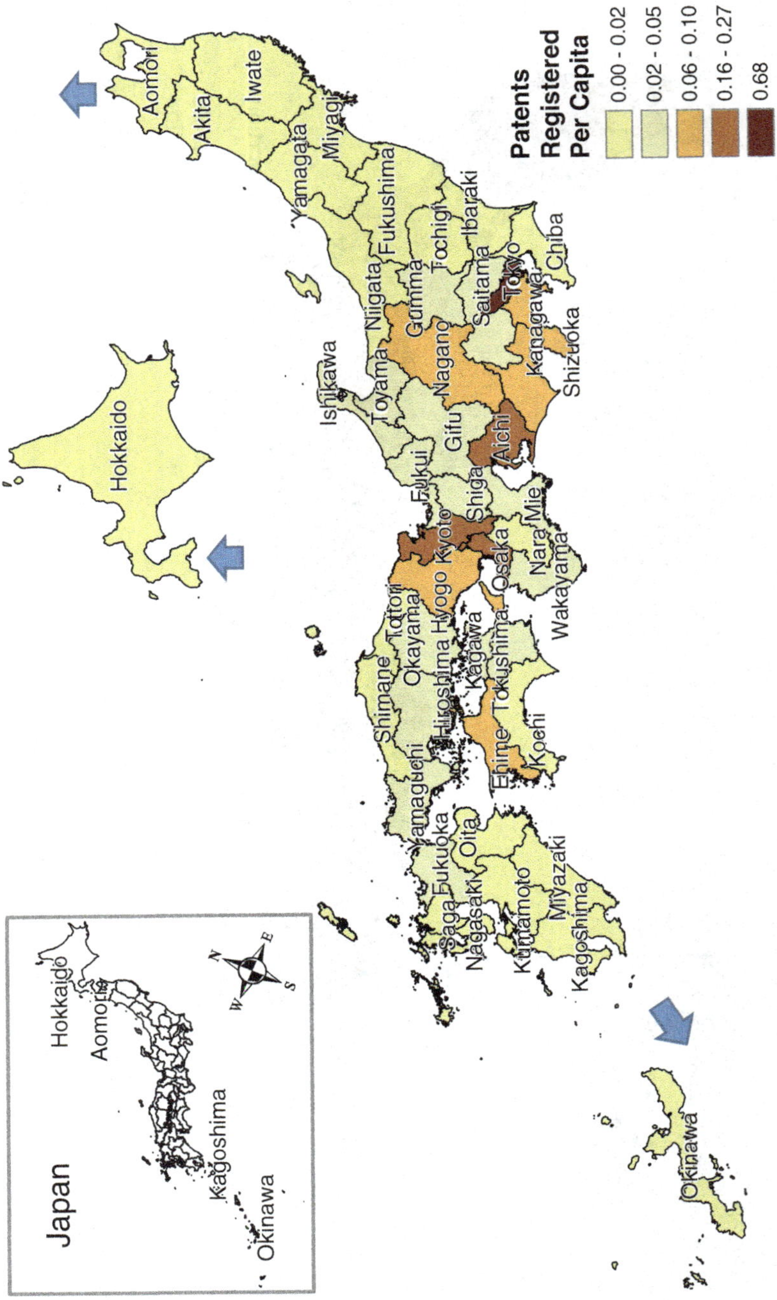

Fig. 7.14 Registered patents per capita, Japan, 2009

through the use of internet, web, and 3-G and 4-G mobile telephones. In the case of Japan, the dependent variables that are significantly correlated to newspaper circulation are content-driven, i.e. the internet and broadband, and mobile IP phones.

Caution must be exercised in interpreting the digital divide as beneficial when governmental and societal measures are taken to try to reduce it. Studies for U.S. technologically underserved populations have questioned whether technological utilization is positive for the individual, for example that additional technological skills for low-income person might not lead to a better job or emergence from poverty (Kvasny and Keil 2006). At the country, state, or prefectural levels, these concerns have greater impact when there is significant proportion of technologically and economically deprived people. In Japan, there is a smaller proportion of such people. However, another caution is that technological knowledge could be put to malicious purposes by individuals (Kvasny and Keil 2006) or by larger societal units. We need to acknowledge these possibilities and state them as a limitation, while recognizing such downside effects have rarely been included in prior studies that examine IT remediation policies and measures. These issues are raised again in Chap. 11, but it is outside the scope of this book to examine this question empirically.

Compared to Japan, China, as seen in Chap. 5, has an even larger and more rapidly growing economy, with a digital divide that has similarities and differences from Japan. Both are influenced by innovation and openness, yet in other respects they differ: Japan's prefectural technology levels are driven more by age, urban structure and education, while China's are propelled more by export commodities value and labor force attributes.

The present findings somewhat correspond to studies for the United States. A small-sample digital divide study of individual members of a disadvantaged public housing community and surrounding neighbors in Pennsylvania examined determinants of e-government services that were in process of being transformed (Sipior et al. 2011), based on technology acceptance. Those findings concur with ours in identifying income, education, age, and employment to be significant determinants of e-government usage. Another study which examined determinants of internet use for a U.S. government national sample, based on two-stage regression (Agarwal et al. 2009) had fairly similar results to the present ones, in that the factors of income, education, age, and children under 18 (i.e. similar to our young dependency ratio) were significant. However, it differed in identifying male gender and number of internet service providers as determinants (Agarwal et al. 2009). A similarity is the recognition of Agarwal et al. (2009) that metropolitan proximity relates to internet use. Case studies of deprived communities in LaGrange, Georgia, and Atlanta (Kvasny and Keil 2006), examined in Chap. 11, suggest that schooling and development of technical skills should be prioritized as most important for the long term.

Recent studies of the European digital divide have considered geography. Since European nations are much more culturally diverse than Japanese prefectures, culture is expected to be more influential in Europe (van Dijk et al. 2009). Based on a model that considered technology motivation, digital skills and usage, and physical access, findings for 25 EU nations revealed geographical gaps in technology indicators (north-south and east-west), and socio-economic gaps of age, education, type of employment, gender, and ethnic minority status (van Dijk 2009). The first three socio-economic gaps correspond to the present Japan findings. The influence of ethnic minority reflects Europe's diversity, but for Japan, ethnic data are unavailable for prefectures and ethnicity's influence nationally is limited given Japan's tiny proportion of ethnic minorities (2 %). The van Dijk et al. (2009) study recommends IT policies somewhat similar to ours for Japan, in particular to reduce deficiencies for the elderly, digitally illiterate, and geographically marginalized.

A recent study investigating the EU's 164 regions for determinants of ICT use (Vicente and Lopez 2011) had a sample somewhat analogous to Japanese prefectures in dropping below the country level, although it encompassed 27 nations. The dependent variable consisted of factor scores based on five ICT components. There was concurrence with Japan on importance of age structure and education, but difference in high importance in the EU of culture, technical workforce and unimportance of population density (in contrast to the present urban/rural importance).

Returning to the present conceptual model, the empirical results suggest that the model can be simplified for Japan by dropping the independent factors of prefectural income (retaining household income), technical workers, and the openness index. Prefectural income and technical workers are of minor importance, and can be better addressed by other variables in their respective categories. The openness index does not influence technology, perhaps because the transparency it measures is of reduced impact for ICT in Japan and also because the differences in prefectural societal openness in Japan are more than offset by geographically uniform central government policies. Since the significant correlates across our dependent variables are largely similar for Facebook and Twitter, a future suggestion is to simplify the Japanese conceptual model by solely including Facebook subscribers or an average of the two.

7.7 Case Studies of ICT Use in Japan

As in other chapters, we now examine case examples related to ICT for Japan with the aim of expounding some of the background discussion as well as our own findings about ICT adoption and usage in Japan.

Case Study 1: Construction of FTTH network in a remote, rural, mountainous village in Hokkaido prefecture

Nishiokoppe Village is a small, remote village in northeast Hokkaido prefecture, constituting a large northern island, one of the most remote areas of Japan that is largely forested and mountainous. Nishiokoppe with a population of 1200 inhabitants in 600 households, as reported by Arai and Naganuma (2010), did not have internet service providers in the early period of internet diffusion in Japan. Although this village was at first outside the range of any internet or television, a fully optical-fiber network to the home was constructed to connect all homes and buildings to cable, phone, and internet services. Almost 60 % of the funding came from the national government and the local government of Hokkaido prefecture (Arai and Naganuma 2010), while another 20 % was borne by a national government scheme. Due to this tremendous support, the village did not have to charge individual home owners for construction of the network. It was successfully deployed with services keyed to the rural and elderly population such as agricultural information and elder care. We assume that ICT training must have been provided, since residents are speculated to largely lack ICT skills and experience.

The role of governments in shaping ICT policy, framing regulations, and providing support for ICT diffusion has been discussed as part of building this book's generic conceptual model in Chap. 3. This case example also serves to reinforce prior discussion about methodical step-by-step ICT policy and strategic initiatives of the Japanese government such as the e- and u-Japan strategies to ensure a ubiquitous networked society in Japan. Furthermore, the case exemplifies prior observation and discussion that Japan's mountainous terrain often imposes infrastructure development constraints for remote, rural areas. The Nishickoppe model thus holds promise, not just for rural areas of Japan but similar geographies in other countries, and highlights the critical role that governments can play in developing networked societies.

Case study 2: Kyoto Poultry Traceability System (Setboonsarng, Sakai, and Vancura 2009)

Over the better part of the last two decades, the world has witnessed several outbreaks of avian flu, popularly known as "bird flu". Each outbreak has tremendous economic implications, causes untold inconvenience to poultry farmers, traders, and consumers alike; and disrupts supply chains worldwide.

In 2004, an outbreak of avian flu caused substantial damage to the poultry industry in Kyoto Prefecture that called for better risk management practices to be put in place. Within a couple of years, with support from the prefectural government, a safety promotion council was set up by producers and distributors of poultry in Kyoto. The council created guidelines for better quality management and traceability of poultry products using networked computer systems. All poultry processing sites acquired a PC with appropriate software, printer, and other hardware, and database server initially. Mitsubishi Electric Corporation was contracted to provide internet connectivity and server maintenance. Retail distributors made similar ICT investments with the aim of tracking poultry products by their identification number, barcode, and receiving date. While rudimentary in terms of use of ICTs, such intervention was critical to track poultry products as they moved along the supply chain, reduce the likelihood of further outbreaks, and decrease financial risk to poultry operators. Tracking information was also uploaded regularly to a website for the benefit of consumers who were able to readily access this information using their mobile phones, thus achieving transparency through the entire traceability system and ultimately restoring public trust. ICTs allowed the Kyoto poultry industry to input, access, and share this critical information much more quickly and broadly than a paper-based system could and therefore ICT was a critical element in the efficiency of this traceability system.

This case highlights the synergy between governments and in this case, the private sector, for a public health issue. It also provides an example of fairly low-level ICT usage at very reasonable costs to all involved. The implications apply not just to the poultry industry in rural Kyoto Prefecture, but to developing countries in general that have significant farming and food production supply chains. This case study also provides context to our regressions findings, specifically the association of farm household population to internet and mobile phone use.

7.8 Policy Implications

First we present policy recommendations for the main geographic regions that emerged from K-means cluster analysis; the findings of which were discussed and clusters characterized earlier in this chapter.

1. "*Tokyo*"—The national government ICT policy contains sections found on Tokyo, in particular to continue and build on policies that have put this zone in world leadership in ICT intensity (IT Strategic Headquarter of Japanese Government 2010). This recommendation is reflected in Japan's ICT policy road-map initiative to move forward ICT R&D to world leading levels in areas such as robotics, 3-D video, and voice translation; to grow its digital content market; to advance in geospatial technologies; and to deploy training through IT advisors and course offerings of the Open University of Japan (IT Strategic Headquarters of Japanese Government 2010). We supplement the national ICT policies by the recommendation to transfer portions of both ICT R&D and training from the Tokyo prefecture to the low-tech areas of the country.
2. "*Other metropolitan areas*"—For Cluster 2, we recommend the Japanese government bring this area up to world-class ICT leadership similar to Tokyo by leveraging expertise available in major national universities located in these regions, while putting in incentives for its world-class industrial and services base to provide ICT improvements.
3. "*Rural-urban mountainous region*"—National government ICT policy appropriate for cluster 3 is to continue to reinforce and move forward more on earlier policies of the 2000s to build nationwide infrastructure (Shimizu et al. 2008). This necessitates overcoming the geographical digital divide by penetrating mountainous regions with infrastructure, to improve mobile connectivity, as well as newer broadband and optical fiber technologies (Arai and Naganuma 2010, p. 13). Consistent with Yuguchi (2008), we recommend subsidies be allocated by the Japanese national government to local governments to alleviate construction costs of base stations and laying of optical fiber especially in

instances where a rural village is located far from the exchange office of a telecom carrier firm. We also recommend subsidies be provided by local municipalities to households for an initial period to mitigate consumer perception of initial tariffs as too prohibitive. Improvements in mobile and broadband connectivity are expected to positively impact adoption and utilization of emerging social media technologies.

An example of an early success in cluster 3 is Nishiokoppe Village, as discussed in the previous section. In summary, for Cluster 3 we recommend that a combination of central government subsidies, local level subsidies, and incentives be provided to educational and digital resource providers of the prosperous clusters 1 and 2, requiring them to address ICT needs in the predominantly rural cluster 3.

4. "*Rural Areas*"—The National ICT policy appropriate for these clusters is to invest funds in developing contemporary communications and IT infrastructure, in technology training with emphasis on the adult and older population, and in incentives such as subsidies for companies (IT Strategic Headquarters of Japanese Government 2010). Although the Japanese government and private companies invested in rural ICT infrastructure in the 2000s based on the e-Japan and u-Japan initiatives (Shimizu et al. 2008), major deficits remain as seen by the lowest levels among all clusters of broadband, mobile phone subscribers, Facebook and Twitter (in cluster 4) and fixed phones and PCs (in cluster 5). The central government ICT plan does contain infrastructure initiatives, which might reflect the considerable progress nationally, but does not address the serious infrastructural deficit in these low-tech areas. The policy challenge is more than just increasing access; it should also prioritize to provide address lifelong skills training and education in these geographic areas possibly through offerings of the Open University of Japan. Although the national plan (IT Strategic Headquarters, Japanese Government 2010) calls for "correcting digital divides," "advancing support of IT-aided lifelong learning," and supporting the Open University of Japan's digital degrees, geography of needs is not mentioned. We supplement these national policies by recommending that they focus concertedly on the rural, low-tech areas for ICT skills training and education.

For central government policymaking in Japan, results imply that additional advancement of technology utilization can be spurred by policies emphasizing internet content generation, as represented by the proxy of newspaper circulation; innovation and R&D; and by education. Findings indicate that newer technologies have lower utilization among the more elderly and more rural Japanese prefectures. The study indicates an inverse association of rural location with technology use for over half of ICT dependent variables, affirming necessity of a Japanese government policy to revitalize agricultural and rural areas through ICT applications (IT Strategic Headquarters of Japanese Government 2010) and pointing to policies to provide specialized support for the older population. Since Japan has among the highest proportion of older citizens worldwide, as seen in Fig. 7.14, policies favoring the elderly would have substantial impact on this large population segment.

For the newer technological areas of broadband, mobile phones, Facebook, and Twitter, innovation is a positive determinant; therefore, the central government should consider strengthening even more its long-standing initiatives and investments in innovation (Shimizu et al. 2008), while adding more geographical balance to these programs.

For practitioners, the study demonstrates that GIS and spatial analysis combined with statistics can provide enhanced insight to managers and planners for analysis and decision-making. This is especially pertinent for practitioners in government, utilities, natural resources, transportation, and marketing. For ICT researchers on the digital divide, the Japan study provides novel findings on technology correlates and geography of ICT for the third largest nation in economic size.

7.9 Conclusions

Japan is one of Asia's and the world's leading developed and industrialized nations. It ranks among the top nations worldwide in terms of subscriber base for mobile phones, internet, and broadband, each exceeding or very close to 100 million in 2013–2014. This has been achieved through systematic policy development and strategic planning by the Japanese Government, especially since the early 2000s. However Japan's rapidly aging population is a demographic reality that impedes ICT adoption, while unfamiliarity with the English language poses some challenges to Japanese society for ICT usage. Overall, findings point to a generational digital divide in Japan that has been documented by anthropological and language studies.

Spatial analysis of ICT patterns reveal disparities between ICT adoption and use between Japan's urban and rural areas, with Tokyo being a very-high ICT outlier, much like Beijing and Shanghai in China, and Delhi in India. Low ICT use prefectures are typically situated at the nation's north-east and south-western peripheries. The urban-rural divide is often exacerbated by Japan's dramatic topography in rural mountainous regions. PC, internet, broadband, and mobile subscribers were found to be spatially autocorrelated, but not fixed phones, alluding to its legacy status. Age, income, education, innovation, and infrastructural variables are the factors that were found to be associated with various forms of ICTs in Japanese prefectures; such results are largely consistent with prior literature and other book findings, and also reinforce the importance of these variables in explaining the digital divide in various nations at the sub-national level, as documented in various chapters of this volume. Findings from cluster analysis of ICT use lead to more geographically detailed policy recommendations that expand on Japan's central government policies, such as providing training to the lowest cluster regions, subsidizing national universities in higher tech prefectures to establishing long-term outreach to deprived regions, and incentivizing high-tech companies to increase infrastructure for the newest technologies and skills training in the low-tech, rural areas. Establishment of innovation centers can be a policy goal coordinated with prefectures, relying less on the central government. Two case examples have been provided that reinforce our findings and support policy recommendations. In conclusion, Japan has accomplished great strides in technology for its citizenry, but substantial digital inequities remain; this chapter demonstrates that more needs to be done to bridge Japan's internal digital divide.

References

Agarwal, R., Animesh, A., & Prasad, K. (2009). Social interactions and the "digital divide": Explaining variations in internet use. *Information Systems Research, 202*, 277–294.

Akiyoshi, M., & Ono, H. (2008). The diffusion of mobile internet in Japan. *The Information Society, 24*, 292–303.

Arai, Y., & Naganuma, S. (2010). The geographical digital divide in broadband access and governmental policies in Japan: Three case studies. *Networks and Communication Studies, NETCOM, 241-2*, 7–26.

Azari, R., & Pick, J. B. (2005). Technology and society: Socioeconomic influences on technological sectors for United States counties. *International Journal of Information Management, 251*, 25–37.

Baliamoune-Lutz, M. (2003). An analysis of the determinants and effects of ICT diffusion in developing countries. *Information Technology for Development, 10*, 151–169.

Chinn, M. D., & Fairlie, R. W. (2007). The determinants of the global digital divide: A cross—country analysis of computer and internet penetration. *Oxford Economic Papers, 591*, 16–44.

Dasgupta, S., Lall, S., & Wheeler, D. (2005). Policy reform, economic growth and the digital divide. *Oxford Development Studies, 332*, 229–243.

Grubesic, T. H. (2006). A spatial taxonomy of broadband regions in the United States. *Information Economics and Policy, 18*, 423–448.

Guillén, M. F., & Suárez, S. I. (2005). Explaining the global digital divide: Economic, political and sociological drivers of cross-national internet use. *Social Forces, 842*, 681–708.

Ida, T., & Sakahira, K. (2008). Broadband migration and lock-in effects: Mixed logit model analysis of Japan's high-speed internet access services. *Telecommunications Policy, 329–10*, 615–625.

Ishii, K. (2004). Internet use via mobile phone in Japan. *Telecommunications Policy, 281*, 43–58.

IT Strategic Headquarters of Japanese Government. (2010). The new strategy in information and communications technology IT. Roadmaps. Report of The Strategic Headquarters for the promotion of an advanced information and telecommunications network society. Tokyo, Japan: IT Strategic Headquarters, Prime Minister of Japan and His Cabinet.

ITU. (2012). *World telecommunication/ICT Indicators 2012*. Geneva, Switzerland: International Telecommunication Union.

Kawaguchi, D. (2006). Are computers at home a form of consumption or an investment? A longitudinal analysis for Japan. *Japanese Economic Review, 571*, 69–86.

Kvasny, L., & Keil, M. (2006). The challenges of redressing the digital divide: A tale of two US cities. *Information Systems Journal, 16*, 23–53.

Lengsfeld, H. B. (2011). An econometric analysis of the sociodemographic topology of the digital divide in Europe. *The Information Society, 27*, 141–157.

Moretti, E. (2012). *The new geography of jobs*. Boston, MA: Houghton Mifflin Harcourt.

Nakayama, S. (2002). From PC to mobile internet—overcoming the digital divide in Japan. *Asian Journal of Social Science, 302*, 239–247.

Omura, T. (1997). Japan's stumbling policy for competition in the telecommunications industry. *Telecommunications Policy, 21*(2), 127–141.

Ono, H., & Zavodny, M. (2007). Digital inequality: A five country comparison using microdata. *Social Science Research, 363*, 116–139.

Pick, J. B., & Azari, R. (2008). Global digital divide: Influence of socioeconomic, governmental, and accessibility factors on information technology. *Information Technology for Development, 142*, 91–115.

Pick, J. B., & Azari, R. (2011). A global model of utilization of technology based on governmental, social, economic, and business investment factors. *Journal of Management Information Systems, 281*, 51–85.

Pick. J. B., Sarkar, A., & Johnson, A. (2015). United States digital divide: State level analysis of spatial clustering and multivariate determinants of ICT utilization. *Socio-Economic Planning Sciences, 49*, 16–32.

Pick. J. B., Sarkar, A., & Johnson, A. (2014a). A multivariate and geospatial analysis of ICT utilization and the digital divide in the African nations. In *Proceedings of the Western Decision Sciences Institute Annual Conference, Napa, CA*.

Pick, J. B., Nishida, T., & Sarkar, A. (2014b). Broadband utilization in the indian states: Socioeconomic correlates and geographic aspects. In J. Choudrie & C. Middleton (Eds.), *Management of broadband technology innovation* (pp. 269–296). Oxford, England: Routledge.

Pick, J. B., Nishida, T., & Zhang, X. (2013). Determinants of China's technology utilization and availability, 2006–2009: A spatial analysis. *The Information Society, 29*(1), 26–48.

Robison, K. K., & Crenshaw, E. M. (2002). Post-industrial transformations and cyber-space: a cross-national analysis of internet development. *Social Science Research, 31*, 334–363.

Setboonsarng, S., Sakai, J., & Vancura, L. (2009). Food Safety and ICT Traceability Systems: Lessons from Japan for Developing Countries. ADBI Working Paper 139. Tokyo: Asian Development Bank Institute. http://www.adbi.org/working-paper/2009/05/28/3012.ict.food.traceability.system/. Accessed 15 Oct 2014.

Shimizu, H., Ogawa, K., & Fujinuma, K. (2008). Information and communication technologies policy in Japan: Meeting the challenges ahead, Chapter 2.3. In S. Dutta (Ed.), *The global technology report, 2007–2008* (pp. 07–116). Geneva, Switzerland: World Economic Forum.

Shiraishi, K. (2004). Is the Japanese press a dinosaur in the 21st century? The IT revolution and newspapers in Japan. In M. Kagami, M. Tsuji & E. Giovannetti (Eds.), *Information technology policy and the digital divide* Chapter 12. Cheltenham, UK: Edward Elgar Publishing.

Simon, J. S. (2004). Critical success factors for electronic services: Challenges for developing countries. *Journal of Global Information Technology Management, 72*, 31–53.

Sipior, J. C., Ward, B. T., & Connolly, R. (2011). The digital divide and t-government in the United States: Using the technology acceptance model to understand usage. *European Journal of Information Systems, 20*, 308–328.

Statistics Japan. (2014). *Statistical handbook of Japan*. Tokyo: Statistics Bureau, Ministry of Internal Affairs and Communications.

Takhteyev, Y., Gruzd, A., & Wellman, B. (2012). Geography of Twitter networks. *Social Networks, 34*, 73–81.

UN. (2014). *World urbanization prospects: The 2014 revision, highlights ST/ESA/SER.A/352*. New York: Department of Economic and Social Affairs, Population Division, United Nations.

van Dijk, J. (2009). One Europe, digitally divided. In A. Chadwick & P. N. Howard (Eds.), *The Routledge handbook of internet politics* (pp. 288–304). London, NY: Routledge.

Vicente, M. R., & Lopez, A. J. (2011). Assessing the regional digital divide across the European Union-27. *Telecommunications Policy, 35*, 220–237.

World Bank. (2014). Data, retrieved November 1, 2014 from http://data.worldbank.org/indicator/SP.POP.TOTL.

Yates, D. J., Gulati, G. J., & Weiss, J. W. (2011). Different paths to broadband access: The impact of governance and policy on broadband diffusion in the developed and developing worlds. In *Proceedings of the 44*th *Hawaiian Conference on System Sciences*. Los Alamitos, CA: IEEE Computer Society.

Yuguchi, K. (2008). The digital divide problem: An economic interpretation of the Japanese experience. *Telecommunications Policy, 325*, 340–348.

Chapter 8
United States Digital Divide

8.1 Introduction

The United States, the world's third largest country with a population of approximately 316 million in 2013 (World Bank 2014), is a leading industrialized nation and is classified by the United Nations as "very high" in terms of human development. The US also has an advanced information society with approximately 277 million internet users in early 2014 (Internet World Stats 2014), 139 million fixed phone lines in 2012, in both cases behind China, and 310 million mobile cellular subscribers in 2012, behind India and China. It leads the world in terms of internet hosts with 505 million in 2012 (Central Intelligence Agency 2014).

Per capita use of ICTs in the US however continues to lag behind several other nations. For example, the US ranks 21st worldwide with 84.2 internet users per 100 people in 2013 behind Nordic nations Denmark, Finland, Iceland, Norway and Sweden, European nations such as the Netherlands, Switzerland, and United Kingdom, and other very high human development peers such as Japan, Canada, and South Korea. In 2013, the US had 96 mobile cellular subscriptions, and 28.54 fixed broadband subscribers per 100 population (World Bank 2014), ranking in the middle segment of nations worldwide. US trade in ICTs presents a mixed picture during the 2010–2012 period; while ICT goods exports and imports both decreased from 10.5 to 9 %, and from 14.2 to 12.8 %, ICT services exports increased from 21.1 to 22.1 % during the same period (World Bank 2014). Overall, while the US is widely recognized as an advanced digital nation, it has lagged several Nordic, European, and other industrialized nations in terms of ICT access, use, and skills, as manifest by its rank of seventeenth in terms of ITU's ICT Development Index (IDI) in 2012 (ITU 2013).

As the demographic, social, political, and economic landscape of the US continues to evolve, we focus in this chapter on the adoption and utilization of ICTs among US states. While the state is certainly not the smallest geographic unit for the United States, examining gaps in ICT adoption and usage at the state level is

J.B. Pick and A. Sarkar, *The Global Digital Divides*, Progress in IS,
DOI 10.1007/978-3-662-46602-5_8

important since U.S. states possess the authority to develop their own digital policy and shape ICT adoption, consumption, and usage within their boundaries, whereas counties have less authority. We begin by studying advancement in the American telecommunications sector and then examine the patterns and regional differences in ICT utilization among US states. Next, we present our conceptual model of ICT utilization among US states in which eight dependent ICT availability and utilization factors are posited to be associated with fourteen independent socio-economic, demographic, innovation, social capital, and societal openness factors. One unique aspect of this chapter's conceptual model is explicit consideration of social capital as an independent correlate of ICT dependent variables.

Our confirmatory analysis findings indicate that social capital, education, societal openness, urbanization, and ethnicities are significantly associated with ICT utilization. Following discussion of regression analysis findings, we present cases of ICT use in US states and conclude this chapter with policy implications of our findings and summary remarks.

8.2 Historical Background

The history of telephony in the United States goes back to the 1800s. By the late 1800s, after Bell's initial patent monopoly lapsed, the company encountered stiff competition from thousands of independent telephone companies. As a result of intense competition, ubiquitous telephone infrastructure developed in the US. Since Bell and the independents refused to interconnect with each other's networks, a race ensued towards universal service that would increase penetration and geographic coverage, even in rural areas (Mueller 1993) at reasonable and affordable rates. However by 1919, the US telecommunications underwent significant policy shift from private competition to regulated monopoly (Schiller 1998). 1934 witnessed the founding of the Federal Communications Commission (FCC) and by the 1934 Communications Act, FCC was entrusted with investigating the telephone industry (Solomon 1978).

Fast forward 62 years, when the 1996 Telecommunications Act overhauled the US telecommunications sector (NTIA 1996). The goal of the law was to let any enterprise, be it local telephone companies, long distance providers, or cable companies enter the communications business, and to let any communications business compete in any geographic market against any other (FCC 2014). The 1996 Act, motivated by deregulation was hailed as a landmark and was expected to catalyze a new era of innovation, investment, inclusion, and declining prices of telecom services (NTIA 2014). However actual outcomes in terms of competition and reduced tariffs paid by American consumers fell far short of anticipated outcomes. The telecommunications and cable industries became largely concentrated together, and there were sharp increases in cable and in-state long-distance services, while local phone and inter-state long distance tariffs stagnated. Only a minority group of intense telecom consumers enjoyed price breaks and competitive options.

Thus, some have argued that rather than bridging the digital divide, the 1996 Telecommunications Act ended up exacerbating the gap between information and telecom haves and have-nots (Kimmelman et al. 2006).

Between 1984 and 1997, telephone penetration rates increased in the US from 91.8 to 94.0 (Belinfante 2014). A handful of states in eastern and Midwestern US (Delaware, District of Columbia, Maryland, Maine, and Illinois) reported declining telephone penetration among households during the same period. More states, such as Colorado, Connecticut, Hawaii, Idaho, Kansas, Louisiana, Mississippi, New Jersey, and others witnessed declines between 1997 and 2000. Since 2000, fixed telephone subscriptions have steadily declined from almost 70 per 100 population to 40 per 100 population in 2013, as shown in Fig. 8.1.

During the same timeframe, mobile cellular subscriptions grew strongly (indicating replacement of fixed telephones) as well as internet use per capita. Broadband use has also increased steadily; in the US, mobile broadband connections numbered 120 million in mid-2011 exceeding fixed broadband connections (Prieger 2013). Broadband has been the focus of attention of federal efforts such as the National Broadband Plan, released in 2010 by the FCC during the latter part of the great economic recession in 2010. The Plan outlined a roadmap for initiatives involving broadband that would stimulate economic activity and job creation in areas such as education, healthcare, homeland security, etc. The National Broadband Map website, available at http://www.broadbandmap.gov/ was launched as a tool to search, analyze, and visualize broadband availability geographically, emphasizing the

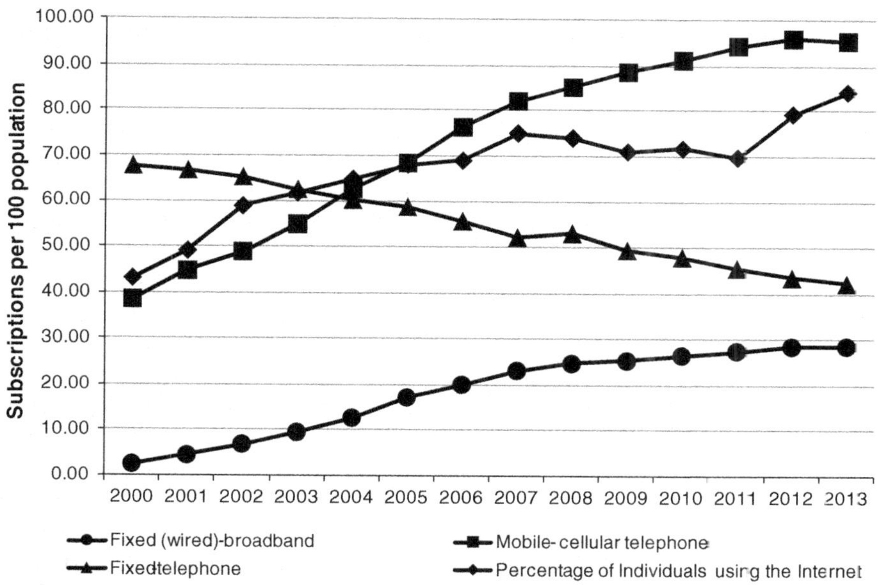

Fig. 8.1 Shift in ICT adoption in the US, 2000–2013 (created with subscription data from ITU)

importance of providing such visualization and analysis capabilities along with robust data to the fingertips of the common citizen.

In the next section, we examine geographic patterns and dimensions of ICT diffusion in the US overall, and also the US states.

8.3 Current Dimensions of Access and Use of ICTs in US States

ICT use in the US, across all forms of technologies has increased over the years, as evident from Fig. 8.1. Starting with a legacy form of ICT, fixed phones, it can be seen that percent of fixed phone use in American households (not subscriptions per 100 population as in Fig. 8.1) did in fact increase slightly between 1997 and 2010; this expansion is indicative of a strong infrastructural foundation that can be leveraged for diffusion of other forms of ICTs. During the same period, use of internet in US households increased four-fold while broadband adoption increased from a mere 4 % in 2000 to 72 % in 2012. Household use of personal computers also doubled between 1997 and 2010. While NTIA (2010) contends that the growth in broadband adoption has occurred among virtually all demographic groups, the rate of growth was uneven among the groups. Broadband adoption expands with increase in family income and education, and varies by race and ethnicity. In 2010, White non-Hispanic and Asian non-Hispanic broadband adoption in households stood at 68.3 and 68.8 % respectively, in comparison to Black non-Hispanic and Hispanic groups in which household adoption rate were 50 and 45 % respectively. Broadband adoption also varies by location, with urban households consistently adopting broadband more than their rural counterparts. The gap in adoption between urban and rural households however decreased from 15 % in 2007 to 11.8 % in 2009 to 10.1 % in 2010. Such evidence has implications for our conceptual model, discussed later in the chapter (Fig. 8.2).

Among US states, broadband leaders in 2010 were almost all states in the Western seaboard and their neighbors including California, Oregon, Washington, Nevada, Arizona, Utah, and Alaska, as well as states in the Boston-Washington DC megalopolis such as Massachusetts, Connecticut, New Hampshire, and Rhode Island. Kansas was a high "outlier" in the central plains (Fig. 8.3). Almost all states that lagged in broadband adoption are in south-central US and include Mississippi, Alabama, Arkansas, Kentucky, Tennessee, each of which have rural percentage population of at least 40 % in 2010; other low broadband adopters are the states of Indiana in the Midwest, and New Mexico in western US. Other midwestern, upper midwestern, and Rocky Mountain States apart from Arizona have moderate, somewhat middling adoption of broadband in 2010. The difference in broadband adoption rate between Utah (highest among 50 states at 79.7 % in 2010) and Mississippi (lowest among 50 states at 51.7 % adoption in 2010) is significant.

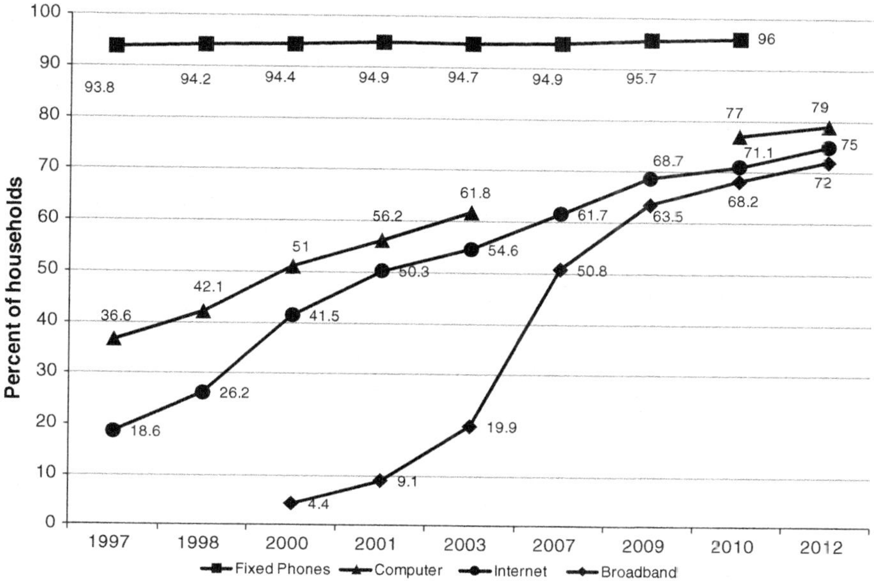

Fig. 8.2 Households with fixed phones, computers, internet, and broadband access, United States, selected years 1997–2012 [recreated and amended with data from NTIA (2010, 2013)]

The geographical patterns of internet use varied for the technology dependent variables. For instance, for Facebook users per capita in 2010 (Fig. 8.4), the highest use states were Washington and Illinois at about 56 %, followed by Georgia, Rhode Island, and Massachusetts at 45–55 %, with the lowest value of 22 % for New Mexico compared to a mean of 38 %, and standard deviation of 7.0. Similar disparity in adoption is also noticed for Twitter (Fig. 8.5).

Lastly, phone usage in US households has been steadily shifting over the past few years (Fig. 8.6). The percentage of households that are wireless-only has been steadily increasing, while the percentage of households that are landline only, or use landline with wireless has been steadily declining since 2009. From a geographic standpoint, adults living in the Midwest (39.6 %), South (41.8 %), and West (39.0 %) were more likely than those living in the Northeast (27.1 %) to be living in households with only wireless telephones (Blumberg and Luke 2013). Figure 8.7 shows a cell-phone user in the high ICT use state of California.

These examples reveal wide technology disparities in ICT adoption among US states and the descriptive mapping shows geographical clustering of states in certain regions, for instance for high levels in New England and low levels in the middle deep South. Fairly wide extent of variation in statewide technological levels is reflected in an average coefficient of variation of 19.1 for the eight ICT dependent variables. Later in this chapter, we examine clustering of states based upon their ICT usage in more detail employing K-means cluster analysis.

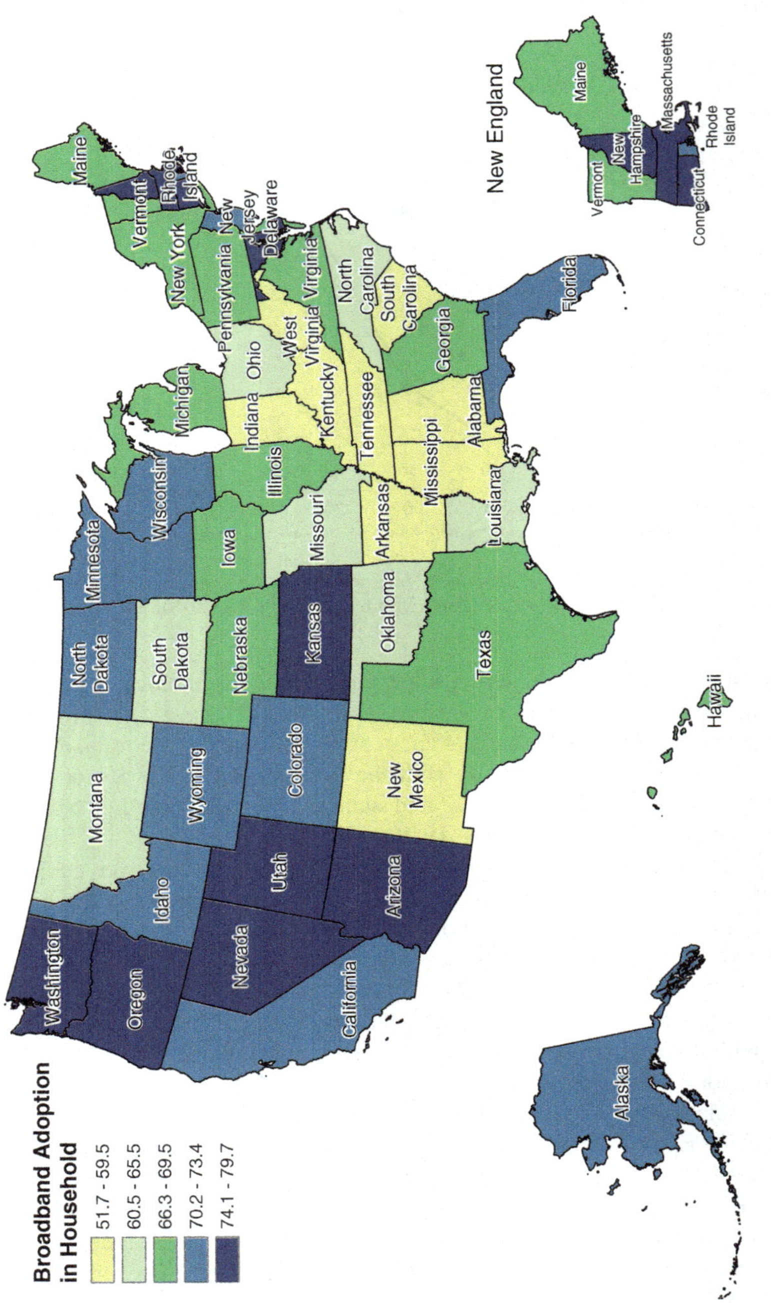

Fig. 8.3 Broadband adoption (subscribers per 100 population) in US States, 2010

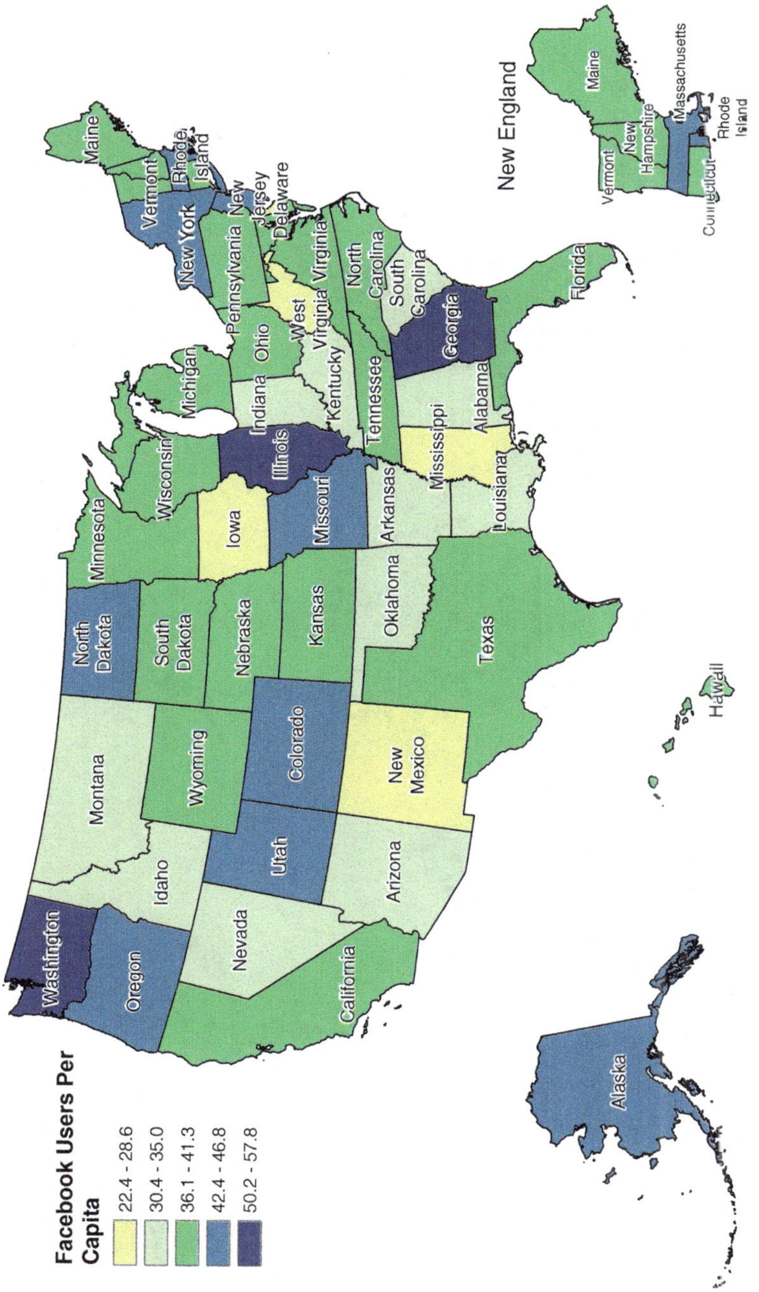

Fig. 8.4 Facebook users per 100 population, United States, 2010

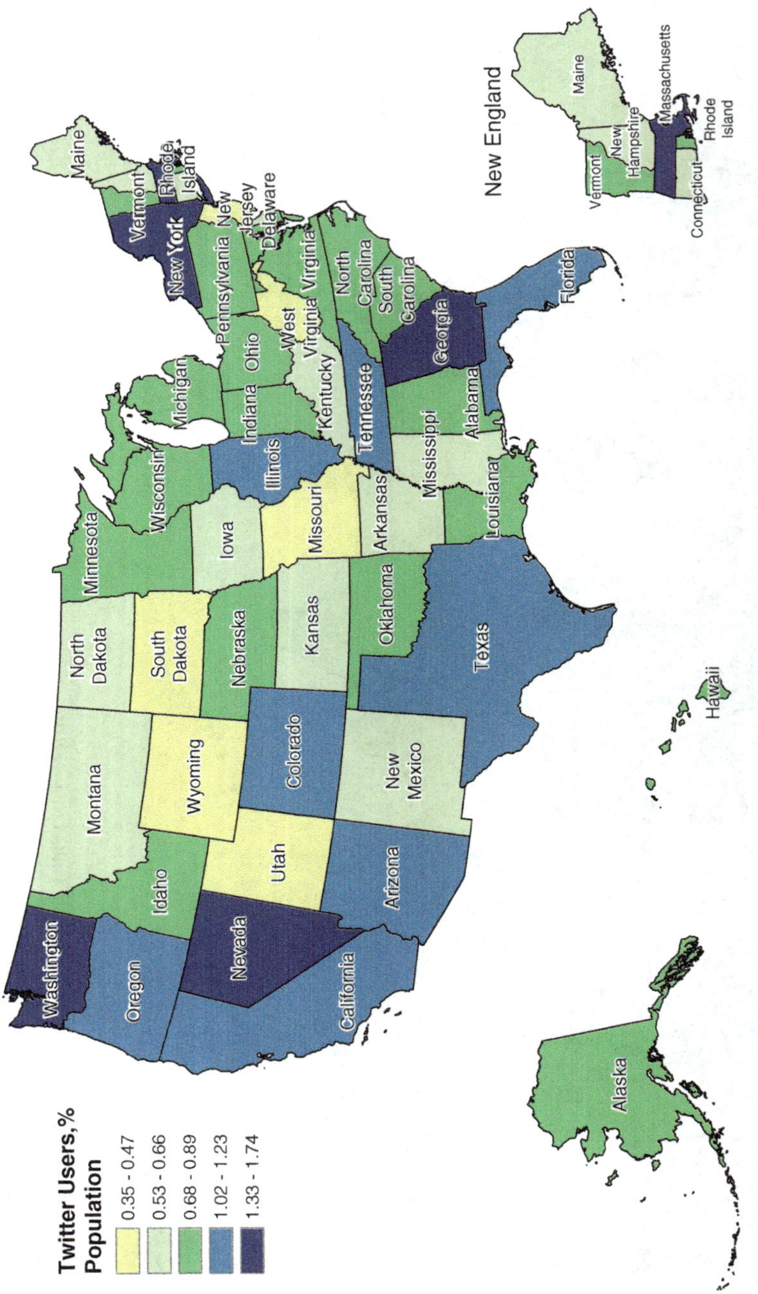

Fig. 8.5 Twitter users, percent of total population, United States, 2010

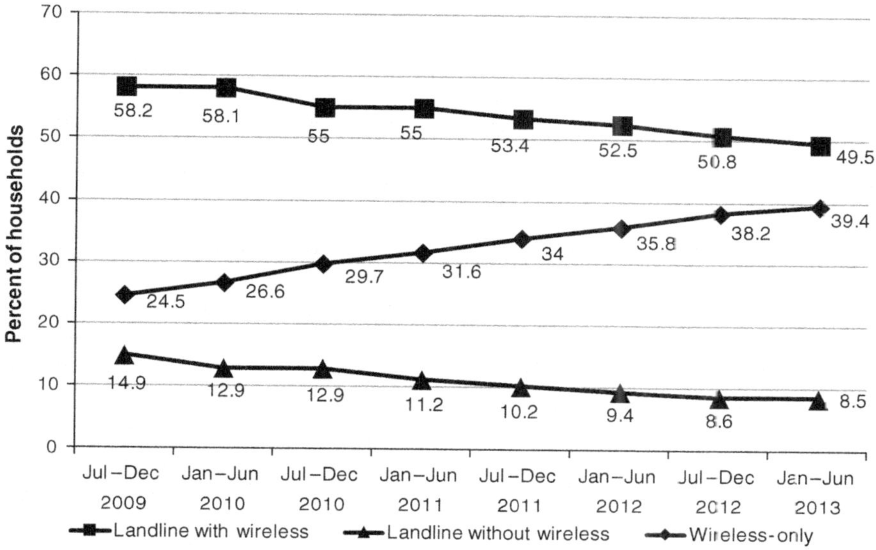

Fig. 8.6 Changing phone use patterns in US households, 2009–13 (data sourced from Blumberg and Luke 2013)

Fig. 8.7 Cell-phone user in California

8.4 Conceptual Model of ICT Utilization in US States

The conceptual model of ICT utilization in US states, depicted in Fig. 8.8 posits eight technology use dependent variables to be associated with fourteen independent correlates that represent demographic, educational, and economic attributes, innovation, societal openness and social capital.

In the book's generic conceptual model of ICT utilization in Chap. 3, the association of demographic variables such as race and ethnicity, extent of urbanization, economic attributes such as income, FDI, and exports, and educational attributes such as level of educational attainment, proportion of students in secondary/tertiary/higher education has been justified. Rationale has also been provided for the inclusion of innovation and societal openness in the generic model.

In the conceptual model for this chapter, we include demographic attributes to account for disparity in technology utilization among various races and ethnicities, as well as urban-rural digital divide that exists in US states. Several US digital divide studies (Chen 2013; Grubesic 2006; Prieger 2013) and well as reports from the National Telecommunications and Information Administration (NTIA 2010, 2013) have documented that urban areas excel in broadband, internet, and other forms of ICT use compared to their rural counterparts. Agarwal et al. (2009) has provided evidence that ICT use in metropolitan areas varies by race and ethnicity. NTIA reports have also pointed out that Black and Hispanic populations tend to use ICTs less than Whites and Asians. We include percent of Asian and Hispanic as independent correlates, and exclude Black because it is highly correlated with Asian (Pearson correlation coefficient of 0.471 at less than 0.001 level).

The association of education with ICT use has been discussed earlier in Chap. 3. For the US, educational attainment has been recognized to be a particularly strong predictor of broadband use (NTIA 2010), and college education has been found to

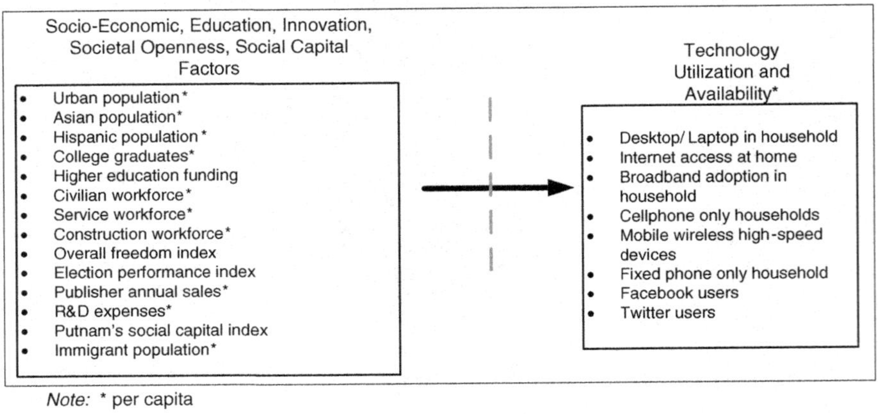

Fig. 8.8 Conceptual model of ICT utilization in US States

be consistently related to internet use (Chen 2013). NTIA (2010) indicated that while almost 85 % of all people aged at least 35 years use broadband at home, the number drops to 51 % for those with a high school diploma in 2010. In a global study (Pick and Azari 2008), public spending on education (as percent of GDP) has been found to be associated with quality of ICT infrastructure. We posit that government funding for higher education will be positively associated with ICT use for the US states.

Chapter 3 has elaborated that free and open societies with effective legal institutions are more likely to exchange information, often digitally, thus enhancing ICT usage. We include direct measures of the societal openness such as the overall freedom index and election performance index for US states. The overall freedom ranking combines personal and economic freedoms and is computed at the state level using scores for fiscal, regulatory, and personal freedom (Sorens and Ruger 2009).

Innovation manifests itself in different forms, and in the digital divide literature, higher R&D expense has been often found to be associated with increased levels of ICT use (Azari and Pick 2005; Pick and Azari 2008; Quibria et al. 2003). For China (Chap. 5), volume of published books has been found to be a significant correlate for ICT use. This has been attributed to the presence of intellectual and innovation centers such as Beijing and Shanghai. Publications such as books, newspapers, magazines, and periodicals tend to generate e-content, websites, and other innovative technologies. Therefore, our conceptual model includes R&D expenditures and publisher annual sales (newspaper, periodical, book, directory) as independent correlates.

Note that the publisher annual sales variable has been posited in this chapter as an independent variable that is not only associated with innovation but also relates to societal openness. The reason is intuitive; citizens in free and open societies are more likely to express and exchange opinions, indulge in public discourse, take positions on matters of social, national, and global interest, and this manifests itself in the form of publications such as newspapers, magazines, books, position papers, periodicals, etc.

Our conceptual model for this chapter is unique for its explicit inclusion of social capital variables. In his seminal book noted political scientist and public policy researcher Robert Putnam has defined social capital to emphasize "not just warm and cuddly feelings, but a wide variety of quite specific benefits that flow from the trust, reciprocity, information, and cooperation associated with social networks" (Putnam 2000). Social capital creates value for the people who are connected and—at least sometimes—for bystanders as well (Bowling Alone 2014). Putnam (2000) provided evidence that Americans have increasingly become disconnected with family, friends, and neighbors. This impoverishes the lives of individual citizens and communities. According to Putnam (2000), the value generated through social networks among homogeneous people, that is, people who share similar ascribed or achieved traits, is bonding social capital and that from heterogeneous people is bridging social capital. Chen (2013) empirically established that bridging social capital is positively associated with internet access in the U.S., but not significantly

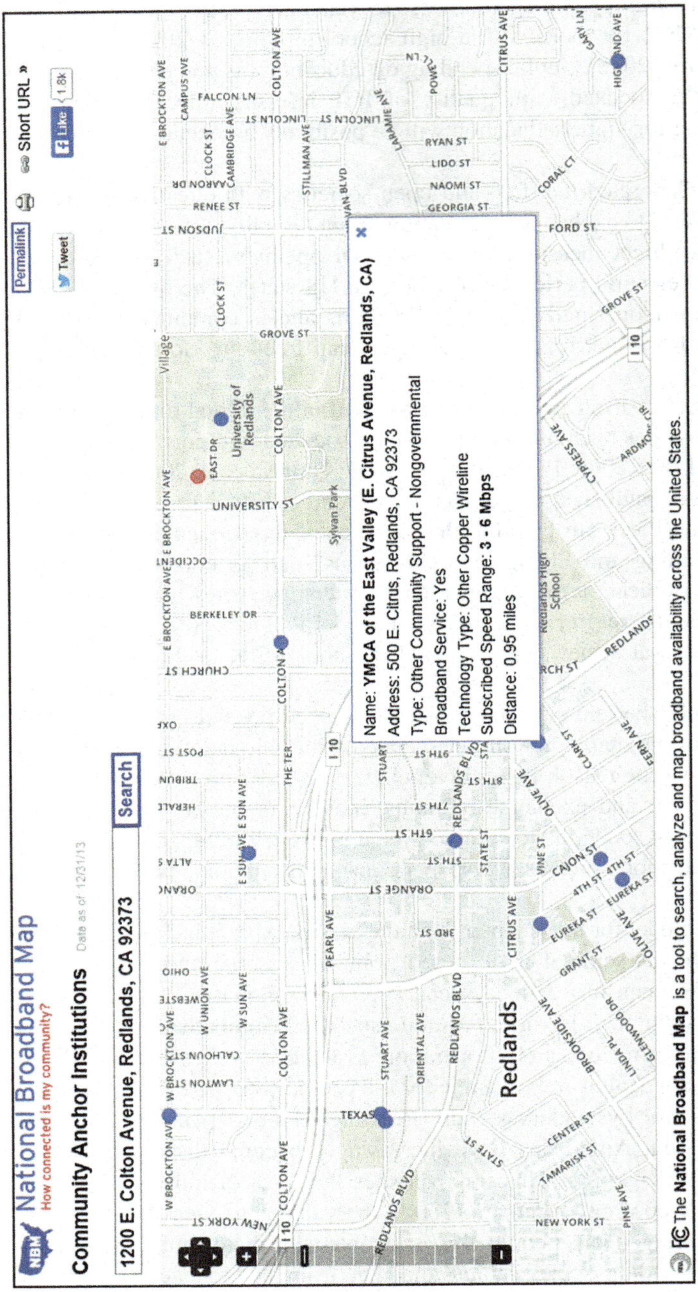

Fig. 8.9 Community Anchor Institutions (*blue dots*) that are closest to the address specified in the search box (*red dot* on map)

associated with internet use, and negatively associated with online communication with close contacts. In contrast, bonding social capital is positively associated with online communication but not with internet access or use. In other words, weak-ties-based bridging social capital helps non-adopters overcome informational, motivational, and technical barriers and gain Internet access. Once this gap is bridged, strong-ties-based bonding social capital takes over and exerts higher peer pressure to maintain social contacts with one's daily contacts. The communal aspect of ICT use is also evident from the National Broadband Map website (http://www.broadbandmap.gov/community-anchor-institutions) in which one can enter an US address and look up the 25 closest community anchor institutions—schools, libraries, medical and healthcare providers, public safety entities, community colleges, and institutions of higher education, and other community support organizations that possess broadband connectivity and might allow an individual to avail of broadband services (National Broadband Map 2014). For an example of the National Broadband Map showing community anchor institutions closest to a user-specified address, see Fig. 8.9.

Research into the association of social capital with ICT use in America is increasingly drawing the attention of digital divide researchers and social scientists; we therefore include Putnam's Social Capital Index as one of two social capital independent variables in our model. A second variable, proportion of immigrant population is also included based upon evidence from the 2000 US Census which showed that the foreign born population was geographically concentrated (refer to maps in Suchan et al. 2007) and the argument that geographically proximate immigrant communities are likely to be cohesive, communal, and thus foster social capital. Note however Chen's (2013) cautionary finding that Internet users, regardless of immigration status, race, employment status, and large income, have no significant differences in online communication with their daily contacts.

8.5 Data Used for This Study

Different sources of data were used to obtain values of various dependent and independent variables. Among independents, variables such as population, employment, workforce, education, income, gross state product, publication estimates such as annual sales revenue of book publishers, and innovation-related variables such as research and development estimates by state, and patents awarded by state were obtained from a variety of federal government sources such as the U.S. Census Bureau's Population Census and American Community Survey (ACS) (U.S. Census Bureau 2007–2009, 2013), National Center for Health Statistics (Blumberg et al. 2012), Federal Communications Commission (FCC 2010) and the National Science Foundation (NSF) (National Science Board 2012). Two of the societal openness variables came from the Mercatus Center of George Mason

Table 8.1 Variable definitions, data sources, and descriptive statistics

	Variable	Definition	Year of data collection	Data source	Mean[a]	SD[a]
Dependent variables						
Personal computers	Desktop/laptop in household	Persons three years or older who live in household with desktop or laptop	2010	CENCPS10	79.99	4.93
Internet users	Internet access at home	Log of persons three years or older who access the Internet at home	2010	CENCPS10	1.82	0.04
Broadband subscribers	BroadbandBroadband adoption in household	Percent of total households with broadband adoption	2010	NTIA	67.76	6.63
Mobile telephone subscribers	Persons in cellphone-only households	Percent of persons 18 years or older living in households with wire-less only phones	2010	NCHS	27.17[b]	6.67[b]
	Mobile wireless high-speed devices	Subscribers to mobile wireless high-speed devices per capita	2008	FCC	20.96	5.38
Fixed phone subscribers	Persons in fixed-phone-only household	Percent of persons 18 years or older living in households with fixed-phones only	2011	NCHS	32.13[b]	7.00[b]
Social media	Facebook users	Facebook users per capita	2010	DCI	38.16	6.98
	Twitter users	Twitter users per capita	2010	DCI	0.80	0.31
Independent variables						
Demographic	Urban	Urban percent of total population	2010	CENPOP10	73.58	14.56
	Asian	Asian percent of total population	2009	CENPOP10	0.04	0.06
	Hispanic	Hispanic/Latino percent of total population	2009	CENPOP10	0.11	0.10
Education	College graduates	Population with any college degree (associate, bachelor, graduate) per population age 25+	2009	CENPOP10	27.42	4.76
	Higher education government funding	Higher education support ($1000s)	2008	SHEEO	0.29	0.10

(continued)

Table 8.1 (continued)

	Variable	Definition	Year of data collection	Data source	Mean[a]	SD[a]
Economic	Employed civilian workforce	Total employment in civilian occupations per population 16+ in labor force	2010	CENPOP10	59.41	4.93
	Service occupation employment	Total employment in service occupations per population 16+ in labor force	2009	CENPOP10	17.48	1.81
	Construction industry employment	Total employment in construction occupations per population 16+ in labor force	2009	CENPOP10	7.05	0.93
Innovation/ societal openness	Overall freedom index	Overall freedom index	2009	MERCATUS	151.31	41.05
	Election performance index	Pew index measuring state elections performance	2008	PEW	66.66	8.03
	Publisher annual sales (newspaper, periodical, book, directory)	Sales revenues of newspaper, periodical, book, and directory publishers/annual sales revenues	2007	CENECON07	2206.92	3937.32
Innovation	R&D expenditures	R&D expenditures/gross state product	2008	NSF	1062.93	810.32
Social capital	Putnam's social capital index	Robert Putnam's Social Capital Index	2000	Putnam	1.45[c]	0.78[c]
	Immigrant population	Immigrant population per capita	2009	CENPOP10	0.09	0.06

[a]n = 50
[b]n = 46
[c]n = 48
DCI Digital America Report, DCI Group; *FCC* Federal Communications Commission; *MERCATUS* Mercatus Center, George Mason University; *NCHS* National Center for Health Statistics; *NTIA* National Telecommunications and Information Administration; *NSF* National Science Foundation; *PEW* Pew Research Center; *CENCPS10* US Census; *CENPOP10* US Census; *SHEEO* State Higher Education Exec Officers; Putnam = Robert Putnam, *Bowling Alone; CENECON07 US Economic Census 2007*

University (Sorens and Ruger 2009) and the Pew Research Center (Pew Charitable Trusts 2013). Most independent data were obtained for the years 2008–2010, although in a couple of instances the sources date back to 2006, to ensure completeness of data for variables such as gross state product and sales revenue estimates of newspapers and periodicals.

On the dependent variable side, in addition to federal sources just mentioned, data were gathered from independent sources such as Internet World Stats (2013) and the DCI Group for Facebook and Twitter Users (DCI Group 2001). Most dependent variable estimates were from the year 2010 or 2011; however, for mobile wireless devices and subscribers, recent complete data were for 2008. Most variables were re-estimated on a per capita basis (percentages or per 100 population) and descriptive statistics of all variables were calculated. Variable definitions, sources, and descriptive statistics appear in Table 8.1.

8.6 Spatial Arrangement of Technologies Within United States, 2010

The presence of agglomeration of states for the eight ICT variables, measured by the Moran's I statistic, is summarized in Table 8.2. Findings show that all the dependent variables have highly significant spatial autocorrelation except Twitter and Facebook, for which geographic distributions are depicted in Figs. 8.3 and 8.4 respectively. We reason the lack of spatial autocorrelation for the social media variables may be due to their appeal to a more youthful base of users/consumers that is not limited or bound in its social media use to ICT-intensive geographies.

The level of spatial autocorrelation for the U.S. can be compared to levels for China and Japan. The similar agglomeration study of ICT in China in 2009 (see Chap. 5) reveals the level of spatial autocorrelation in China is three quarters that of the U.S., and half the U.S. level if the municipalities of Beijing and Shanghai are excluded. The reason the U.S. is more agglomerated is unknown, but we speculate it may be due to China's relatively more centrally planned economy and urban residential and migration restrictions which might lower the potential for technological agglomeration of provinces. In China, out of eight technology factors, web page volume is least clustered, which may reflect less government control over web pages, which tend to concentrate virtually, rather than geographically. As seen in Chap. 7 for Japan, similar testing of agglomeration by Moran's I indicates mostly high agglomeration with a different set of randomly distributed variables.

For broadband in the U.S., at the zip code level there is a high level of agglomeration, due in part to socio-economic and demographic differences (Grubesic 2006). When broadband providers were mapped by zip code, they were predominantly located in the large metropolitan areas, a finding confirmed by very high Moran's I values (Grubesic 2006, Fig. 3).

Table 8.2 Spatial autocorrelation of dependent variables, U.S. 2010 and China 2009, as measured by Moran's I Statistic

Moran's I values for United States, N = 50, 2010									
	Desktop/ laptop in household	Log of internet access at home	Broadband adoption in household	Persons in cellphone-only household	Mobile wireless high-speed devices	Persons in fixed-phone-only household	Facebook users	Twitter users	Average
	0.547^{c}	0.471^{c}	0.457^{c}	0.621^{c}	0.230^{c}	0.648^{c}	−0.004	0.069	0.381
Moran's I values for China, N = 31, 2009									
PCs per 100 urban families	PCs per 100 rural families	Internet users per 100 pop.	Broadband subscribers per 100 pop.	Mobile telephone subscribers per 100 pop.		Urban fixed phone subscribers per capita	Number of domain names per 100 pop.	Number of web pages per Capita	Average
0.346^{c}	0.206^{a}	0.264^{b}	0.272^{b}	0.205^{a}		0.252^{b}	0.615^{c}	0.086^{a}	0.285
Moran's I values for China, N = 29, Excluding Beijing and Shanghai, 2009									
0.343^{c}	0.239^{b}	0.137	0.258^{b}	0.061		0.143	0.236^{b}	0.177^{a}	0.199

Source for China: Pick et al. (2013)
[a]= Signif. at 0.05
[b]= Signif. at 0.01
[c]= Signif. at 0.001

8.6.1 Cluster Analysis Findings

K-means cluster analysis reveals distinctive geographical clustering of states, based on all eight dependent variables. K-means selection of 4 clusters is mapped in Fig. 8.10, with cluster characteristics (see Table 8.3) summarized as follows:

Cluster 1. "Northeast, California, Hawaii, Alaska." Technology levels are high and resemble Cluster 2. Most states are in the Boston-Washington megalopolis and California, which are regarded politically as "blue states."

Cluster 2. "Western, Sunbelt Cluster." This highest cluster overall in ICT access use is similar to Cluster 1, but higher in cell-phone-only, fixed-phone-only households, and Twitter use. It comprises twenty states mostly in the Rocky Mountain region, and some in the upper and western Midwest, as well as Georgia, and the Sunbelt states of Arizona, Texas, and Florida, while only Maine is in the East. The states tend to be large in land area. Figure 8.11 shows an example of technology use in a state in this cluster—researchers using laptops on the sidelines of a conference in Tampa, Florida.

Cluster 3. "Selected non-Metropolitan." Technology access/use levels are intermediate. Most of the states are more rural than urban, and are low to medium in their proportion metropolitan.

Cluster 4. "Middle to Far South, Indiana, New Mexico." The cluster has the lowest technology access/use levels, with broadband adoption in 57 % of homes, and computers in 72 % of households. These states have lower educational and income levels, and few large metropolitan areas.

Overall, the four clusters identified are mostly agglomerated as geographical regions. This is consistent with Tobler's Law, which states that geographic units that are similar in characteristics will tend to be in proximate locations (see Chap. 1). The higher technological clusters (2 and 3) tend to be in the Northeast, prosperous Sunbelt states, Pacific Northwest, and Rocky Mountain areas, while the lower technology clusters (1 and 4) are in more rural, mostly interior parts of the country.

The present U.S. findings on agglomerated geographical areas of technology utilization, although partly supported by prior studies in the literature, reveal new insights. For the U.S. the clusters are somewhat even in size, ranging from 8 to 20 states each, whereas in China (Pick et al. 2013) and India (Pick et al. 2014), there were unique 1-unit clusters for the highest-technology provinces/municipalities/states. For China as discussed in Chap. 5, the Beijing and Shanghai municipalities stood out as unique clusters, with technology levels 3-fold to 20-fold higher than for the lowest clusters. For India, Delhi was a one-state "cluster" with technology levels 4-times to 14-times higher than states located in the north central/northeast part of the country belonging to the lowest technology cluster, as discussed in Chap. 6. In contrast, for the U.S., ratios of highest-tech to lowest-tech clusters vary from 1.2-fold to 1.7-fold. The difference is partially due to Beijing and Shanghai being heavily metropolitan municipalities. More important as a cause is that the U.S. states are generally more even in ICT use compared to the Chinese provinces and Indian states

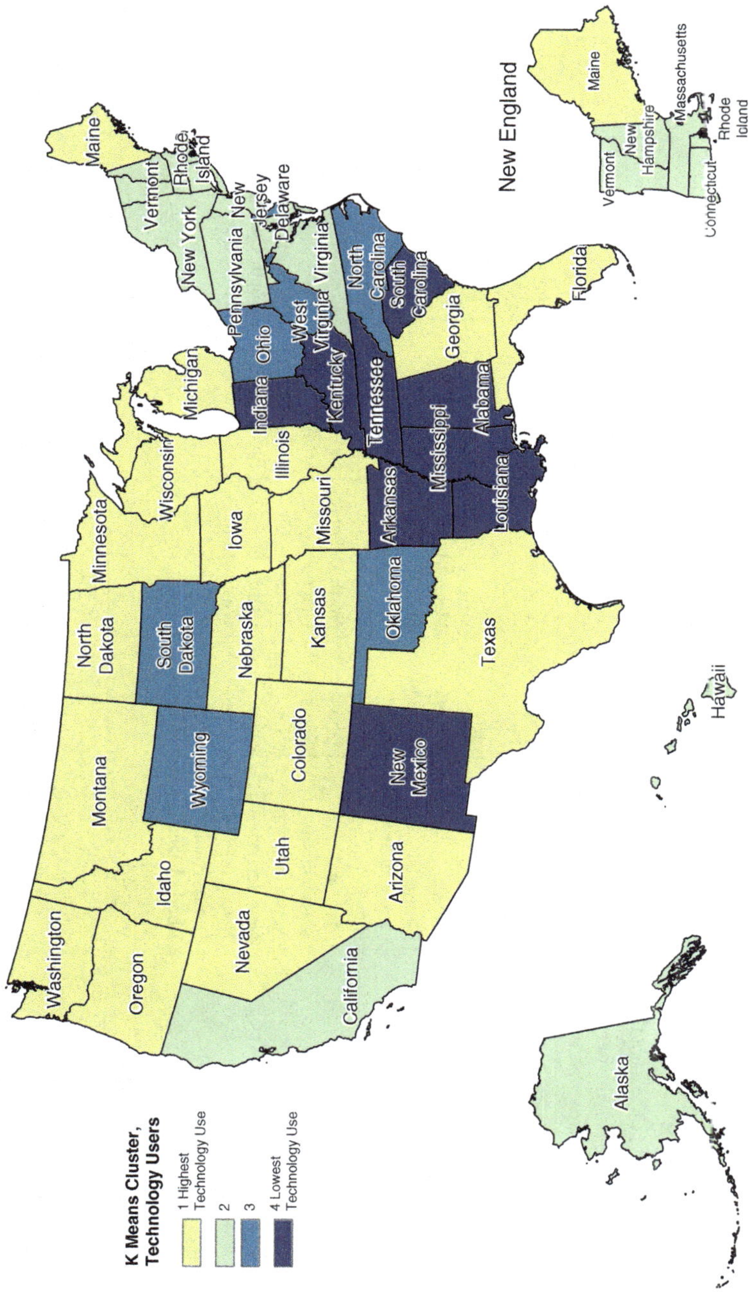

Fig. 8.10 K-means cluster of ICTs in US States, 2010

Table 8.3 K-means cluster composition and characteristics for US, 2010

ICT dependent variable	Cluster 1	Cluster 2	Cluster 3	Cluster 4	Ratio of highest to lowest
Desktop/laptop in household	83.30	82.20	78.30	71.90	1.16
Internet access at home	69.00	68.10	63.30	56.70	1.22
Broadband adoption in household	72.12	71.20	64.16	57.06	1.26
Persons in cellphone-only household	18.76	30.90	26.18	31.89	1.70
Mobile wireless high-speed devices	23.60	22.5	17.20	17.00	1.39
Persons in fixed-phone-only household	23.43	36.15	31.22	36.54	1.56
Facebook users	40.80	41.90	32.40	31.10	1.35
Twitter users	0.82	0.89	0.64	0.72	1.39
	Alaska California Connecticut Hawaii Maryland Massachusetts New Hampshire New Jersey New York Pennsylvania Rhode Island Vermont Virginia	Arizona Colorado Florida Georgia Idaho Illinois Iowa Kansas Maine Michigan Minnesota Missouri Montana Nebraska Nevada North Dakota Oregon Texas Utah Washington Wisconsin	Delaware North Carolina Ohio Oklahoma South Dakota West Virginia Wyoming	Alabama Arkansas Indiana Kentucky Louisiana Mississippi New Mexico South Carolina Tennessee	

Fig. 8.11 Researchers using laptops on the sidelines of a conference in Tampa, Florida

and is a commentary on the more even spatial pattern in the evolution of ICT adoption- and diffusion-related policies in the US in contrast to China and India.

These comparisons also may be partially explained by the fact that unique metropolitan high-tech states and provinces in China and India skew the high-low ratios more than for the U.S. states, even though the U.S. states overall are more agglomerated as shown earlier. Greater U.S. technological evenness might be due to fewer economic barriers in provision of technological capacities among U.S. states, whereas in China the distribution of ICT capacities is centrally planned and controlled, yielding very high emphasis on its two most important metropolises. In India, we reason that Delhi as the national capital and a locus of political and economic activity has received higher per capita technology investment and infrastructure.

8.7 Determinants of ICT Utilization in US States

The conceptual model testing by OLS regression indicates that societal openness and urban factors have the greatest association with ICT dependent variables for the U.S. For the full sample of 50 states, as seen in Table 8.4, social capital variables are significantly associated with five technology factors, while societal openness variables are associated with the two factors of cell-phone-only and fixed-phone-only households. Education is related to desktop/laptop in household, internet access, and broadband, and inversely related to cellphone-only household and Twitter.

Table 8.4 Standardized regression results, 50 states, 2010

		Desktop/ laptop in household	Log of internet access at home	Broadband adoption in household	Persons in cell-phone only household	Mobile wireless high-speed devices	Persons in fixed phones-only household	Facebook users	Twitter users
Demographic	Urban	0.415^{**}	0.455^{***}	0.524^{***}					
Ethnicity	Asian					0.488^{***}			
Ethnicity	Hispanic	−0.254	-0.329^{**}				0.245^{*}	-435^{*}	
Education	College graduates	0.295^{*}	0.311^{**}	0.279^{**}	-0.354^{**}		−0.202	300	-0.265^{*}
Education	Higher ed government funding				0.247^{*}		0.189		−0.207
Economic	Employed civilian workforce								
Economic	Service occupation employment					−0.261			
Economic	Construction industry employment								
Societal openness	Overall freedom index				0.440^{***}		0.485^{***}		0.224
Societal openness	Election performance index								
Innovation/ societal openness	Publisher annual sales (newspaper, periodical, book, directory)								
Innovation	R&D expenditures						-0.301^{*}		
Social capital	Putnam's social capital index	0.493^{***}	0.486^{***}	0.446^{***}				0.321^{*}	

(continued)

Table 8.4 (continued)

		Desktop/ laptop in household	Log of internet access at home	Broadband adoption in household	Persons in cell-phone only household	Mobile wireless high-speed devices	Persons in fixed phones-only household	Facebook users	Twitter users
Social capital	Immigrant population							0.487*	0.764***
Regression adjusted R squared and significance level		0.515***	0.562***	0.608***	0.422***	0.163**	0.494***	0.294***	0.372***
Sample size		50	50	50	50	50	50	50	50
OLS regression tests									
Joint wald statistic		74.326***	88.058***	139.435***	66.681***	15.443***	89.609***	39.005***	10.834*
Koenker (BP) statistic		7.636	6.288	2.712	2.439	0.995	4.164	2.265	5.498
Jarque-Bera statistic		1.644	1.620	1.042	0.596	1.036	0.078	5.125	3.401
Spatial autocorrelation of residuals									
Moran's index		0.410***	0.288***	0.236**	0.042	0.289***	−0.016	−0.010	0.070

*Signif. at 0.05
**Signif. at 0.01
***Signif. at 0.001

Asian ethnicity is strongly linked to mobile wireless high-speed devices, with the caveat that the model does not account for geographic agglomeration, while Hispanic is related inversely to internet access and Facebook, and positively linked to fixed phones. Other significant factors are urban, significant for three of eight technology factors (desktop/laptop, internet access, and broadband) and inverse for fixed-phone only households.

Findings meet the OLS diagnostic tests entirely for all of eight regressions, while Moran's Index is significant for desktop/laptop, internet access, broadband, and mobile wireless high-speed devices, signaling that the model does not account for geographical agglomeration.

In studies of China and India in Chaps. 5 and 6, populous metropolitan states/provinces had very large impact on findings; hence another regression is conducted that excludes the eight U.S. states with 90 % or higher urban population. For the less-urban sample (results in Table 8.5), social capital is again a significant correlate for 4 of 8 technology factors, while in the societal openness category, freedom index is associated with fixed phones and election performance index with Facebook users.

A striking difference in the less-urban sample is that in the demographic category the effect of ethnicity is more profound than urban location, which disappears. Ethnicity's greater impact for the less-urban sample is unexplained and points to need for further research. Directionally, ethnicity is positive for Asian and mixed for Hispanic. Findings are identical to the full sample, except for the addition of an inverse association for fixed phones. The OLS diagnostic tests and Moran's I tests for this sample indicate no problems. In short, the testing that excluded highly urban states yields somewhat consistent findings to the entire sample, a result contrary to the profound result differences for China and India when heavily metropolitan provinces/states were excluded (Pick et al. 2013, 2014).

8.7.1 Discussion on Findings of Regression Analysis

The present results on the digital divide in the U.S. reveal that the major variable groups associated with higher ICT levels are social capital, societal openness and ethnicity.

Social capital

Social capital dominates as a correlate, relating to desktop/laptop in household, internet access at home, broadband adoption in household, and Twitter users. This finding corresponds to a limited set of empirical studies. In a U.S. study of very large surveys of individuals, bonding social capital (i.e. social capital with strong personal ties) to access resources was moderately associated with internet access, internet use intensity, and intensity of online communication (Chen 2013), while in another U.S. survey study, size of interpersonal networks was associated with strength of web use, e-mail use, and chat (Zhao 2006). For Chinese high school students, peer influence had a strong influence on enjoyment and curiosity factors,

Table 8.5 Standardized regression result, 42 less urban states, 2010

		Desktop/ laptop in household	Log of internet access at home	Broadband adoption in household	Persons in cell-phone only household	Mobile wireless high-speed devices	Persons in fixed phones-only household	Facebook users	Twitter users
Demographic	Urban		0.257						
Ethnicity	Asian	0.417^{***}	0.288^{*}	0.235		0.513^{***}		0.635^{***}	
Ethnicity	Hispanic	-0.268^{*}	-0.416^{***}		0.353^{**}		0.377^{**}		
Education	College graduates	0.314^{**}	0.303^{**}	0.296^{**}	-0.464^{***}		-0.405^{**}		-0.373^{*}
Education	Higher ed government funding								
Economic	Employed civilian workforce								
Economic	Service occupation employment								
Economic	Construction industry employment								
Societal openness	Overall freedom index				0.225		0.248^{*}		
Societal openness	Election performance index							0.341^{*}	
Innovation/ societal openness	Publisher annual sales (newspaper, periodical, book, directory)								
Innovation	R&D expenditures				-0.340^{*}		-0.394^{*}	-0.306	
Social capital	Putnam's social capital index	0.500^{***}	0.513^{***}	0.491^{***}					

(continued)

Table 8.5 (continued)

		Desktop/ laptop in household	Log of internet access at home	Broadband adoption in household	Persons in cell-phone only household	Mobile wireless high-speed devices	Persons in fixed phones-only household	Facebook users	Twitter users
Social capital	Immigrant population								0.606^{***}
Regression adjusted R squared and significance level		0.614^{***}	0.644^{***}	0.647^{***}	0.509^{***}	0.244^{***}	0.512^{***}	0.315^{***}	0.254^{***}
Sample size		42		42	42	42	42	42	42
OLS regression tests									
Joint wald statistic		81.618^{***}	117.745^{***}	66.745^{***}	58.003^{***}	14.652^{***}	47.184^{***}	25.287^{***}	14.828^{***}
Koenker (BP) statistic		8.486	9.099	4.637	2.890	0.002	2.153	6.051	4.423
Jarque-bera statistic		1.085	1.378	0.554	1.009	0.629	1.077	1.179	1.063
Spatial autocorrelation of residuals									
Moran's index		0.182	0.259	−0.083	−0.033	0.147	−0.011	0.021	−0.002

*Signif. at 0.05
**Signif. at 0.01
***Signif. at 0.001

which in turn influenced Internet use (Zhao et al. 2011). These technology effects are explained by Chen (2013) as follows. For weaker social capital ties, there is advantage to being exposed to diverse contacts to build cultural capital and develop flexibility, while strong social capital ties "facilitate people's socialization into the technoculture" (Chen 2013; referring to Selwyn 2004).

Societal openness

Societal openness variables have been included in studies at the national level (Robison and Crenshaw 2002; Baliamoune-Lutz 2003; Guillen and Suarez 2005; Pick and Azari 2011), although findings are mixed. For a sample of developing nations, Baliamoune-Lutz hypothesized that social openness and democracy impact diffusion of mobile phones, internet hosts, internet use and PC use, but the findings supported only limited, mixed effects for civil liberties index and political rights index on internet use. Another study (Guillén and Suárez 2005), tested a sample of mostly developed nations, demonstrating only slight positive effects for democracy.

In a more convincing study mentioned earlier, "political openness" was hypothesized and empirically proven to foster more acceptance of technology (Robison and Crenshaw 2002). Correspondingly, for a world sample and for developed nations (Pick and Azari 2011), freedom of press, property rights, and government ICT prioritization strongly influenced a socio-economic level factor, which in turn influenced technology utilization. Accordingly, our societal openness finding corresponds to several literature studies for developed countries.

For the developing nations of China and India, societal openness variables are not available to directly measure such dimensions as freedom, democracy, openness of elections, and judicial independence. In European countries in 2008–2009, judicial independence stood out as the dominant technology determinant (Pick and Nishida 2015). Likewise, in a case study analysis of Denmark, its law on personal information protection and its Data Protection Agency were pinpointed as enabling use of ICT (Igari 2014).

Urban location

Urban population extent is positively correlated with broadband, mobile wireless, and Facebook usage. This finding is in accord with NTIA (2010), for which an urban-rural differential exists for broadband adoption in households. In 2011, urban household residents exceeded their rural peers in terms of broadband usage by 10 %. The urban bias in broadband use is also supported by Grubesic (2004, 2006) and more recently by Weiner et al. (2012) and Prieger (2013). Interestingly, Prieger (2013) pointed out that if the most inclusive definition of broadband (200 kbps upload or download speed) provided by the Federal Communications Commission (FCC) is adhered to, the urban/rural divide, as measured by number of broadband providers ceases to exist. However Prieger (2013) also observed that an urban-rural divide begins to appear when higher speed broadband (at least 3 mbps download and 768 kbps upload, specified by the National Broadband Plan) is considered. The implication is that urban usage of broadband is possibly spurred by the elevated quality of service as measured by higher speeds of uploading and downloading.

Grubesic (2004) had contended that at a very basic level more populous urban areas provide a critical mass of users to broadband service providers. Residents in urban areas, characterized by higher levels of income, median home value, and education generate a significant demand for broadband services. A subtle aspect is that business demand for broadband is elevated in large urban areas (Grubesic 2004). In other words, broadband demand and infrastructure densities tend to be higher in metropolitan areas due to the presence of business districts. This allows broadband service providers to "spill over" their services to adjacent residential areas thereby extending urban access of broadband.

Percent of urban population is not linked to social media use, even though a recent study determined that the use of Twitter is concentrated in large urban areas in the U.S. (Takhteyev et al. 2012). The study identified 25 "tweet" clusters worldwide of which 6 among the top 10, and 15 out of the 25 clusters are large U.S. cities such as New York, Los Angeles, San Francisco, New Jersey, Chicago, Washington, DC. The inconsistency of our findings with Takhteyev et al. (2012) underscores the need for further empirical testing of the relationships between social media forms of ICT and urbanization.

Ethnicity

The positive ethnic correlations of Hispanic ethnicity on fixed-phone and cell-phone only in the household correspond to findings for a national sample of U.S. counties that percent Latino is related to both receipts and payrolls for motion-picture/sound (Azari and Pick 2005). The strong relationship of Asian ethnicity to mobile wireless high-speed devices is supported by a Pew Foundation finding that mobile wireless connectivity of Asian Americans is 77 %, 20 % higher than for the total U.S. population (cited by Nielsen Company 2012). The negative association of internet access with Hispanics is supported by Tolbert et al. (2007) and NTIA (2011) which determined that Hispanic households lagged behind Asian non-Hispanics and White-non-Hispanics although they were marginally better than Blacks. According to the same report, Asian households exhibited the highest rates of home computer ownership (86 %) and internet access (79 % when dialup and broadband are combined) among all race and ethnic groups thus supporting our finding of the positive association of Asians with computers in the household. It is worthwhile to note that computer usage in Asian households remains almost the same for urban and rural areas (86 % vs. 85 %); yet, significant positive association was found in the less urban sample, not the full sample of 50 states.

Our ethnicity findings also correspond to studies of the epicenters of creativity and innovation in the U.S. (Florida 2012). While Asians comprise nearly one half on the creative class in the U.S., followed by whites (almost one-third), Hispanics constitute only 18 % (Florida 2012). At the other end of the spectrum, the working class (in other words, blue collar workers) is comprised of nearly 40 % Hispanics, while ethnic groups are more evenly divided in the service class (Florida 2012). The implication is that increased levels of creativity and innovation catalyze the use of computers and internet, especially broadband. Therefore, the present positive and negative associations of Asians and Hispanics respectively with general utilization of ICTs can be inferred from Florida's study.

Europe has multiple ethnic differences, which so far have been little studied regarding digital divide. A study of 162 small EU regions in 27 nations did find that Nordic and Native-English-speaking regions were correlated with higher intensity of an ICT factor, results demonstrating that certain ethnicities can be influential for ICT utilization in the European context.

Education and R&D

Other significant correlates for ICT factors are education and R&D. As mentioned earlier, literature has consistently supported the relation of education to levels of ICTs, for national, state/provincial, and individual units of analysis. Regarding sub-national investigations of Indian states in Chap. 6, Chinese provinces in Chap. 5, U.S. counties (Azari and Pick 2005), and economic units within European nations (Vicente and Lopez 2011), higher education generally was significantly associated with ICT measures, a relationship ascribed to greater need of educated people to use information in their work and personal lives; their greater income to afford technology; and their better English-language skills advantageous for technology use.

The present inverse relationship of education with cellphone-only households is surprising. We speculate it reflects that less-educated households are more prevalent per capita in rural and remote states in the U.S., where cellphone-only households are more commonplace, since fixed phone services from providers is less available, aging, and/or unreliable. For less-urban states, the inverse effect with fixed-phone only households might reflect generally less fixed-phone use among the less educated households away from large cities. Although outside the scope of this book, future research might investigate further the relationships of educational levels and comparative phone rates in U.S. metropolitan, urban, and rural areas.

The finding that greater R&D is associated with reduced number of persons in fixed-phone-only households can be explained that in a state with high R&D, fixed phones have been displaced by cell phones. This is consistent with research that R&D and innovation are related to higher technology utilization including studies for U.S. counties (Pick and Azari 2005), nations worldwide (Quibria et al. 2003), European sub-national regions (Vicente and López 2011), and Japanese prefectures (Nishida and Pick 2014).

Regarding the correlates of social media variables, for Facebook, social capital attributes were most important, along with ethnicities (positive for Asian and inverse for Hispanic), while for Twitter, immigrant population was dominant, and college education was inverse. The positive relationships with social capital correspond to an investigation of Chinese high school students in which the social capital factor of peer influence was most important in exploratory internet use (Zhao et al. 2011), but differ from Chap. 7 findings for Facebook and Twitter subscribers in Japanese prefectures, for which urban status, education, and innovation were significant. For ethnicities, studies have found Asian population generally has higher rates of use of technologies, while Hispanic population sometimes lags in use (NTIA 2011). Social media variables are new to digital divide research, so the present findings call for further detailed studies.

Returning to this chapter's U.S. conceptual model, it appears largely robust for the empirical results. For the labor force factors, findings suggest that service labor force is important, while construction workforce is insignificant and could be dropped. This reflects that the 21st century U.S. economy is much more a service and knowledge economy. Likewise, newspaper, magazine and book publishers, important proxies for innovation in prior studies, could be dropped and replaced by an indicator of electronic content.

The model's exploratory spatial analysis shows distinctive clustering of states to be evident for the U.S., a finding further confirmed by positive, significant measures of spatial autocorrelation for the technology variables. The model's socio-economic factors are strongly supported by OLS regression analysis tests, both for the entire country sample (N = 50) and less-urban subsample (N = 42). Socio-economic categories especially associated with the dependent variables are societal openness and demographic, which are supported by empirical findings in the literature, particularly for developed nations. Overall, spatial influences are partly accounted for by the model, as seen by the random spatial autocorrelation of the residuals for half of the variables. An example of a non-random spatial distribution of regression residuals (significantly clustered) for dependent variable, desktop/laptop in households, found in Fig. 8.12, reinforces the need of spatial screening of prediction errors.

In summary, the conceptual model of ICT utilization proposed for U.S. states is largely supported by the empirical results. In the next section, we provide case examples of ICT adoption and utilization in two US states, Georgia and Nebraska.

8.8 Case Studies of ICT Utilization in US States

8.8.1 Georgia Enterprise Technology Services (GETS)

Our K-means cluster analysis of ICT utilization in US states has revealed the state of Georgia in the South to be a high-ICT-use state, yet it is surrounded in almost all directions by low-tech states such as Alabama, Mississippi, South Carolina, Louisiana, Arkansas, Kentucky, and Indiana, that are part of the lowest ICT use cluster in the US. A related observation is Georgia's higher urbanization (only one quarter of Georgia's population is rural) compared to that of the low-tech neighboring states.

To trace Georgia's rich modern history of telecommunications excellence, one needs to go back to the mid-nineties when fiber investments were made in Atlanta in anticipation of the Summer Olympics of 1996 and Georgia's Statewide Academic and Medical System for distance learning, healthcare, and teleconference facilities were launched. While evolutionary at that time, the state IT system

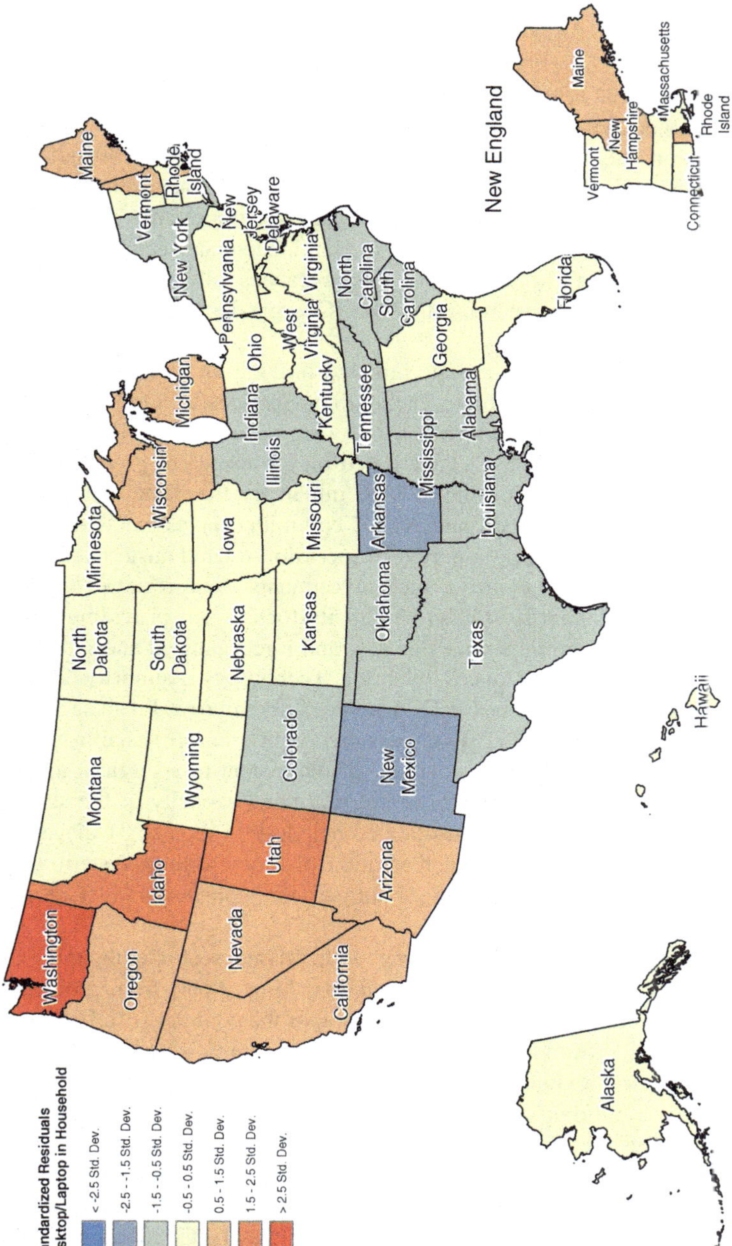

Fig. 8.12 OLS regression Residuals (N = 50, with Putnam) for desktop, laptop in household, United States, 2010

subsequently became non-integrated, and difficult to understand, use, and maintain. Various state agencies maintained their individual applications, and even simple citizen requests such as a change of address necessitated changes across multiple systems. This was neither cost effective, nor efficient for the state as well as for the concerned citizen (Georgia Technology Authority 2011). Compounding the problem were multiple other challenges due to aging equipment with almost no funding for refreshment, inconsistent operating standards, failure to comply with industry best practices, underfunding of security aspects to prevent data breaches, inadequate power backup for critical IT systems, failure to back up important data due to broken servers, and skilled employee turnover. In fact, inadequate security resulted in a major loss of face for the state with data breaches affecting more than 4.5 million citizens of the state beginning in 2005 (Georgia Technology Authority 2009).

To respond to these challenges, the state initiated in 2008 an IT modernization project known as the Georgia Enterprise Technology Services (GETS) initiative, and in April–May 2009, IT infrastructure services were outsourced to IBM, while network services were taken over by AT&T based on a transparent private sector vendor procurement process. With the GETS initiative, the state committed $1.2 billion over 10 years, while IBM and AT&T committed investments of $188 and $99 million respectively. During a period of recession-related austerity, it was virtually impossible for the state to make such investments. As a result of GETS, the Georgia Technology Authority (GTA) shifted its focus from providing technology services to management of service delivery and it reorganized and radically downsized from 660 to 175 employees (Georgia Technology Authority 2009). Major accomplishments as a result of GETS have been (a) improved data security, (b) establishment of the state's first IT disaster recovery program, (c) establishment of an ongoing equipment refreshment program, (d) movement to a services model for hardware rather than equipment purchase, (e) expanded capacity of the state's wide area network by 100 %, (f) one consolidated helpdesk to replace 21 disparate helpdesks or analogous entities, (g) implementation of a new billing system with web-invoicing, and detailed information about service consumption (Georgia Technology Authority 2009, 2011).

At the time, the GETS model of engaging with private sector companies to overhaul a state's ICT was innovative and groundbreaking. Apart from projected cost savings exceeding $200 million over the duration of the contract with IBM and AT&T, the state benefitted from industry best practices, winning trust back from its citizens, especially those previously impacted by data breaches, and most importantly progress towards building a sustainable model for long-term investments in critical technology infrastructure.

In 2013, GETS moved beyond the 75 % transformation threshold and completed 95 % modernization in 2014. Since 2012–13, the state has embarked on an enhanced GETS+ initiative that assist government agencies better manage GETS resource consumption and their IT budget (Georgia Technology Authority 2013). Based upon modernization of IT in the state, Georgia has produced an IT strategic plan for 2020 that enables government workers to be mobile and closer to their "customers," improve citizen access to government by shifting appropriate services to mobiles, encourage technology use innovation in government, manage data as an asset with security at the core, and develop/procure enterprise solutions to the extent feasible when multiple agencies have overlapping needs (Georgia Technology Authority 2014).

The Georgia case study highlights the critical role of careful planning in shaping IT strategy and policy at the state level. While government support and prioritization of ICT has not been considered as a factor in our US conceptual model due to lack of systematic data at the state/provincial level, this case study reinforces observations made by prior ICT and digital divide case studies about the important role that governments can play in facilitating the adoption and diffusion of ICTs. It exemplifies the importance of the inclusion of government prioritization of ICT which has been included as an independent correlate in the book's generic conceptual model in Chap. 3.

8.8.2 ICT Adoption Challenges Encountered by Micro-Enterprises in Nebraska

In this case study, we focus on a specific context of ICT use and related challenges. Micro-enterprise specifications in terms of number of employees and revenue generated vary across the United States. In 2011, approximately 80 % of all firms (4.5 million firms out of a total of 5.6 million) in the United States employed fewer than 10 employees (considered as the employment threshold for micro-enterprises), and their payroll accounted for 9 % of the total payroll of firms with fewer than 500 employees, generally accepted as small and medium-sized enterprises (SMEs) (US Census Bureau 2011). Selective use of ICTs by micro-enterprises can help these businesses gain access to better information and expertise, innovate, reach more customers, administer the business more efficiently, and grow in knowledge and skills required to run the business better (Qureshi 2005). However ICT adoption and utilization by micro-enterprises, whether in developing or developed nations, is fraught with challenges.

Wolcott et al. (2008) examined ICT adoption or lack thereof in eleven micro-enterprises in underserved communities of Omaha, Nebraska, a majority of which delivered a variety of services, and found attitude-related challenges including resistance to technology about fundamental concepts such as data back-up, lack of awareness, and lack of confidence, sometimes stemming from a micro-enterprise business owner's entire time and energy being devoted to running the business, to be significant impediments for ICT use. Other challenges are deficits in education and lack of trust in circumstances when external support was made available (in the case of Wolcott et al. (2008) by students of a local university). Operational challenges such as the utter lack of ICT support services due to budget constraints and ICT skill challenges were also common. ICT access challenges were perpetuated by the lack of adequate hardware and software and poor infrastructural issues such as poor computing infrastructure and limited to no internet connectivity.

While these findings reported by Walcott et al. (2008) largely reinforce ICT adoption challenges in micro-enterprises previously reported in the literature (Furuholt and Orvik 2006; Duncombe and Heeks 2003), they are also consistent with this chapter's finding that education is as an essential facilitator of ICT adoption and use in the US. The importance of affordability and infrastructure as a catalyst for ICT adoption among microenterprises supports the inclusion of similar factors posited to be associated with ICT utilization in this book's generic conceptual model in Chap. 3. Wolcott et al. (2008) have further conjectured that when offered training and external communal support, a lack of trust or mismatch of expectations with the support service system impedes ICT use, a situation that may have its roots in ethnicity and cultural background of the micro-entrepreneurs. This lends support to the inclusion of ethnicity variables in our conceptual model for this chapter.

Wolcott et al. (2008) prescribe "IT therapy" as a potential solution for overcoming attitudinal challenges that hinder ICT adoption in microenterprises, but also caution that shortage of time, money, information, and infrastructure will not be remediated as a result.

In conclusion, both cases reinforce certain independent correlates included in the generic conceptual model of ICT utilization developed in this book, as well as the US state variant developed in this chapter. We now discuss policy implications of our findings.

8.9 Policy Implications

A central finding of our work is the strong positive association of social capital with personal computer use in household, internet access at home, and broadband adoption in household. The role of communal engagement that fosters social interactions thus becomes crucial. Extending the argument of Agarwal et al. (2009), it is fair to state that individuals who are geographically proximate to others who have adopted and are using ICTs are more likely to adopt as well; versus those who are surrounded by people who choose to not adopt and use ICTs are less likely themselves to adopt ICTs. This has implications for policymakers in US states.

Communities in the US develop social ties through interactions mostly at the local level. This can be at churches, libraries, local volunteer organizations, clubs, or at the grocery store. Therefore, while enacting policy and regulations, the federal as well as state governments should encourage and if possible, incentivize local governments at the city and county level as well as organizations to act as ambassadors of ICT adoption and use. This can be achieved by communal events such as workshops, training events, demonstration sessions, and rewarding referrals that may encourage and promote adoption and diffusion of ICTs. The same is applicable for ICT service providers which are expected to be more in touch with specific needs of local communities, and can organize designated "Technology Day(s)" that bring in local communities, community leaders, government officials, educators and gradually build an ecosystem that empowers individuals and households at the bottom of the pyramid and encourages adoption and use of ICTs. Agarwal et al's (2009) word of caution about half-hearted efforts realizing no benefits is especially pertinent here. Also, as indicated in Chen (2013), it is important to recognize that peer effects in the absence of resources are also unlikely to alleviate technology divide problems. In other words, non-adopters may be encouraged to access ICTs through bridging social capital in which a person knows someone from other occupations only through weak ties. However, for the same person to actually use ICTs, it is important to foster resource-rich bonding capital, where the individual has strong ties with people in several other occupations who have access to resources.

Community based efforts (Agarwal et al. 2009), encouraged by the state, and hosted by local governments, organizations, and service providers can thus provide a forum for ICT- impoverished individuals and households to connect with peers who access and use ICTs as well as individuals and entities which possess resources to facilitate ICT adoption, and is therefore outlined as a policy recommendation. Contextualizing such local efforts to the demographic, socio-economic, and cultural attributes of the local population is also suggested since people on the wrong side of digital divides are often also found to be on the wrong side of social inequalities (Chen 2013).

Other important U.S. findings relate to the association of societal openness, ethnicity, urbanization, and college education to use of ICTs. US states shape their own fiscal and regulatory policies among which tax burden, employment opportunities, government debt, labor market laws, property protection, health insurance freedom, cable and telecom freedom, along with interference in peoples' personal lives have a direct impact on the quality of life of its citizens and the economic vitality of its people and businesses; the aforementioned factors along with many others influence peoples' perceived sense of freedom (Ruger and Sorens 2013). It is worthwhile to note that a free, open society is more likely to engender freedom of expression, exchange of ideas and information, and enable a truly empowered information society. Therefore, we recommend states to pay attention to their fiscal and regulatory policies that largely determine economic freedom, and policy recommendations by Ruger and Sorens (2013) may serve as useful guidelines.

NTIA (2010) has clearly showed that adoption of broadband increases with the level of educational attainment as there is almost a 20 % point difference in broadband adoption between people with just a high school diploma and those who have some college education. The gulf widens to almost 35 points with people who possess an undergraduate degree. In light of the significant positive association of college education with PC, internet, broadband, and cellphone use, we recommend state governments along with regional and local governments to support public education especially at the college level. Support may manifest itself in tangible forms of funding, tax relief, special programs, work study, and subsidies for government workers who want to pursue college education (Azari and Pick 2005). Intangible forms of support such as political backing of various programs and initiatives of colleges and universities are also valuable in this regard. Figure 8.13

shows a university staff member in California using a tablet to take photos at a commencement.

To bridge urban-rural divides, a variety of policies have been suggested in the literature. Weiner et al. (2012) has suggested resource-based initiatives, such as strategic commercial pricing or underwriting, direct government subsidies, or cooperation between the government and private sector for rural areas where affordability is often an issue. Grubesic (2006) has provided examples of smaller "peripheral" communities that have launched their own broadband companies to bring access to their regions, thereby suggesting that local level initiatives and policy can supersede free-market mechanisms that resulted from the 1996 Telecommunications Act. Collaborations between national and state policymakers with telecommunications service providers and local government have been suggested (Sylvester and McGlyn 2010) to provide internet access as well as training to areas and groups that are digitally disadvantaged. Often telecom service providers shy away from providing service in rural areas due to lack of economic activity and a critical mass of household demand. Examining a rural region's economic structure and information infrastructure has been recommended (Grubesic 2002) as a first step towards the remediation of digital inequalities in such instances.

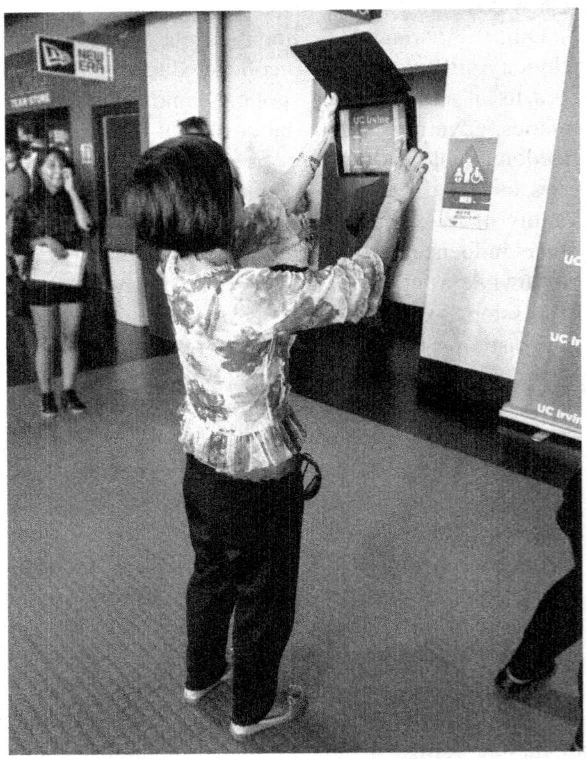

Fig. 8.13 Staff member using a tablet to capture photos at a university commencement

Earlier in the chapter, the proportion of rural population among several low-ICT states in Cluster 4 was found to be at least 40 % in 2010; for these states, several of the policy recommendations outlined in prior literature are applicable. As suggested in the National Broadband Plan (FCC 2010), state level policies should also facilitate demand aggregation (pooling demand for broadband) which can be especially beneficial in rural America and use state, regional and local networks when they offer the most cost-efficient solution for anchor institutions such as rural schools, hospitals, and government buildings to meet their connectivity needs.

In conclusion, while the digital divide between urban and rural America has persisted, Grubesic (2002) has contended based upon historical evidence that ICT use presently exhibited in urban areas will eventually diffuse to more rural areas, just like such spatial diffusion of fixed telephone infrastructure and service historically took place over many years.

8.10 Conclusions

Despite being a developed nation with an advanced information society, the use of ICT in US states continues to remain uneven. Given this background, this chapter has examined patterns of ICT utilization in US states. We find that utilization of six ICT dependent variables—personal computers, internet, broadband, mobile wireless, fixed phones, and mobile wireless devices, is highly spatially agglomerated, although both Facebook and Twitter users are randomly distributed. Cluster analysis with all eight ICT variables reveals the presence of four clusters; most US states (34 out of 50) are in the top two ICT utilization clusters while a few are in a moderate ICT-use cluster in the central and eastern US. The lowest ICT use cluster is predominantly comprised of states in the South extending up to the lower Midwest, as well as the state of New Mexico.

We also determine that social capital, college education, and ethnicities are associated with ICT use in the US states, followed by urban location and R&D. Despite high spatial autocorrelation in six out of the eight ICT variables, it is greatly reduced following regression modeling, in which half have random residuals. This is indicative of the fact that our conceptual model of ICT utilization is well-specified and accounts endogenously for geographic influences affecting the diffusion and use of ICTs.

Policies recommended to improve information technology utilization and access are to give state support of higher education; make efforts to increase social capital using locally organized and culturally appropriate education and training programs; build a fiscal and regulatory environment and instill a culture of moderated societal intervention at the state level that fosters openness and transparency; support ICT initiatives for ethnic groups in society; and develop public-private partnerships in ICT with accompanying financial incentives and investments.

References

Agarwal, R., Animesh, A., & Prasad, K. (2009). Social interactions and the 'digital divide': explaining variations in internet use. *Information Systems Research, 20*(2), 277–294.

Azari, R., & Pick, J. (2005). Technology and society: Socio-economic influences on technological sectors for United States counties. *International Journal of Information Management, 25*(1), 25–37.

Baliamoune-Lutz, M. (2003). An analysis of the determinants and effects of ICT diffusion in developing countries. *Information Technology for Development, 10*, 151–169.

Belinfante, A. (2014). Telephone penetration by income by state, report of the federal communications commission. Retrieved November 1, 2014, from http://www.fcc.gov/ccb/stats.

Blumberg, S., & Luke, J. (2013). *Wireless substitution: Early release of estimates from the national health interview survey, January–June 2013*. Atlanta: Center for Disease Control and Prevention.

Blumberg, S. J., Luke, J. V., Ganesh, H., Davern, M. E., & Boudreaux, M. H. (2012). *Wireless substitution: State-level estimates from the National Health Interview Survey, 2010–2011* (National health statistics report no. 61). Washington, DC: National Center for Health Statistics, U.S. Department of Commerce.

Bowling Alone. (2014). *Social capital primer*. Retrieved from http://bowlingalone.com/?page_id=13.

Central Intelligence Agency. (2014) *The world fact book*. Retrieved November 1, 2014, from https://www.cia.gov/library/publications/the-world-factbook/geos/us.html.

Chen, W. (2013). The implications of social capital for the digital divides in America. *The Information Society, 29*, 13–25.

DCI Group. (2001). *Digital America, report*. Washington, DC: DCI Group.

Duncombe, R., & Heeks, R. (2003). An information systems perspective on ethical trade and self-regulation. *Information Technology for Development, 10*(2), 123–139.

FCC. (2014). *Telecommunications Act of 1996*. Retrieved 2014, from http://transition.fcc.gov/telecom.html.

FCC. (2010). *14th annual report and analysis of competitive market conditions with respect to mobile wireless, May* 20. Washington, DC: U.S. Federal Communications Commission.

Florida, R. (2012). *The rise of the creative class—Revisited*. New York: Basic Books.

Furuholt, B., & Orvik, T. U. (2006). Implementation of information technology in Africa: Understanding and explaining the results of ten years of implementation effort in a Tanzanian organization. *Information Technology for Development, 12*(1), 45–62.

Georgia Technology Authority. (2009). *State IT report*. Available at http://gta.georgia.gov/state-it-report.

Georgia Technology Authority. (2011). *Georgia broadband and mapping analysis* (Semi-annual report, Vol. 1, January 2011). Atlanta Georgia, Georgia Technology Authority, State of Georgia.

Georgia Technology Authority. (2013). *Technology transformation overview*. Atlanta, GA: State of Georgia. Available from http://gta.georgia.gov/technology-transformation-overview.

Georgia Technology Authority. (2014). *Georgia enterprise IT strategic plan*. Available at http://gta.georgia.gov/state-it-report.

Grubesic, T. H. (2002). Spatial dimensions of internet activity. *Telecommunications Policy, 26*, 363–387.

Grubesic, T. H. (2004). The geodemographic correlates of broadband access and availability in the United States. *Telematics and Informatics, 21*, 335–358.

Grubesic, T. H. (2006). A spatial taxonomy of broadband regions in the United States. *Information Economics and Policy, 18*, 423–448.

Guillen, M. F., & Suarez, S. L. (2005). Explaining the global digital divide: Economic, political, and sociological drivers of cross-national Internet use. *Social Forces, 84*(2), 681–708.

Igari, N. (2014). How to successfully promote ICT usage: A comparative analysis of Denmark and Japan. *Telematics and Informatics, 31*(1), 115–125.

Internet World Stats. (2013). *United States of American Internet and Facebook users stats*. Available from www.internetworldstats.com.

Internet World Stats. (2014). *Usage and population statistics*. Retrieved November 1, 2014, from http://www.internetworldstats.com/list2.htm.

ITU. (2013). *Measuring the information society, report*. Geneva, Switzerland: International Telecommunication Union.

Kimmelman, G., Cooper, M., & Herra, M. (2006). The failure of competition under the 1996 telecommunications act. *Federal Communications Law Journal, 58*(3), 511–518.

Mueller, M. (1993). Universal service in telephone history: A reconstruction. *Telecommunications Policy, 17*(5), 352–369.

National Broadband Map. (2014). *National broadband map: How connected is my community?* http://www.broadbandmap.gov/.

National Science Board. (2012). *Science and engineering indicators 2012*. Arlington, VA: National Science Foundation.

Nielsen Company. (2012). *State of the Asian American consumer* (Report, quarter 3). New York, NY: Nielsen Company.

Nishida, T., Pick, J.B., & Sarkar. A. (2014). *Japan's prefectural digital divide: A multivariate and spatial analysis. Telecommunications Policy, 38*(11), 992–1010.

NTIA. (1996). *The United States Telecommunications Act of 1996*. Retrieved 2014, from http://www.ntia.doc.gov/legacy/opadhome/overview.htm.

NTIA. (2010). *Digital nation: Expanding internet usage*. Washington, DC National Telecommunications and Information Administration, U.S. Department of Commerce.

NTIA. (2011). *Exploring the digital nation: Computer and internet usage at home*. Washington, DC: National Telecommunications and Information Administration, U.S. Department of Commerce.

NTIA. (2013). *Exploring the digital nation: Computer and internet usage at home*. Washington, DC: National Telecommunications and Information Administration, U.S. Department of Commerce.

NTIA. (2014). *National broadband map*. Washington, D.C.: National Telecommunications and Information Administration. Available at http://www.broadbandmap.gov.

Pew Charitable Trusts. (2013). *Measuring state elections performance*. Washington, DC: Pew Charitable Trusts. Available at www.pewstates.org.

Pick, J. B., & Azari, R. (2008). Global digital divide: Influence of socio-economic, governmental, and accessibility factors on information technology. *Information Technology for Development, 14*(2), 91–115.

Pick, J. B., & Azari, R. (2011). A global model of utilization of technology based on governmental, social, economic, and business investment factors. *Journal of Management Information Systems, 28*(1), 51–85.

Pick, J. B., Nishida, T., & Zhang, X. (2013). Determinants of China's technology utilization and availability 2006–2009: A spatial analysis. *The Information Society, 29*(1), 26–48.

Pick, J. B., Nishida, T., & Sarkar, A. (2014). Broadband utilization in the Indian states: Socio-economic correlates and geographic aspects. In J. Choudrie & C. Middleton (Eds.), *Management of broadband technology innovation* (pp. 269–296). Oxford, England: Routledge.

Prieger, J. E. (2013). The broadband digital divide and the economic benefits of mobile broadband for rural areas. *Telecommunications Policy, 37*, 483–502.

Putnam, R. (2000). *Bowling Alone: The collapse and revival of American community*. New York: Simon and Schuster.

Quibria, M. G., Ahmed, S. N., Tschang, T., & Reyes-Macasaquit, M.-L. (2003). Digital divide: Determinants and policies with special reference to Asia. *Journal of Asian Economics, 13*(6), 811–825.

Qureshi, S. (2005). How does information technology effect development? Integrating theory and practice into a process model. In *Proceedings of the 11th Americas Conference on Information Systems*, Omaha, NE.

Robison, K. K., & Crenshaw, E. M. (2002). Post-industrial transformations and cyber-space: A cross-national analysis of internet development. *Social Science Research, 31*, 334–363.

Ruger, W., & Sorens, J. (2013). *Freedom in the 50 states: An index of personal and economic freedom*. Arlington, VA: Mercatus Center, George Mason University.

Schiller, D. (1998). Social movement in Telecommunications: Rethinking the public service history of US telecommunications 1894–1919. *Telecommunications Policy, 22*(4–5), 397–408.

Selwyn, N. (2004). Reconsidering political and popular understandings of the digital divide. *New Media and Society, 6*, 341–362.

Solomon, R. J. (1978). What happened after Bell spilled the acid?: Telecommunications history: a view through the literature. *Telecommunications Policy, 2*(2), 146–157.

Sorens, J., & Ruger, W. (2009). *Freedom in the 50 states: An index of personal and economic freedom*. Arlington, VA: Mercatus Center, George Mason University.

Suchan, T., Perry, M., Fitzsimmons, J., Juhn, A., Tait, A., & Brewer, C. (2007). *Census Atlas of the United States, Series CENSR-29*. Washington, D.C.: U.S. Census Bureau.

Sylvester, D. E., & McGlynn, A. J. (2010). The digital divide, political participation, and place. *Social Science-Computer Review, 28*(1), 64–74.

Takhteyev, Y., Gruzd, A., & Wellman, B. (2012). Geography of Twitter networks. *Social Networks, 34*, 73–81.

Tolbert, C., Mossberger, K., King, B., & Miller, G. (2007). Are all American women making progress online? *African-Americans and Hispanics Information Technologies and International Development, 4*(2), 61–88.

U.S. Census Bureau. (2007–2009). American Community Survey. Washington, DC: U.S. Census Bureau.

U.S. Census Bureau. (2013). *U.S. population census of 2010*. Washington, DC: U.S. Census Bureau.

U.S. Census Bureau. (2014). *Statistics of U.S. Businesses, 2011*. Retrieved November 1, 2014, from http://www.census.gov/econ/susb/.

Vicente, M. R., & Lopez, A. J. (2011). Assessing the regional divide across the European Union-27. *Telecommunications Policy, 35*, 220–237.

Wolcott, P., Kamal, M., & Qureshi, S. (2008). Meeting the challenges of ICT adoption by micro-enterprises. *Journal of Enterprise Information Management, 21*(6), 616–632.

Weiner, M. D., Puniello, O. T., Noland, R. B., Ciemnecki, D., & Turakhai, C. (2012). Consider the non-adopter: Developing a prediction model for the adoption of household-level broadband access. *Socio-Economic Planning Sciences, 46*, 183–193.

World Bank. (2014). *Data*. Retrieved November 1, 2014, from http://data.worldbank.org/indicator/SP.POP.TOTL.

Zhao, L., Lu, Y., Wang, B., & Huang, W. (2011). What makes them happy and curious online? An empirical study on high school students' internet use from a self-determination theory perspective. *Computers and Education, 56*, 346–356.

Zhao, S. (2006). Do internet users have more social ties? A call for differentiated analyses of internet use. *Journal of Computer-Mediated Communication, 11*, 844–862.

Chapter 9
Digital Divide in Africa

9.1 Introduction

Africa is the second largest continent behind Asia in terms of its population, which is estimated to be 1.14 billion in 2014 (Population Reference Bureau 2014). However, a large majority of nations in the African continent, especially nations in sub-Saharan Africa are classified as low income or lower middle income nations (The World Bank 2012). Exceptions are the northern African nations of Algeria and Libya, as well as South Africa and its neighbors Botswana and Namibia, which are classified as upper middle income nations. African nations also fare poorly in terms of social, health, and economic indicators such as life expectancy, undernourishment, participation in education, unemployment, poverty. Most African nations are also ranked at the bottom end of the United Nations Human Development Index (HDI), which is a composite measure that averages achievement in three basic dimensions of human development—a long and healthy life, knowledge and a decent standard of living. This is especially true for a vast majority of nations in sub-Saharan Africa.

ICT adoption and utilization in many African nations have lagged far behind those of their more industrialized counterparts in other continents such as North America, Europe, and Asia. At the turn of the 21st century, Africans had the use of one fixed landline telephone per 100 people (one per 200 people excluding South Africa) against the worldwide average of 50 telephones per 100 (Cooper 2002). This considerable lack of availability of what many might argue to be a legacy form of ICT must be viewed as part of an overall infrastructural malaise comprised of deteriorating roads, unreliable electricity, and arbitrary regulatory and judicial systems (Cooper 2002). In 2005, Africa, with an estimated 14 % of the world's total population, accounted for less than 2 % of the world's internet usage. However this has improved noticeably, and at end of 2013, Africa with an estimated 15.7 % of the world population accounted for 8.6 % of the world's internet users (Population Reference Bureau 2014; Internet World Stats 2014). Africa's internet penetration

J.B. Pick and A. Sarkar, *The Global Digital Divides*, Progress in IS,
DOI 10.1007/978-3-662-46602-5_9

rate is 19 % (of total population) in 2014 (ITU 2014), lagging behind the worldwide rate of 40 % and far behind the rates of Europe (75 %), Americas (65 %), and Asia and Pacific (32 %). Only six African nations—Morocco, Seychelles, Egypt, South Africa, Kenya, Mayotte, and Tunisia report internet penetration rates (Internet World Stats 2014) that exceed the worldwide rate of 40 % (ITU 2014). Perhaps unsurprisingly, internet usage in Africa has leapfrogged by 3607 % during 2000–2012, compared to moderate rates of growth in all other continents or regional agglomerations such as the Middle East. On the other hand, a barrier for African internet access is the high cost barrier for internet access. Average dial-up internet access for 20 hours a month costs US$68 in addition to internet subscription fees (Gebremichael and Jackson 2006). Warf (2013) indicates that while cybercafés are commonly found in commercial districts in African cities, they are mostly frequented by young tourists, students, and business executives; since most cafes derive their profits from non-Africans, their fees are too high to make them accessible to the low income citizenry residing in the outskirts of the urban areas.

Onyeiwu (2002) has observed that the digital divide literature has often tended to focus on the differences between Africa and the industrialized world, possibly at the cost of paying inadequate attention to the widely varying levels of digitization and ICT availability and utilization among the African nations. While countries such as South Africa, Namibia, Morocco, Tunisia, Botswana, Cape Verde, and Seychelles have made significant progress in ICT adoption and access, many other nations have lagged far behind, several of which are in sub-Saharan Africa. Onyeiwu (2002) has further commented that the digital divide in Africa is made even more complicated since countries with similar socio-economic foundations have tended to have substantially different access to ICT. For example, Angola and Cameroon, classified in terms of the UN's human development index (United Nations 2013) at low ranks of 148 and 150 respectively, and with comparable 2012 estimated populations of 20 million possess considerably different internet penetration rates of 14.8 and 5 % respectively (Internet World Stats 2014).

Given this context, in this chapter we focus on ICT availability and utilization and the resulting digital divide for the entire African continent at the country unit of analysis. Several studies have focused on sub-regions of Africa, most notably sub-Saharan Africa (Buys et al. 2009; Rhine 2006; Ojedokun and Lumande 2005; Gyamfi 2005), or on ICT development, policies, practices, and impacts of specific African nations or provinces within a nation as examples (Mbatha et al. 2011; Bankole et al. 2011; Warschauer 2003; Alemna and Sam 2006); but only a handful of studies such as Fuchs and Horak (2008) and Onyeiwu (2002) have examined aspects of Africa's digital divide at the country level. To examine ICT utilization and the digital divide in Africa, we consider five different forms of ICT—broadband, fixed/landline telephones, mobile phones, internet, and virtual social networks. Onyeiwu (2002) is the only study which has considered multiple forms of ICTs for Africa; however it excluded modern ICTs such as broadband and social networking in lieu of use of personal computers and internet hosts.

Next this chapter focuses on the historical evolution of ICT adoption and diffusion in Africa. We then examine ICT utilization in African nations along with spatial patterns and disparities that exist at the nation level using descriptive mapping and cluster analysis techniques employed in previous chapters. Subsequently we build a conceptual model of ICT utilization for Africa based upon induction from prior literature. Our model posits that with eighteen demographic, socio-economic, education, technology tariff, societal openness, and business competitiveness variables are associated with five dependent ICT utilization variables. Regression analysis reveals that laws that relate to the use of ICT are a dominant predictor of all forms of ICT use for African nations, a novel finding. The effectiveness of a national parliament/congress as a lawmaking institution is significantly associated with modern (broadband) as well as legacy forms (fixed telephones) of ICT. Next, case studies of ICT utilization in African countries are presented and discussed. These case examples provide context to the discussion on ICT utilization in Africa and reinforce findings from the regression analysis. We conclude this chapter by discussing policy implications of our findings that are positioned within the broader ICT policy framework for Africa, and a brief summary of the chapter.

9.2 Historical Background for Africa

Prior to gaining independence from colonialism, Africa's telecommunications infrastructure existed primarily to serve administrative purposes of colonial rulers and also economic objectives to some extent. As a majority of nations became independent in the sixties, Africa's telecommunications landscape did not undergo any significant change and was in fact hamstrung by antiquated hardware infrastructure, as well as an obsolete organizational structure. The first major change catalyst was the Pan-African Telecommunications network (Panaftel) project.

Panaftel was conceptualized in the early 1960s towards establishing intra-African telecom links that would facilitate regional cooperation, trade, and development. As a result of Panaftel, African nations were to be connected by copper wire and microwave links. However, the project encountered bureaucratic hurdles that spanned years and caused substantial delays. Despite initial funding from organizations such as the International Telecommunications Union (ITU), United Nations Development Program (UNDP), UNESCO and others, language barriers between French and English speaking countries and an acute shortage of telecom-related skills slowed cooperation. Panaftel's implementation began around 1975 and progress was achieved through the remainder of the decade and the 1980s. By 1990, progress made as a result of this project has resulted in 39,000 km of radio-relay links, 8000 km of submarine cable, and 43 international switching stations (Noam 1999). As a result of the RASCOM (Regional African Satellite Communications Organization) project, a key component of the Panaftel network, 42 out of the 45 member countries had international satellite stations (ITU 2003; Noam 1999). Overall, it is fair to say that Panaftel was the first post-colonial telecom infrastructure and development initiative and laid the foundation for digital connectivity in the continent. The project was discontinued in 1992 due to low revenues from existing links and political inertia in the central sub-Saharan Africa, a sub-region where connectivity was missing. The United Nations Development Program, the project's original funding agency felt Panaftel's problems could not be overcome (Noam 1999).

Following Panaftel, efforts towards telecommunications and ICT revival in Africa led to the New Partnership for African Development (NEPAD) initiative, which coincided roughly with the formation of the African Union. While NEPAD's primary focus was on socio-economic development, ICT infrastructure development was considered to be an important aspect to foster economic cooperation and integration among African nations. Currently NEPAD's e-Africa program has multiple e-initiatives, for example in education, and aims to connect all African countries with each other and the rest of the world using existing and new submarine and terrestrial cable systems as part of a larger broadband infrastructure development network (NEPAD 2014).

Due to insufficient infrastructure coupled with affordability issues, landline telephones started making way to mobile telephony in Africa since the mid-late 1990s. This was despite the fact that in the preceding decade, mobile provision was practically non-existent in Africa (Hamilton 2003). Uganda became the first country in Africa to have more mobile than fixed telephone users, and by the year 2000, the same was applicable across the entire African continent. As the African population increasingly adopted mobile phones through the 2000s due to affordability resulting from competition and privatization in the mobile sector, convenience and prestige, the growth of fixed telephony was gradually stunted. By 2003, virtually all countries in the continent has access to mobile services, many with two service providers, one of which usually belonged to the private sector (Hamilton 2003).

Despite high cell phone penetration, teledensity remained abysmal in many nations due to the low starting base of telephone users. In many countries, telecommunications operators, often nationalized, diverted investments to the lucrative mobile sector giving less priority to the expansion of fixed services. The downside of this approach was that mobile service providers by the early 2000s did not provide sufficient transmission rates for good internet connectivity and data communication; as a result, services that increase business efficiency, deliver e-applications, and generally foster wealth creation and societal progress could begin to be impeded (ITU 2003). In the next section, we focus on the evolution and leapfrogging that Africa experienced in various forms of ICTs since 2005, and explore the landscape of ICT adoption and utilization in Africa.

9.3 Current Dimensions of Access and Use of Technologies for Africa

ICT use in Africa continues to remain very low. Among major world regions, as defined by the International Telecommunications Union, Africa's ICT Development Index (IDI) of 2.0 (on a 7.0 scale) is the lowest among continents. Excluded from this computation are the 10 African Arab nations. The world average was 4.35

in 2012 and the average for developing nations was approximately 3.75. Also, Africa had the second highest coefficient of variation in its IDI behind Asia and the Pacific in 2012; in fact, the coefficient of variation increased from 2011 to 2012 indicating a further widening of the digital divide among the African nations (ITU 2013). ICT laggards worldwide as measured by the IDI are predominantly African countries which occupied the bottom 19 places in IDI rankings in 2012 (ITU 2013).

This is hardly surprising because ICT use in Africa got off to a very slow start with negligible penetration of fixed phones and internet. For penetration rates in 2005, there were 1.5 fixed phone subscribers, 12.4 mobile-cellular subscribers, 0.03 fixed (wired)-broadband subscribers, 2.9 households with a computer, 1.0 household with internet access at home, 2.4 individuals using the internet, and in 2010, there were 1.8 active mobile-broadband subscribers in Africa, per 100 population (ITU 2014). At the same time, the corresponding numbers for developing nations were 12.7, 22.9, 1.3, 14.6, 8.1, and 7.8 respectively, while the world per capita usage was higher in each category. Since then, the use of all forms of ICTs in Africa shown in Fig. 9.1 increased rapidly, surpassing those of all other major world regions, including Asia and the Pacific and Arab states. The only world region that has surpassed Africa's growth in fixed broadband and household computer use is the Commonwealth of Independent States (CIS) which is not shown in Fig. 9.1 due to the small number of nations that comprise CIS. This leapfrogging between 2005 and 2014 is understandable given worldwide growth in use of the figure's ICTs and the low starting base of adoption and use in Africa.

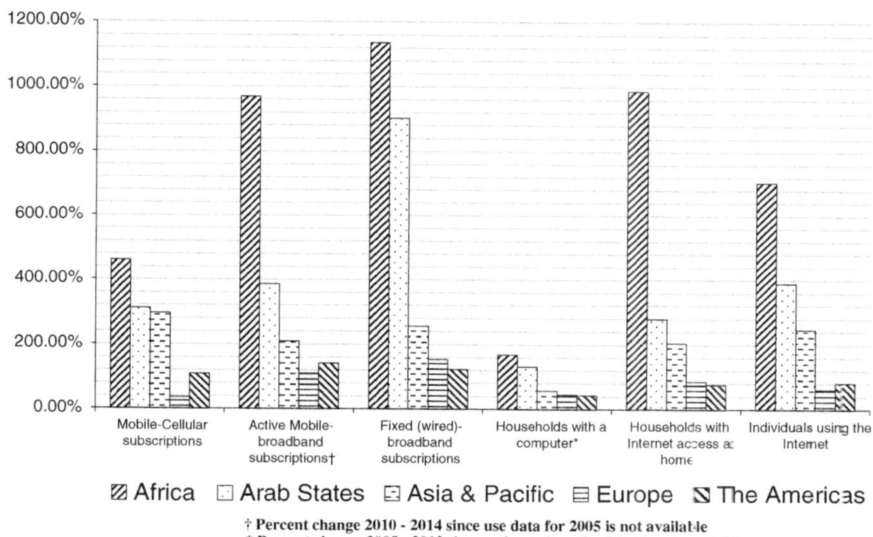

Fig. 9.1 Percent change (estimated) in use of ICTs in world regions, 2005–2014 (Data retrieved November 11, 2014 from http://www.itu.int/en/ITU-D/Statistics/Pages/stat/default.aspx)

By contrast, fixed telephone use in Africa along with all other world regions declined between 2005 and 2014.

It is important to note that the world regions depicted in Fig. 9.1 are defined by the ITU's Telecommunication Development Bureau, which classifies the African nations of Algeria, Cameroon, Comoros, Djibouti, Egypt, Libya, Mauritania, Morocco, Sudan, and Tunisia, most of which are in North Africa, as part of the Arab states regional group. Nonetheless, these nations also witnessed significant leapfrogging, with average increase of 98.11, 143.32, and 402.41 % in mobile cellular, internet, and fixed broadband use between 2008 and 2013.

Despite this leapfrogging, the discrepancy in ICT use between African nations continues to be significant. Among African nations, ICT leaders, as measured by ITU's ICT Development Index (IDI) (ITU 2013) are Seychelles, Mauritius, Egypt, Morocco, Tunisia, and South Africa with IDI values exceeding the average for developing nations. This is corroborated later in this chapter by results of our K-means cluster analysis of ICT use. These nations are usually among the top five nations in Africa for fixed telephone, mobile cellular (Fig. 9.2), fixed broadband, and internet subscriptions—both in terms of aggregate subscribers as well as per capita use.

Conversely, ICT laggards are mainly in sub-Saharan Africa, where a majority of countries are classified as low income or lower middle income (The World Bank 2012).

In addition to deficits in the UN Human Development Index mentioned earlier, ICT progress in several of these nations has also been severely impeded by decades of civil war, post-independence conflict, and general political instability. In fact, national governments in many of these nations have further impeded miniscule ICT penetration rates by using censorship in various ways. In some instances, such as Sudan and Kenya, censorship has been used to protect and sustain the government's telecommunications monopoly, while in some countries such as Zimbabwe, multiple laws were enacted by the government to limit freedom of expression by the media, and curtail democratic and political activities of the citizenry and opposition (Warf 2013). On the other end of the spectrum, South Africa has negligible interference in cyberspace (Warf 2013) that has clearly benefitted the rapid strides made by the country towards ICT progress and development since the end of apartheid. This provides credence to our inclusion of societal openness factors, manifested as different context-specific independent variables as determinants of ICT utilization in various empirical studies of digital divides discussed in this book. Lastly, infrastructural malaise also plagues sub-Saharan countries as is evident from the low penetration of fixed telephones in 2000 and virtually no change during 2000–2013 (Fig. 9.3) even though leading ICT nations in the continent followed the global pattern of increasing per capita use of fixed telephones followed by a period of gradual decline marked by the emergence and rapid adoption of mobile phones.

Given the landscape of significant disparities in ICT adoption and utilization among the African nations, we now induce our conceptual model of ICT utilization for Africa.

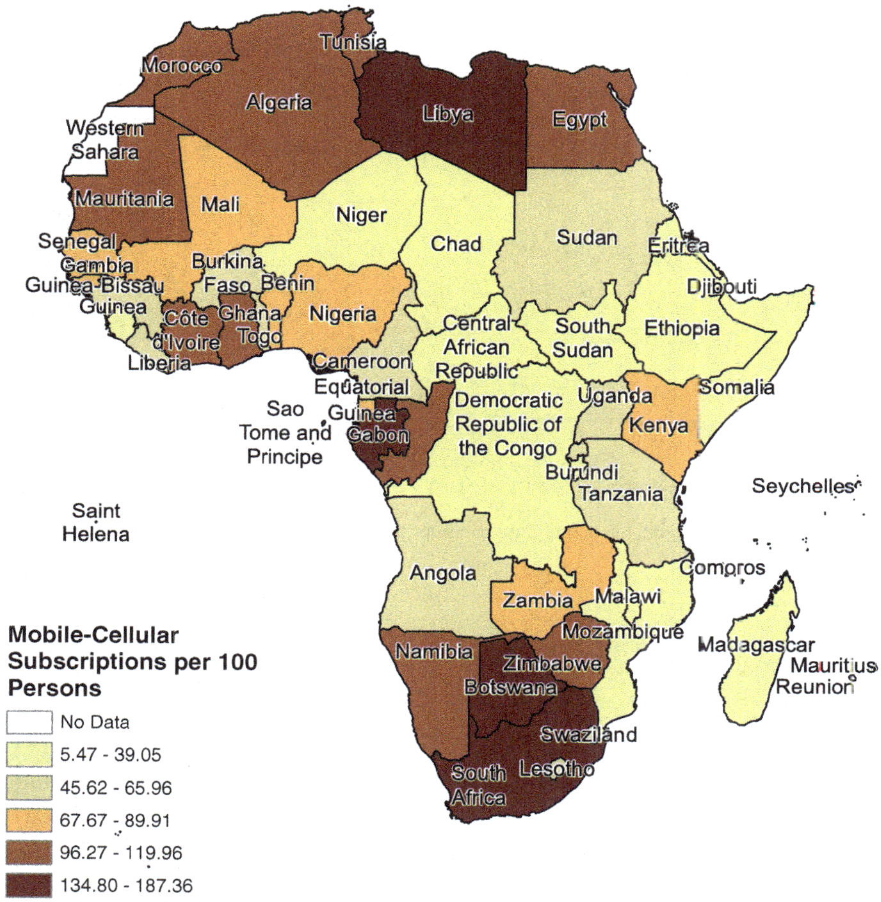

Fig. 9.2 Mobile-cellular subscriptions per 100 population, Africa, 2012

9.4 Conceptual Model of ICT Utilization in African Nations

In our conceptual model of ICT utilization for Africa (Fig. 9.4), five dependent variables are posited to be associated with eighteen demographic, socio-economic, education, technology tariff, societal openness, infrastructure, and ICT services competitiveness variables.

The conceptual model has justifications from prior studies. Justifications for including *demographic* factors such as age structure, and urban location, *education and training* factors such as literacy, *economic* factors such as income, manufacturing and agricultural output or exports, and FDI, all included as potential correlates of ICT utilization for Africa have been presented earlier in Chap. 3.

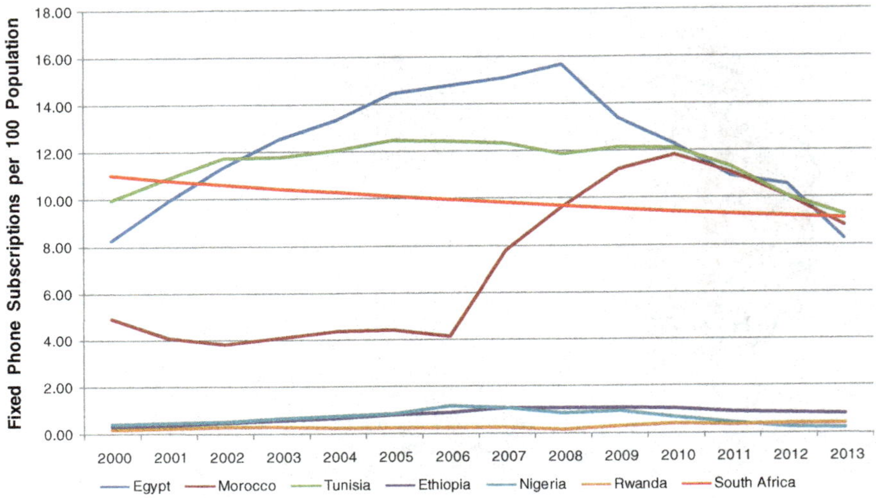

Fig. 9.3 Change in fixed telephone subscriptions per 100 population in a sample of African nations, 2000–2013

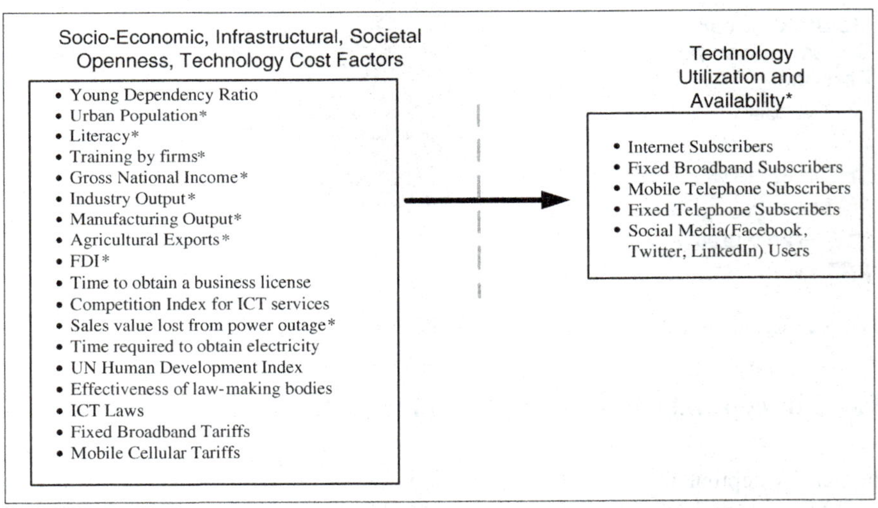

Note: * per capita

Fig. 9.4 Conceptual model of ICT utilization for Africa

Socio-economic factors involve a wide variety of dimensions that include categories of life expectancy, health, income, and education. Although income and education are known to increase technology use, the health-related dimensions have not been studied. However, we reason that people who are healthier have greater opportunity to take advantage of technologies. Life expectancy does not directly link to technology, but serves as another indicator of health. In the African context, Roycroft and Anantho (2003) investigated the influence of economic development, estimated by the United Nations human development index, along with other factors such as international internet bandwidth, domestic internet hosts, ISP market structure, the cost of a local telephone call, and the use of English as an official language on internet subscription in Africa, and determined that ISP market structure, overall economic development, and the amount of international bandwidth are the strongest predictors of internet subscription. Another study (Andoh-Baidoo et al. 2013) has investigated the degree to which economic, human development, political, and population conditions impact ICT capacity in African nations, and determined that investment in telecommunications is motivated mainly by human development factors such as life expectancy, enrollments in tertiary education, expenditures per student, and health expenses as percent of GDP. We posit that apart from traditional economic factors such as gross national income, industrial, manufacturing, agricultural outputs, and FDI, the UN human development index (HDI) is associated with technology use.

Infrastructure is an important factor in technology use, whether in poor developing nations or advanced ones, and its inclusion in our conceptual models has been justified earlier in Chap. 3. For Africa, lagging infrastructure was described earlier in the chapter. Infrastructural correlates have manifested themselves as investment variables (for telecom, in Oyelaran-Oyeynka and Lal 2005), installation and maintenance costs for cellphone towers (Buys et al. 2009), and supply of electrical power as well as other infrastructure that was crucial for ICT development and adoption in Ghana (Ewusi-Mensah 2012). *We posit that sales value lost from electrical outages and time delay in obtaining electricity reduce use of technologies.* Figure 9.5 shows a shopkeeper of a small mobile phone shop busy charging cellphones.

Business efficiency and competition have rarely appeared in the digital divide literature for nations and states/provinces. This might be due to limited data availability until recently on these dimensions. Pick and Azari (2011) posited a factor, business and technology utilization, which included technological readiness and venture capital, which relate to competition; that factor was consistently influential in a structural equation model of use of technologies in a multi-country global study. African digital divide studies have advocated deregulation of telecom markets (Onyeiwu 2002; Roycroft and Anantho 2003) and simplifying licensing regimes (Buys et al. 2009) to encourage competition. We reason that business efficiency and greater competition lead to a more vibrant private sector which stimulates technology use and therefore posit that *the delay in obtaining a business license is associated with reduced technology use, while competition, measured by a national competition index, increases ICT use.*

Note that a study by Fuchs and Horak (2008) has cautioned that neoliberal policies of market liberalization and privatization are not a panacea for the digital inequities in African nations; the authors corroborate this by providing an example of Ghana where such policy shifts have hardly contributed to increased access to ICTs due to prevalent social and poverty related problems that involve fundamental quality of life issues such as access to food, clean drinking water, shelter, education,

Fig. 9.5 Shopkeeper charging cellphones in small mobile shop in Nairobi, Kenya. *Source* Tom Spender

and lack of reliable electricity. This provides further justification for inclusion of UN's Human Development Index and infrastructural variables related to electricity in our conceptual model.

Surprisingly, *technology cost* has not been frequently included in digital divide studies. While discussing the proliferation of internet cafes in urban areas and central business districts in African nations, Warf (2013) commented that such cafes derive a substantial proportion of their profit from non-African tourists and the relatively well-off; as a result, their fees are often too high and hardly affordable for low income populations in African nations. We reason that at the low income level of most African nations, many with annual incomes under $1500 per capita, costs of technologies constitute very important factors. Many African citizens have to weigh technology use against basic necessities of life, and would reject use unless cost is very low. *We posit that fixed broadband and mobile cellular internet tariffs reduce use of technologies.*

Societal openness refers to societies that have greater freedoms, legal system, effective and transparent government, and has been found to foster ICT use in prior chapters. In Africa, censorship in sub-Saharan countries has often been employed as a tool to curb competition, maintain the state's telecommunications monopoly, and/or curb democratic rights of citizens and opposition groups (Warf 2013). Hamilton's (2003) empirical results on substitution of fixed phones by mobile telephony supports the theory that credible institutions that can provide stable and sound legal structures will foster modernization and a surge in Africa's telecom sector. Based upon evidence from prior literature, we reason that an autocratic or highly controlling society that lacks credible institutions and ICT laws is likely to limit ICT use since cyberspace could threaten the government; hence we posit that *laws related to ICT* and *the effectiveness of lawmaking bodies are positively related to use of technologies*.

We now describe various data sources for the dependent and independent variables included in our conceptual model along with their definitions, and descriptive statistics.

9.5 Data Use for This Study

Data for various dependent and independent variables were obtained from a variety of sources such as World Bank's World Development Indicators database (The World Bank 2012), World Telecommunications/ICT Indicators Database of the International Telecommunications Union (ITU 2012, 2013), the World Economic Forum's Global Information Technology Report (WEF 2012), and the International Human Development Indicators database produced by the United Nations Human Development (The United Nations 2013). To control for population size, variables were converted to per capita (in terms of per person or per 100 persons) wherever possible. A handful of variables are expressed as indices and the young dependency ratio independent variable is expressed as a ratio. Missing data were a significant challenge during data compilation. Due to widespread missing data for a number of variables—mainly independent ones, seven nations had to be eliminated from the study (out of a total of 58 nations). Nations excluded are Mayotte, Reunion, Saint Helena, Sierra Leone, Somalia, South Sudan, and Western Sahara, which account for approximately 2.6 % of Africa's 2012 estimated population. Time simultaneity of the variables was honored to the extent possible and descriptive statistics for all 5 dependent and 18 independent variables were computed.

Table 9.1 shows the five dependent variables as well as 18 independent variables. Many of the variables for nations are converted to per capita; however, henceforth in the chapter text, per capita shown in the table will not be repeated.

The descriptive statistics show that technology use in Africa is at a low level by worldwide standards. For instance, in 2012 the average percent of individuals using the Internet and fixed broadband for the African nations was 13.8 and 1.0 % respectively, which is much lower than the world average of 41.3 and 9.8 % (ITU 2012). Within Africa, there is large variation among nations, as seen by high coefficients of variation for Internet Users (102.3), Fixed Broadband Subscriptions (232.0), Mobile-cellular Subscriptions (57.1), and Fixed Telephone Subscriptions (166.3), while Social Network Use (17.6) is moderate. The variation implies large differences geographically throughout Africa, as seen for example in Fig. 9.2 for mobile-cellular subscriptions. This chapter concerns analyzing and explaining this variation. The moderate differences for social network use can be ascribed to its use predominantly by more educated people in metropolitan areas, which would imply a more consistent level of use throughout the continent.

The dependent technology variables, with the exception of fixed phones, are highly correlated, as seen in Table 9.2. Even though four of them are closely

Table 9.1 Variable definitions and descriptive statistics

Variable category	Variable name	Definition	Year of data collection	Source	Mean (N[a])	SD
Dependent variables						
ICT use	Internet users per 100 people	Percent of individuals using the internet	2012	ITU	71.76 (54)	41
ICT use	Fixed broadband internet subscriptions per 100 people	Fixed (wired) broadband subscriptions per 100 persons	2012	ITU	13.83 (53)	14.15
ICT use	Mobile subscriptions per 100 people	Mobile-cellular telephone subscriptions per 100 persons	2012	ITU	3.77 (54)	6.27
ICT use	Fixed telephone subscriptions per 100 people	Fixed telephone subscriptions per 100 persons	2012	ITU	1.00 (52)	2.32
ICT use	Use of social networks per 100 people	How widely virtual social networks are used (e.g., Facebook, Twitter, LinkedIn) for professional and personal communication per 100 persons	2010–11	WEF	4.43 (54)	0.78
Independent variables						
Demographic	Young dependency ratio	Ratio of population aged 0–14 to population aged 15–64	2010	WB	73.06 (45)	16.59
	Urban	Urban population as percent of total population	2010	WB	41.43 (46)	18.4
Education	Literacy	Adult literacy rate per 100 people	2005–10	WB	62.55 (42)	18.41
Training	Firms offering formal training	Percent of firms offering formal training	2005–10	WB	30.54 (42)	10.8
Economic	Gross national income per capita	Gross national income per capita ($)	2010	WB	3194.08 (49)	6673.95
	Industry output	Industry output as a percent of GDP	2010	WB	32.92 (38)	17.34
	Manufacturing output	Manufacturing output as a percent of GDP	2010	WB	10.50 (45)	7.65
	Agricultural exports	Agricultural exports as percent of total exports	2010	WB	7.65 (37)	12.23
	Foreign direct investment	Foreign direct investment per person	2010	WB	681.46 (45)	792.89

(continued)

Table 9.1 (continued)

Variable category	Variable name	Definition	Year of data collection	Source	Mean (N[a])	SD
Business efficiency/ competition	Time to obtain business license in days	Time required to obtain operating license (in days)	2005–10	WB	26.33 (40)	16.33
	Level of competition for internet, long distance fixed phone, and cell phone	Level of competition index for Internet services, international long distance services, and mobile telephone services on a 0–2 (best) scale	2011	WEF	1.42 (34)	0.52
Infrastructure	Sales value lost from electrical outages	Percent of sales value lost due to electrical outages	2005–10	WB	6.17 (42)	4.86
	Time required to get electricity in days	Time required (days)	2011	WB	133.40 (48)	109.72
Socio-economic	Human development index of UN	Composite index for long and healthy life, knowledge and a decent standard of living.	2010	UN	0.80 (51)	0.401
Societal openness	Effectiveness of lawmaking bodies	How effective is the national parliament/ congress as a law-making institution (scale 1 lowest to 7 highest)	2010–11	WEF	3.49 (32)	0.784
	Laws related to ICT	Laws relating to ICT (scale 1 lowest to 7 highest)	2010–11	WEF	3.35 (34)	0.72
Technology cost	Fixed broadband internet tariffs	Subscription charge for fixed (wired) broadband internet service (in PPP $/month)	2010	WEF	225.44 (33)	385.1
	Mobile cellular tariffs	Average per-minute cost of different types of mobile cellular calls in PPP ($/min)	2010	WEF	0.41 (34)	0.18

Data sources ITU International Telecommunications Union, World Telecommunication/ICT Indicators 2012; *UN* United Nations; *WB* World Bank World Development Indicators, 2012; *WEF* World Economic Forum
[a]N Sample size without imputation

related, prior research has indicated considerable variation in their regression correlates (see Chap. 4, as well as Guillén and Suárez 2005; Pick and Azari 2008), justifying keeping them separate rather than merging them into an index. The lack of correlation of fixed phone subscriptions can be explained by their displacement by mobile phones in most parts of Africa (Rashid and Elder 2009). Fixed phones are fading due to their expensive infrastructure, reliability problems, limited functionality compared to mobile phones, and advantages of the mobility itself.

Table 9.2 Correlations of dependent variables

ICT dependent variables	Internet users	Broadband subscriptions	Mobile phone subscriptions	Fixed phone subscriptions
Internet users				
Broadband subscriptions	0.658***			
Mobile phone subscriptions	0.573***	0.477***		
Fixed phone subscriptions	−0.108	−0.016	−0.127	
Social network use	0.738***	0.488**	0.617***	−0.174

Note: *Indicates levels of significance–already indicated at the bottom of the table
*Signif. at 0.05 level
**Signif. at 0.01 level
***Signif. at 0.001 level

For the independent variables, Africa's lower worldwide development level is evident. For instance, the average literacy is 63 % and gross national income per capita (GNIPC) is only $3194. Manufacturing is low throughout the continent, comprising only 10.5 % of gross domestic output, while the continent has high agricultural production including substantial export of agricultural raw materials (excluding food) of 7.6 % of merchandise exports. Incoming foreign direct investment has an average of $681 per capita, with a high coefficient of variation of 116.

On average African nations have low societal openness, as gauged by effectiveness of lawmaking bodies and extent of development of laws relating to ICT. The high costs of fixed broadband and mobile cellular usage serve as constraints to wider adoption and use of technologies, e.g. broadband monthly subscription is US$225 on average, which equates of 84 % of average income, so that its use would be restricted to more prosperous nations and wealthier users. Likewise average mobile-cellular cost of 41 cents per minute is a barrier in poorer African nations. At this level of cost, it is unsurprising that level of competition for internet, mobile phone, and long distance fixed phone services is relatively high (1.42 on a 0–2 worldwide scale) (WEF 2012). Another facet of Africa is business inefficiency, as seen by an over 4 month average delay in obtaining electricity and about a month delay for getting a business license.

9.6 Spatial Arrangement of Technologies Within Africa

Cluster analysis reveals distinctive technology-use clusters within Africa, which are mostly spatially contiguous. K-means cluster analysis was performed for four of the five dependent variables, namely, internet users, broadband subscriptions, mobile-cellular subscriptions, and fixed telephone subscriptions. Social network use was not included in the cluster analysis due to its multiple missing values in the West Coast, Central-North, and Central parts of Africa, and to its high correlation with

Internet users, broadband subscriptions and mobile phone subscription, implying its addition to the analysis would have limited influence on the overall cluster findings. The clusters are seen geographically in Fig. 9.6, with their characteristics given in Table 9.3.

Clusters 1 and 2, which stand out as much higher than the others, have averages of internet users at 44.5 and 33.5 and broadband at 4.5 and 6.4 respectively; internet-use values correspond to world average and broadband corresponds to half of world average. There is a strong geographic agglomeration, as expected from Tobler's Law, introduced in Chap. 1 (Goodchild and Janelle 2004; Longley et al. 2011). Mobile phone subscriptions are 40 % above world average reflecting leapfrogging and lower relative cost relative to their more prosperous economies. These more

Fig. 9.6 K-means cluster analysis, use of 4 technologies, 5 clusters, N = 51, Africa 2012

Table 9.3 Characteristics of five K-means clusters derived from 4 dependent variables, 51 African Nations, 2012

Dependent ICT use variable	Cluster 1	Cluster 2	Cluster 3	Cluster 4		Cluster 5		Ratio of high to clusters
Internet users	33.47	44.58	10.06	14.9		5.12		8.71
Broadband subscriptions	6.38	4.47	0.55	0.62		0.13		49.38
Mobile telephone subscriptions	153.41	120.57	168.72	83.85		38.36		4.4
Fixed telephone subscriptions	22.85	13.02	4.45	2.96		0.97		23.5
	Libya	Egypt	Botswana	Algeria	Mauritania	Angola	Lesotho	
	Seychelles	Mauritius	Gabon	Benin	Namibia	Burkina Faso	Liberia	
		Morocco		Cameroon	Nigeria	Burundi	Madagascar	
		South Africa		Cape Verde	Republic of Congo	C. African Rep.	Malawi	
		Tunisia		Côte d'Ivoire	Sao Tome and Principe	Chad	Mozambique	
				Equatorial Guinea	Senegal	Comoros	Niger	
				Gambia	Sudan	Dem. Rep. of Congo	Rwanda	
				Ghana	Swaziland	Djibouti	Tanzania	
				Guinea-Bissau	Zambia	Eritrea	Togo	
				Kenya	Zimbabwe	Ethiopia	Uganda	
				Mali		Guinea		

economically advanced African nations in clusters 1 and 2, which historically established more mature fixed telephone bases, have fixed phone subscriptions that are seven-fold higher than for clusters 3 and 4. The very high level of mobile phone subscriptions in these nations points to the mobile phone as the primary widespread mode for information technology in these countries. These African "high-tech" countries are all coastal, which reflects Africa's historical greater urbanization/industrialization on its coasts, a topic discussed later.

Clusters 3 and 4 tend to be located in the northwest, central west, and south parts of the country. About half are coastal and they are agglomerated geographically, with the exception of Sudan and Kenya. These nations have low internet use and almost no broadband use by world standards, while mobile phones are somewhat higher than world average, reinforcing that the cellphone is Africa's most widespread and primary ICT. Kenyans in Nairobi streets using their cell phones are seen in Fig. 9.7.

At the low end for the continent is Cluster 5, comprising mostly central and eastern countries. Here technology use is among the world's lowest. The reasons include history of less urbanization/industrialization, poor economies with lower productivity, less infrastructure including technological, and lower literacy and education (Davidson 1994; Gebremichael and Jackson 2006; Warf 2013).

Overall, the cluster analysis points to a continent with sharp inequalities in use of technologies, agglomerations of contiguous nations at similar technology levels, mobile phone "leapfrogging" occurring at a high rate but especially in the middle to high level clusters, and coastal location as beneficial to higher technology levels. Our results are largely consistent with Onyeiwu (2002) which classified African nations into six classes based upon their extents of digitization as measured by an IT index.

The amount of agglomeration for each of the five technology dependent variables is first analyzed for spatial autocorrelation using the Moran's I statistic. Moran's I was calculated for the full study sample of 51 countries, as well as for 44 "lower tech" nations that excluded the seven nations in Clusters 1 and 5, namely Egypt, Libya, Mauritius, Morocco, Seychelles, South Africa, and Tunisia. Four of them are Arab-speaking Mediterranean nations, while South Africa is Africa's most advanced economy, and the island nations of Madagascar and Seychelles are

Fig. 9.7 Cell phone users on the streets of Nairobi. *Source* Tom Spender

Table 9.4 Moran's I values for African Nations, 2010–2012, U.S. States, 2010, and Chinese Provinces, 2009

Moran's I values for African Nations, N = 51, 2010									
		Internet users	Fixed-broadband internet subscriptions		Mobile subscriptions	Fixed telephone subscriptions	Use of virtual social networks		Average
		0.059	0.288***		0.277**	0.324***	0.001		0.223
Moran's I values for less-urban African Nations, N = 44, 2010									
		Internet users	Fixed-broadband internet subscriptions		Mobile subscriptions	Fixed telephone subscriptions	Use of virtual social networks		Average
		0.034	−0.034		0.264**	0.133	−0.013		0.077
Moran's I values for United States, N = 50, 2010									
	Desktop/ laptop in household	Log of internet access at home	Broadband adoption in household	Persons in cellphone-only household	Mobile wireless high-speed devices	Persons in fixed-phone-only household	Facebook users	Twitter users	Average
	0.547***	0.471***	0.457***	0.621***	0.230***	0.648***	−0.004	0.069	0.381
Moran's I values for China, N = 31, 2009									
PCs per 100 urban families	PCs per 100 rural families	Internet users per 100 pop	Broadband subscribers per 100 pop	Mobile telephone subscribers per 100 pop		Urban fixed phone subscribers per Capita	No. of domain names per 100 pop	No. of web pages per Capita	Average
0.346***	0.206*	0.264**	0.272**	0.205*		0.252**	0.615***	0.086*	0.285
Moran's I values for China, N = 29, Excluding Beijing and Shanghai, 2009									
0.343***	0.239**	0.137	0.258**	0.061		0.143	0.236**	0.177*	0.199

*Signif. at 0.05
**Signif. at 0.01
***Signif. at 0.001

among the continent's most prosperous. The 44-nation sample, which in this section allows initial comparison of spatial agglomeration, will be repeated in comparing regression findings for the whole continent and its lower tech countries.

As seen in Table 9.4, for the full 51-nation sample, broadband subscriptions, mobile subscriptions and fixed telephone subscriptions are highly agglomerated, whereas internet users and virtual social network use have random spatial patterns. These significant effects are explained by the first three variables tending to have sharp contrast between high value agglomerations of the North African, Mediterranean nations plus southern tip ones, compared to the lower-valued agglomerated central swathe of Africa. Mobile phone subscriptions and social network users are more random throughout the continent (e.g., for mobile phones as seen in Fig. 9.2 and social networking in Fig. 9.8). For the 44-nation lower-tech sample, spatial autocorrelation

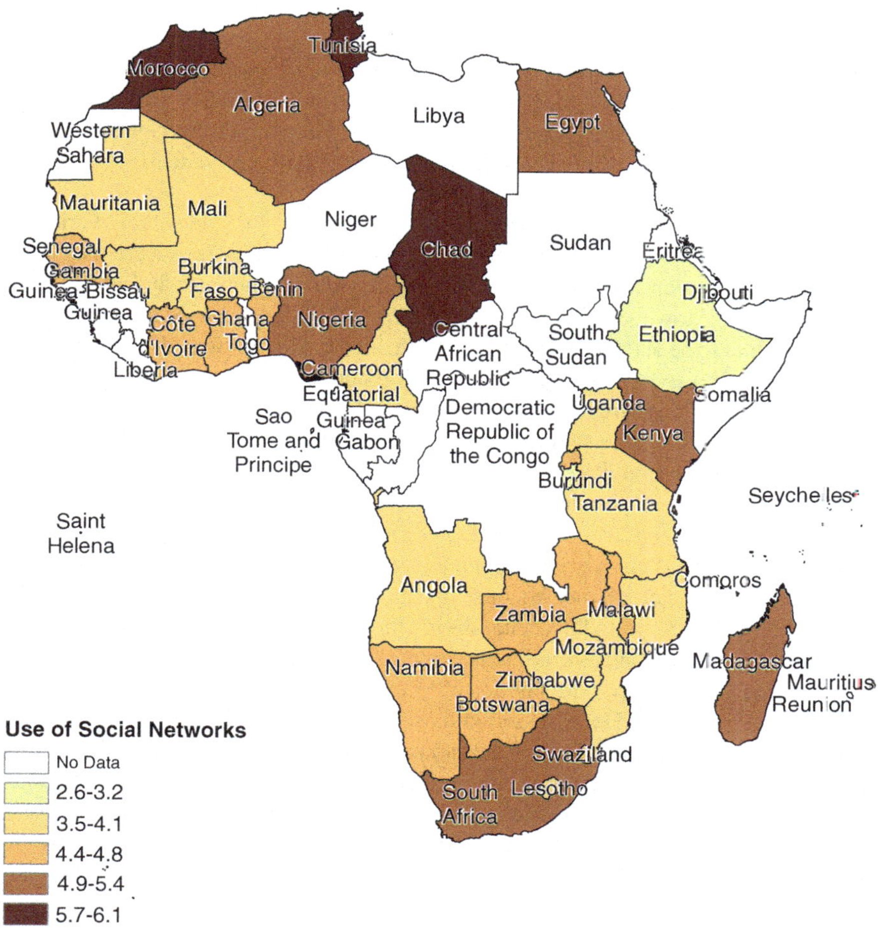

Fig. 9.8 Use of social networking in Africa per 100 population, 2010–11

results are weaker, present only for mobile-cellular subscriptions. This is understandable since the juxtapositions of high-value with low-value agglomerations of nations are lessened by elimination of North Africa and the far south.

Comparing the full samples which are predominantly agglomerated, the main difference is that Africa is more random in geographic patterns for Internet use than for the United States and China. The random pattern for African social media resembles that for the U.S. For the less urban samples for Africa (N = 44) and China (N = 29), there is less significant agglomeration, which is understandable because the samples' dependent variables become more evenly balanced geographically.

9.7 Determinants of ICT Utilization for African Nations

9.7.1 Background on OLS Regressions

The conceptual model is tested with five dependent technology use variables of internet users (as percent of population), broadband subscriptions per 100 persons, mobile-cellular subscriptions per 100 persons, fixed telephone subscriptions per 100 persons, and social network use, which is a Likert variable on a scale of 1 (lowest level of use to 7 highest level).

Ordinary least squares (OLS) regression is performed for five dependent variables, based on the eighteen independent variables. Two sets of regressions were conducted for each ICT dependent variable, one for the full set of N = 51 nations, and another for a slightly reduced, "low tech" set of N = 44 nations. The reduced set excluded countries that were found to be at the high end of ICT utilization spectrum from K-means cluster analysis, namely Egypt, Libya, Mauritius, Morocco, Seychelles, South Africa, and Tunisia. Two sets of regressions were performed to examine if moderate-low ICT utilization nations have the same correlates of ICT utilization as the entire continent. Also, for each of the samples, we further conducted two sets of regressions, one with Gross National Income per capita included as an independent variable but extent of urban population excluded as an independent variable, and in the other, Gross National Income per capita excluded but extent of urban population included.

This was necessitated since income and urbanization are often significant correlates of ICT utilization in the digital divide literature; however these two variables also tend to be significantly correlated with each other. In our study, GNI and urban are moderately positively correlated with a correlation coefficient of 0.461 that is significant at 0.01 level. Hence to avoid multi-collinearity problems, GNI and urban were not allowed to enter into regressions simultaneously. Missing values for a handful of independent variables were imputed by the mean of the variable since such a function is readily available in any statistical software package. We chose to impute missing values and not shrink the sample size for variables to preserve the richness of regression analysis. However imputation by means may be considered to be a limitation of this analysis.

Variables were pre-tested for multi-collinearity, stepwise regression methods were employed, and the regression findings were tested for conformance to regression assumptions by the Joint Wald Statistic, Koenker (BP) Statistic, and Jarque-Bera Statistic. OLS regression residuals were further tested for spatial autocorrelation using Moran's I. Further details and insights into the research methodology adopted in this chapter have been outlined earlier in Chap. 1.

9.7.2 Results of Regression Analysis

Results of the regression analysis provide support for the conceptual model and validate the benefit of screening for spatial autocorrelation of error terms. As seen in Table 9.5, for the full sample with GNIPC included, factors in the societal openness and economic categories dominate. The most important correlate, consistent across the five dependent variables, is ICT laws. Countries with well-developed laws for ICT were likely to have higher ICT levels. We reason that a country with well-developed ICT laws is encouraging to vendors, content providers, infrastructure developers, trainers/educators, and other service entities.

On the other hand, effectiveness of law-making bodies reduced broadband subscriptions and fixed telephone subscriptions. We reason that ineffectiveness of congresses and legislatures characterize periods of turmoil, instability, and change, which might reduce regulation at times and necessitate better information exchange, and provide openings and opportunities for deployment of technologies. Economic factors were influential selectively across all five dependent variables. GNIPC was associated with broadband and fixed telephone subscriptions, while industrial and/or manufacturing output was associated with internet users, mobile subscriptions, and social networks. The inverse association of agricultural exports on use of virtual networks is explained by countries with heavily agricultural economies being less amenable to social networking, which appears to be more present in Africa's metropolitan and urban areas.

For the full sample with urban included, ICT laws again was the most important correlate across all five regressions, while effectiveness of law-making bodies was significant only in reducing broadband. Economic influences disappeared for broadband and fixed telephone subscriptions and urban was of secondary importance for mobile and social networks.

For the "lower tech" 44-nation sample (see Table 9.6), with GNIPC included, economic categories dominate while societal openness is reduced. GNIPC, industrial output, manufacturing output are important for reasons previously given. Export of agricultural materials reduces use of virtual networks, with explanation similar to the full sample. Since ICT laws are much more developed in the seven "high tech" nations, eliminating them reduces ICT laws' effect except for mobile subscriptions and social networks. An explanation is that for lower-tech nations, the latter two technologies remain sensitive to legislative ineffectiveness for the reason stated previously of rapid change/turmoil and can remain favored for the general citizenry by ICT laws. For the 44-nation sample, with urban included, urban, ICT laws, and manufacturing/industrial output account together for the effects.

Table 9.5 Standardized regression results for dependent variables, 2012, 51 African Nations

Dependent variable											
		GNI included, urban excluded					Urban included, GNI excluded				
Category	Independent variable	Internet users per 100 persons	Fixed broadband internet subscriptions	Mobile subscriptions per 100 persons	Fixed telephone subscriptions per 100 persons	Use of virtual social networks	Internet users per 100 persons	Fixed broadbands internet subscriptions	Mobile subscriptions per 100 persons	Fixed telephone subscriptions per 100 persons	Use of virtual social networks
Demographic	Dependency ratio										
	Urban							0.24	0.258*		0.267*
Education	Literacy										
Training	Firms offering formal training (percent)										
Economic	Gross national income per capita (Atlas method)		0.350**		0.407**						
	Industrial output as percent of GDP	0.227*		0.494***			0.23		0.438***	0.234	
	Manufacturing output as percent of GDP	0.384***				0.337**	0.384***				0.278**
	Agricultural exports					−0.342**					−0.223*
	FDI										
Business efficiency	Time to obtain business license in days										
Competition	Competition index	0.233*					0.233*				
Infrastructure	Sales value lost from electrical outages										
	Time required to get electricity in days										

(continued)

Table 9.5 (continued)

Dependent variable											
		GNI included, urban excluded					Urban included, GNI excluded				
Category	Independent variable	Internet users per 100 persons	Fixed broadband internet subscriptions	Mobile subscriptions per 100 persons	Fixed telephone subscriptions per 100 persons	Use of virtual social networks	Internet users per 100 persons	Fixed broadbands internet subscriptions	Mobile subscriptions per 100 persons	Fixed telephone subscriptions per 100 persons	Use of virtual social networks
Socio-economic	Human development index of UN					0.233					
Societal openness	Effectiveness of lawmaking bodies (scale 1 lowest to 7 highest)		−0.406**		−0.301*			−0.411**			
	Laws related to ICT (scale 1 lowest to 7 highest)	0.589***	0.311*	0.461***	0.265*	0.641***	0.589***	0.392**	0.451***	0.353*	0.580***
Technology cost	Fixed broadband internet tariffs										
	Mobile cellular tariffs ($/min) PPP		−0.241*					−0.22			
Regression adjusted R square and significance level		0.428***	0.314***	0.305***	0.245***	0.484***	0.409***	0.235**	0.357***	0.099*	0.522***
Sample size (N)		51	51	51	51	51	51	51	51	51	51
OLS regression tests											
Joint Wald statistic		65.4*	33.6*	41.8	22.1*	88.7*	65.4*	12.7*	48.6*	10.3*	72.8*
Koenker (BP) statistic		2.498	3.840	1.732	3.714	5.463	2.498	5.182	1.655	1.002	1.902
Jarque-Bera statistic		10.218*	475.006*	2.094	333.964*	1.751	10.218*	327.960*	1.269	346.620*	0.059
Spatial autocorrelation of residuals											
Moran's index		0.061	0.213**	0.126	0.236**	0.111	0.06**	0.218**	0.102	0.298***	0.115
p-value		0.379	0.003	0.116	0.002	0.156	0.379	0.003	0.188	0.00008	0.146

*Signif. at 0.05
**Signif. at 0.01
***Signif. at 0.001

Table 9.6 Standardized regression results for dependent variables, 2012, 44 African Nations

Dependent variable											
		GNI included, urban excluded					Urban included, GNI excluded				
Category	Independent variable	Internet users per 100 persons	Fixed broadband internet subscriptions per 100 persons	Mobile subscriptions per 100 persons	Fixed telephone subscriptions per 100 persons	Use of virtual social networks	Internet users per 100 persons	Fixed broadband internet subscriptions per 100 persons	Mobile subscriptions per 100 persons	Fixed telephone subscriptions per 100 persons	Use of virtual social networks
Demographic	Dependency ratio										
	Urban							0.374*	0.306*	0.458**	0.392**
Education	Literacy										
Training	Firms offering formal training (percent)										
Economic	Gross national income per capita (Atlas method)	0.529***	0.750**		0.725**						
	Industrial output as percent of GDP			0.537***					0.436***		
	Manufacturing output as percent of GDP					0.376**	0.324***				0.291*
	Agricultural exports					−0.388**					
	FDI										
Business Efficiency	Time to obtain business license in days	−0.249	−0.174								
Competition	Competition index						0.272				
Infrastructure	Sales value lost from electrical outages										
	Time required to get electricity in days				−0.177					−0.268	
Socio-economic	Human development index of UN	−0.368**				0.294*					
Societal openness	Effectiveness of lawmaking bodies										

(continued)

Table 9.6 (continued)

Dependent variable											
		GNI included, urban excluded					Urban included, GNI excluded				
Category	Independent variable	Internet users per 100 persons	Fixed broadband internet subscriptions per 100 persons	Mobile subscriptions per 100 persons	Fixed telephone subscriptions per 100 persons	Use of virtual social networks	Internet users per 100 persons	Fixed broadband internet subscriptions per 100 persons	Mobile subscriptions per 100 persons	Fixed telephone subscriptions per 100 persons	Use of virtual social networks
	(scale 1 lowest to 7 highest)										
	Laws related to ICT (scale 1 lowest to 7 highest)			0.393**		0.632***	0.459**		0.346*		0.589***
Technology cost	Fixed broadband internet tariffs										
	Mobile cellular tariffs ($/min) PPP		−0.216	−0.297*							−0.263
Regression adjusted R square and significance level		0.351***	0.583***	0.278**	0.550***	0.385***	0.284**	0.140*	0.280***	0.218**	0.705***
Sample size (N)		44	44	44	44	44	44	44	44	44	44
OLS regression tests											
Joint Wald statistic		146.2*	268.5*	20.3*	167.9*	74.6*	7.398931	3.139397	14.9*	4.009672	48.7*
Koenker (BP) statistic		1.563	5.730	6.903	0.661	11.988*	7.336052	7.570*	5.278	14.635*	3.347
Jarque-Bera statistic		28.803*	190.497*	1.063	28.681*	5.196	2.608925	82.385*	1.684	31.694*	0.23
Spatial autocorrelation of residuals											
Moran's index		−0.020	−0.043	0.225*	0.086	−0.028	0.078	−0.014	0.152	0.086	−0.093
p-value		0.975	0.82	0.0102	0.24	0.96	0.3	0.92	0.07	0.24	0.48

*Signif. at 0.05
**Signif. at 0.01
***Signif. at 0.001

The somewhat greater influence of urban in the "lower tech" nations versus the full sample is explained by urban standing out for technological uses more in this sample, which has overall a lower urbanization level.

9.7.3 Discussion: Implications of Findings

The study findings demonstrate that Africa's relatively high-tech countries are located in the North Africa adjoining the Mediterranean and in the far south of the continent, while the lowest tech nations are agglomerated in the middle and middle south of Africa. The reasons for this relate to the greater industrial and urban development of the high-tech parts of the continent due to their proximity to ocean transport and earlier history of economic development. The more central sections of the continent did not have the transport advantages and were too remote to industrialize and develop large cities. It can be expected in the 21st century that more of central Africa will industrialize and urbanize, especially since population is growing rapidly, a trend implying greater use of technologies. Presently, these central nations can be compared to the far western provinces of China (see Chap. 5) or northeastern states of India (see Chap. 6), regions that are more remote from transport, trading, commerce, and industries and which trail considerably in technological growth.

The spatial randomness for internet users and social network users relates to the greater evenness of their distribution throughout the continent. They agglomerate less since these forms of technology are relatively inexpensive and hence not tied to the economic disparities of the continent, as is seen for fixed broadband and fixed telephones. For mobile phones, location does matter, which might relate to the cell-tower and other mobile-phone infrastructure which is tied to the economic level of nations.

For the regression model, the categories of variables most strongly supported are economy, societal openness, demographic factors particularly urban, and technology cost. Education and literacy, so influential in the extensive literature for the national unit, in Africa have no impacts as measured by literacy and training. This implies that at the very low standard of living for most of Africa, education and training are supplanted by more basic factors of income, urbanization, and industrialization. It is surprising that societal openness is highly influential, but less so if its recent influence at the national level of change in the Arab Spring and its often profound effects in advanced nations are considered.

The absence of effects for infrastructure (electricity) and business efficiency/competitiveness can again be explained as less important, on the scale of things, in very basic and poor environments of most of Africa. Those factors also can be expected to grow in importance as Africa industrializes/urbanizes in this century. The model's weakness in incorporating spatial errors for certain dependent variables stems from strong starting spatial autocorrelation of the variable for broadband and fixed telephones in the full sample, although the spatial errors diminish for the "lower technology" subsample. Consequently, for broadband, standardized residuals are not randomly distributed, as shown in Fig. 9.9.

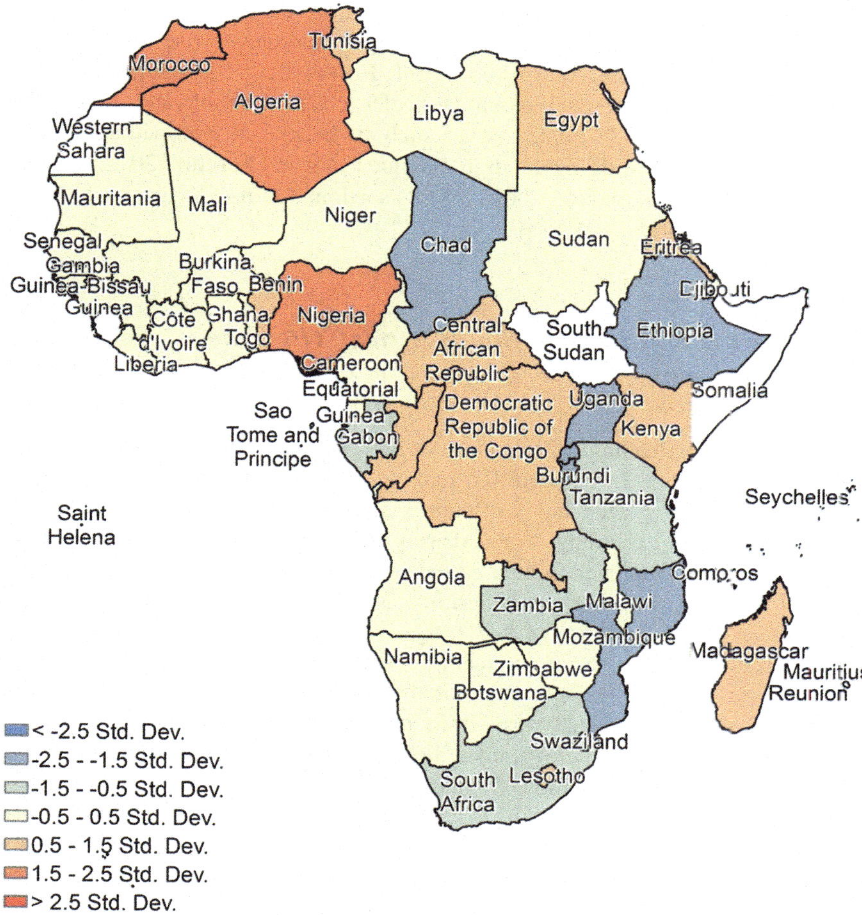

Fig. 9.9 Standardized residuals of estimate of fixed broadband subscriptions, 2012 (percent urban included, gross national income excluded)

9.8 Cases Studies of ICT Utilization in Africa

The literature on digital divide in Africa is comprised of several detailed case studies and vignettes that focus on ICT utilization in specific countries such as Egypt (Warschauer 2003; Wheeler 2007), Ghana (Alemna and Sam 2006; Ewusi-Mensah 2012; Fuchs and Horak 2008), Liberia (Best et al. 2010), Nigeria (Jagun et al. 2008; Bankole et al. 2011), South Africa (Chigona and Licker 2008; Mbatha et al. 2011; Fuchs and Horak 2008), Tanzania (Furuholt and Kristiansen 2007; Sein and Furuholt 2012), Tunisia (Danowitz et al. 1995), and Uganda (Martin and Abbott 2011). While a few of these studies focus on the adoption, diffusion, and use of a specific ICT within a particular context, for example cellular phones for mobile banking, agricultural development, and microenterprise development. (Bankole et al. 2011; Martin and Abbott 2011; Best et al. 2010; Jagun et al. 2008) and internet for social empowerment (Wheeler 2007; Furuholt and Kristiansen 2007), others examine ICT-related issues such as the role of intermediaries manifested as telecenters in bridging the digital divide (Sein and Furuholt 2012; Wheeler 2007; Furuholt and Kristiansen 2007). In this section, we discuss two case studies related to ICT in Africa.

9.8.1 Internet Diffusion in South Africa: The Role of Government

South Africa is a leading ICT nation in the African continent with an ICT Development Index of 3.95 (world IDI ranking of 84 among 157 nations) in 2012 compared to an average IDI index of approximately 3.75 and 2.00 for developing nations and Africa (excluding Arab African nations) respectively. In our cluster analysis of ICT utilization presented earlier in this chapter, South Africa belonged to cluster 2 along with Egypt, Mauritius, Morocco, Tunisia, and just behind Seychelles and Libya in terms of ICT usage.

Liberalization of the telecommunications sector in South Africa started in 1993 and continued to gain steam throughout the 1990s into the early 2000s. While providing fixed telephone lines was within the purview of the state-owned Telcom, mobile service providers were granted licenses with conditions to serve a vast majority (70 %) of the population. Similarly Telcom was mandated to build almost 3 million new telephone lines by 2002 with more than sixty percent in underserved areas such as townships and rural markets. Penalties were to be imposed for underachievement. As the provider of fixed telephone services, Telcom also controls fixed broadband infrastructure in South Africa. In 1997, the Universal Service Agency was established by the South African government; this agency was tasked with financing and setting up telecenters, especially in disadvantaged areas. Furthermore in 2002, the government enacted the Telecommunications Amendment Bill that allowed licenses to be issued to small telecom businesses in the highly underserved areas.

This case study illustrates that market liberalization along with laws and regulations enacted by the government creates the potential for higher access to ICTs (Fuchs and Horak 2008), especially among underserved, disadvantaged communities and is consistent with this chapter's central finding that laws relating to ICTs and effectiveness of law-making bodies foster and encourage adoption and use of ICTs. Fuchs and Horak (2008) however caution that privatization and liberalization may ensure access to fixed and mobile phones, and by extension to the Internet, but by themselves are not a panacea for the digital divide; essential underlying social, ideological, and economic factors such as affordability are equally or more important and should be considered in conjunction with infrastructural factors and the regulatory and legal environment while analyzing digital divides. This is consistent with independent factors incorporated in this chapter's conceptual model of ICT utilization.

In this context, Warf (2013) has highlighted the role of South African regional governments which have promoted the development of internet cybercafes in slums under the aegis of the Universal Service Agency. The role of community computing centers (CCFs), located in selected libraries in the city of Cape Town, and developed as part of an initiative of the Cape Town City Council has been discussed by Chigona and Licker (2008). CCFs provided access to computers and the internet free of cost for 45 min a day to the urban poor in the city of Cape Town. While acknowledging that the time limit and lack of training in using computers and accessing information on the web were impediments, CCFs provided computer access to those in need, those who were unable to pay, and also to the unemployed, who went to a CCF to draft resumes and apply for jobs. From a censorship perspective, South Africa, unlike many other nations in the African continent has negligible governmental interference in cyberspace (Warf 2013) which promotes societal openness, democratic free exchange of ideas, and freedom.

This case study is consistent with our overall conceptual model of ICT utilization which posits that a free and open society will increase ICT use. Overall the case highlights the role of governments at national, provincial, as well as city level to draft legislation, create laws, and mandate compliance for service to communities in need, thus catalyzing the adoption and use of ICTs.

9.8.2 Mobile Telephony for Nigerian Micro-enterprise Development

In this case study, we summarize findings of a case study (Jagun et al. 2008) that examined the impact of mobile telephony on the supply chain of a cloth-weaving microenterprise in Nigeria, a low human development nation, as classified by the United Nations, and the world's seventh most populous nation with a population of 173.6 million in 2013. This contrasts with the previous case which described the state of internet in South Africa, a medium human development nation, and an ICT leader among African nations. Figure 9.10 shows the discrepancy in 2012 in ICT adoption between Nigeria, ranked 122nd by the ITU's ICT Development Index, and South Africa, ranked 84th (ITU 2013).

Jagun et al. (2008) described the impact of mobile phone intervention on constituents of the supply chain in Nigeria's cloth-weaving sector. In Nigeria, in this sector, trading is slow, costly, and often risky. This is because data gathering about pricing, potential trading partners, suppliers, weavers, and customers is almost impossible without physical intervention requiring time-intensive travel. Further compounding matters are risks of opportunistic overpricing, or agreeing to contracts

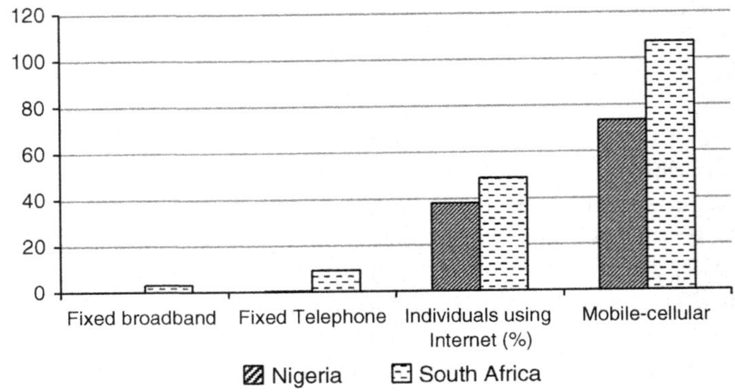

Fig. 9.10 ICTs in Nigeria and South Africa in 2013 (broadband, fixed phone, mobile cellular, in subscriptions per 100 population; all data from ITU)

by constituents of the supply chain (buyers, producers, intermediaries) with knowledge that they cannot be fulfilled.

Mobile phone calls were used in this sector to realize two main process-related benefits, improvement in time and financial costs of transactions. Mobile phones enabled checking identities and possibly of ordering, seeking and confirming orders, searching raw materials suppliers, confirming credit arrangements, and arranging for payment and subsequent pickup of raw materials (Jagun et al. 2008). Such calls often saved travel time, otherwise necessitated at a cost of US$8 per journey on the average, much higher in comparison to cheap call rates of 40 cents U.S. per minute; calls would typically last about 5 min. This underscores the importance of mobile phone tariffs which was found to be significantly associated (inversely) with mobile phone usage for our regression analysis of the 44-nation smaller sub-sample of low-tech ICT nations in Africa.

While mobile phone usage engendered process efficiencies, their impact on the structure of supply chains in the weaving sector was minimal. Intermediation was not eliminated; in fact it became entrenched by the emergence of a new category of intermediaries, the coordinator-weaver. Essentially when a weaver invested in a mobile phone, other weavers aligned themselves with the cellphone-owning weaver, thus enlarging the community of weavers (Jagun et al. 2008). In such cases, the coordinator-weaver was able to act on behalf of his peers to procure larger orders and allocate work much more efficiently than before. From a competitive advantage standpoint, weavers without cellphones became marginalized, thus raising the prospect of a competitive divide arising from the digital divide.

Unlike the previous case, this case study (Jagun et al. 2008) illustrates the use of mobile telephony, a form of ICT growing very rapidly in Africa. It has implications from an entrepreneurial standpoint and reinforces the importance of tariffs in relation to ICT usage. Next we discuss African policy implications of our findings.

9.9 Policy Implications for Africa and African Nations

Gillwald (2010) has argued that ICT policy outcomes in Africa, especially as they relate to the deployment of ICTs for poverty alleviation remain poor. This is attributed to the paucity of research that places policy reform and economic regulations in light of political dimensions existent in Africa.

One of the central findings of this chapter relates to the positive association of laws related to ICT with each of the five ICT dependent variables for our 51-nation full sample, and with three out of the five dependent variables for our 44-nation reduced sub-sample. The correlate "Laws relating to ICT" is measured by the effectiveness of a country's national parliament/congress as a law-making institution (WEF 2013). However, a related variable, effectiveness of lawmaking bodies, measured by the assessment of a country's laws regarding the use of ICTs (for example, in e-commerce, digital signatures, consumer protection) has been found to be negatively associated with selected ICT dependent variables.

While these findings do not relate directly to the use of ICTs for poverty alleviation, a policy implication is to reduce and ultimately remove all forms of censorship that are presently pervasive in African nations with a few notable exceptions. While levels of censorship vary widely across the continent (Warf 2013), they are detrimental to the adoption, diffusion and use of ICTs. National governments in Africa have sometimes censored cyberspace activities that challenge autocratic regimes, while at other times, censorship has been used as a garb to protect monopolistic tendencies of a country's nationalized telecommunications provider, for example, in Kenya (Warf 2013). This is a complex geopolitical problem with no easy solution. In extreme instances, such as in Sudan which openly censored the internet, Sudatel, the nation's telecom monopoly was blacklisted by the U.S. as part of larger efforts to resolve the violence in Darfur. Ultimately a nation's willingness to provide unfettered access to ICTs for its citizens is part of a commitment to honor basic human rights. In the absence of such a commitment, citizens' political mobilization is evolving into revolutions. Cases in point are the recent Arab Spring revolutions, first in Tunisia, and subsequently in Egypt, which saw significant use of new forms of ICTs, particularly social media to vent against socio-economic malaise, corruption, and censorship.

Once censorship is alleviated or removed, ICT laws need to be enacted that ensure free and fair access to a population. ICT programs and initiatives in post-apartheid South Africa were largely successful because various laws at the national, provincial, and even city levels were put in place to mandate compliance with ICT coverage requirements. Private as well as public sector telecom providers were mandated to provide their services and coverage to poverty-stricken, underserved, rural communities and penalties were imposed in the event of non-compliance (Fuchs and Horak 2008). This is especially recommended as a policy outcome for nations in Africa that are emerging from major social, political upheavals such as civil war, for example, in Liberia.

Other policy implications of our findings relate to economic development that results in increased income of local populations and alleviate concerns about affordability. In this context, our empirical analysis revealed significant inverse relationship between mobile cellular tariffs and cell-phone use for both the full sample as well as the reduced sub-sample. Several of the African nations in sub-Saharan Africa are classified as low human development nations; their ICT Development Indices also are among the lowest worldwide. It is hardly surprising that an individual with a daily earning of a dollar per day would not be willing to spend a third or half of that income for ICT usage. Privatization of telecommunications, coupled with deregulation has fostered competitiveness and reduced prices of ICT related services in several instances in Africa (for example, in Seychelles, Onyeiwu 2002) and are recommended as policies. However, as cautioned by Fuchs and Horak (2008), market liberalization may increase the potential of ICT access, but the digital divide is still likely to persist if income inequality, educational and ICT skill deficiencies, all part of the broader tapestry of socio-economic development, are not remediated.

Our cluster analysis findings indicate that African ICT clusters largely conform to Tobler's Law of geography, introduced in Chap. 1, which states that "everything is related to everything else, but near things are more related than distant things" (Goodchild and Janelle 2004). An ICT leading nation of Africa, such as South Africa, is flanked by moderate to high ICT use nations of Botswana and Seychelles in the south, while another leader Libya in the north is closely proximate to other high ICT peers of Egypt, Tunisia, and Morocco. Conversely, large swathes of sub-Saharan Africa (42 out of our 51 nation sample) are low-tech countries that are surrounded by other low ICT nations.

Fig. 9.11 University of Cape Town, South Africa. *Source* Ian Barbour

All of the high-tech nations also have extensive coastlines, which ensure unrestricted access to trade, tourism, and geographical proximity to locations where submarine cables enter into land from the seas. Increasing levels of cooperation with regional high to moderate ICT allies, especially ones whose people speak the same language is recommended as a policy direction. This can be possibly facilitated by programs in higher educational institutions, for example the University of Cape Town in South Africa, depicted in Fig. 9.11. It is worth recalling at this point, that language was a key barrier to successful and timely implementation at various stages of the Panaftel initiative. Increased cooperation with regional ICT leaders will ensure exchange of ICT know how, innovation, and best practices, increase trade in the telecommunications sector, bridge skills gaps, and foster increased adoption and use among ICT laggard nations.

9.10 Conclusion

This chapter examined utilization of various forms of ICT including broadband and social media technologies in the countries of Africa and how they have shaped the digital divide in the world's second most populous continent. Our model of ICT utilization posits five ICT dependent variables to be associated with eighteen socio-economic, education, income, ICT price, legal effectiveness, and infrastructure independent variables. Spatial analysis of ICT dependent variables reveal clusters of high utilization in North Africa (Egypt, Libya, Morocco, and Tunisia) and Southern Africa (Mauritius, South Africa, Seychelles) while west, central, and large parts of sub-Saharan Africa are at the low end of ICT utilization spectrum. Our finding on high-ICT nations has been rationalized by Onyeiwu (2002); that study has argued that a higher degree of economic openness in Africa is characteristic of nations that are either popular tourist destinations (such as Seychelles, Mauritius, South Africa, Egypt, and Morocco) or are located along the oceanic coastlines with unfettered access to the global economy and trade flows.

The leading determinants of ICT utilization for the full sample are laws that relate to the use of ICTs and the effectiveness of national parliament/congress as a lawmaking institution. This is supported by previous literature which has emphasized that national government in Africa must play a significant role to shape economic and telecommunications policies for public and private sector investments in ICT to build infrastructure, deregulate telecom markets to encourage competition, simplify licensing regimes, and encourage adoption of ICTs among individuals and organizations in education, government and business to increase ICT penetration and bridge the digital divide (Andoh-Baidoo et al. 2013; Buys et al. 2009; Oyelaran-Oyeynka and Lal 2005; Onyeiwu 2002; Roycroft and Anantho 2003). For a reduced set of 44 nations that excludes high ICT countries, the effect of ICT laws is moderated, but gross national income remains a significant predictor for broadband, fixed landline phones, and internet users. Also, mobile tariffs are found to reduce mobile phone subscription. Combined, this makes sense, and is consistent with contentions in existing literature that higher tariffs act as a barrier for ICT penetration for low ICT-intensity nations.

Two case studies on internet use in South Africa and use of mobile telephony in Nigeria, an ICT leader and an ICT laggard respectively reinforce our findings about the importance of societal openness, effectiveness of law-making bodies, and tariffs as important correlates of ICT adoption in Africa. Policy recommendations follow from empirical findings of this study and include removal of censorship, development of institutions of credibility that draft ICT laws and support privatization and deregulation with an eye on socio-economic development, and fostering of partnerships between Africa's high ICT leading nations and their significantly trailing regional neighbor countries.

References

Alemna, A. A., & Sam, J. (2006). Critical issues in information and communication technologies for rural development in Ghana. *Information Development, 22*(4), 236–240.

Andoh- Baidoo, F. K., Osatuyi, B., & Kunene, K. (2013). ICT capacity as the investment and use of ICT: Exploring its antecedents in Africa. *Information Technology in Development*, forthcoming.

Bankole, F., Bankole, O., & Brown, I. (2011). Mobile banking adoption in Nigeria. *The Electronic Journal of Information Systems in Developing Countries, 47*(2), 1–23.

Best, M. L., Smyth, T. N., Etherton, J., & Wornyo, E. (2010). Uses of mobile phones in post-conflict Liberia. *Information Technologies & International Development, 6*(2), 91–108.

Buys, P., Dasgupta, S., Thomas, T., & Wheeler, D. (2009). Determinants of a digital divide in sub-Saharan Africa: A spatial econometric analysis of cell phone coverage. *World Development, 37* (9), 1494–1505.

Chigona, W., & Licker, P. (2008). Using diffusion of innovations framework to explain communal computing facilities adoption among the urban poor. *Information Technologies & International Development, 4*(3), 57–73.

Cooper, F. (2002). *Africa since 1940: The past of the present*. New York: Cambridge University Press.

Danowitz, A. K., Nassef, Y., & Goodman, S. E. (1995). Cyberspace across the Sahara: Computing in North Africa. *Communications of the ACM, 38*(12), 23–28.

Davidson, B. (1994). *Modern Africa: A Social and Political History* (3rd ed.). Harlow, England: Pearson Education Ltd.

Ewusi-Mensah, K. (2012). Problems of information technology diffusion in sub-Saharan Africa: the case of Ghana. *Information Technology for Development, 198*(3), 247–269.

Fuchs, C., & Horak, E. (2008). Africa and the digital divide. *Telematics and Informatics, 25*, 99–116.

Furuholt, B., & Kristiansen, S. (2007). A rural-urban digital divide? Regional aspects of internet use in Tanzania. *The Electronic Journal of Information Systems in Developing Countries, 31*(6), 1–15.

Gebremichael, M. D., & Jackson, J. W. (2006). Bridging the gap in Sub-Saharan Africa: A holistic look at information poverty and the region's digital divide. *Government Information Quarterly, 23*(2), 267–280.

Gillwald, A. (2010). The poverty of ICT policy, research, and practice in Africa. *Information Technologies and International Development*, *6*(Special Edition), 79–88.

Goodchild, M. F.,& Janelle, D. G. (2004). Thinking spatially in the social sciences. In M. Goodchild & D. G. Janelle (Eds.), *Spatially Integrated Social Science*. New York: Oxford University Press.

Guillen, M. F., & Suarez, S. I. (2005). Explaining the global digital divide: Economic, political and sociological drivers of cross-national internet use. *Social Forces, 84*(2), 681–708.

Gyamfi, A. (2005). Closing the digital divide in Sub-Saharan Africa: Meeting the challenges of the information age. *Information Development, 21*(1), 22–30.

Hamilton, J. (2003). Are main lines and mobile phones substitutes or complements? Evidence from Africa. *Telecommunications Policy, 27*, 109–133.

Internet World Stats (2014). Internet usage statistics for Africa. Retrieved October 10, 2013 from http://www.internetworldstats.com/stats1.htm.

ITU. (2003). Assessment of telecommunication and ICT infrastructure in Africa. BDT Paper. Retrieved November 15, 2014 from http://www.itu.int/ITU-D/afr/events/arusha-ITU-NEPAD/Documents/doc7(elotu-english).pdf.

ITU. (2012). *World Telecommunication/ICT Indicators 2012*. Geneva, Switzerland: International Telecommunication Union.

ITU. (2013). *Measuring the Information Society 2013*. Geneva, Switzerland: International Telecommunications Union.

ITU. (2014). The world in 2014: ICT facts and figures. Retrieved November 15, 2014 from https://www.itu.int/en/ITU-D/Statistics/Documents/facts/ICTFactsFigures2014-e.pdf.

Jagun, A., Heeks, R., & Whalley, J. (2008). The impact of mobile telephony on developing country micro-enterprise: A Nigerian case study. *Information Technologies & International Development, 4*(4), 47–65.

Longley, P., Goodchild, M., Maguire, D., & Rhind, D. (2011). *Geographical Information Systems and Science*. Hoboken, NJ: Wiley.

Martin, B. L., & Abbott, E. (2011). Mobile phones and rural livelihoods: Diffusion, uses, and perceived impacts among farmers in rural Uganda. *Information Technologies & International Development, 7*(4), 17–34.

Mbatha, B. T., Ocholla, D. N., & Le Roux, J. (2011). Diffusion and adoption of ICTs in selected government departments in KwaZulu-Natal, South Africa. *Information Development, 27*(4), 251–263.

NEPAD. (2014). NEPAD e-Africa Programme, Retrieved November 15, 2014 from http://nepad.org/regionalintegrationandinfrastructure/infrastructure/ict.

Noam, E. M. (1999). *Telecommunications in Africa*. UK: Oxford University Press.

Ojedokun, A., & Lumande, E. (2005). Cooperative electronic networks of academic libraries in southern Africa. *Information Development, 21*(1), 66–73.

Onyeiwu, S. (2002). Inter-country variations in digital technology in Africa. In *Proceedings of the UNU/WIDER Conference on the New Economy in Development*, Helsinki Finland.

Oyelaran-Oyeyinka, B., & Lal, K. (2005). Internet diffusion in sub-Saharn Africa: A cross-country analysis. *Telecommunications Policy, 29*, 507–527.

Pick, J. B., & Azari, R. (2008). Global digital divide: Influence of socioeconomic, governmental, and accessibility factors on information technology. *Information Technology for Development, 14*(2), 91–115.

Pick, J. B., & Azari, R. (2011). A global model of utilization of technology based on governmental, social, economic, and business investment factors. *Journal of Management Information Systems, 28*(1), 51–85.

Population Reference Bureau. (2014). *2014 World Population Data Sheet* Washington, D.C.: Population Reference Bureau.

Rashid, A. T., & Elder, L. (2009). Mobile phones and development: An analysis of IDRC-supported projects. *The Electronic Journal on Information Systems in Developing Countries, 36*(2), 1–16.

Rhine, L. (2006). The impact of information technology on health information access in Sub-Saharan Africa: The divide within the divide. *Information Development, 22*(4), 242–250.

Roycroft, T. R., & Anantho, S. (2003). Internet subscription in Africa: Policy for a dual digital divide. *Telecommunications Policy, 27*(1–2), 61–74.

Sein, M., & Furuholt, B. (2012). Intermediaries: Bridges across the digital divide. *Information Technology for Development, 18*(4), 332–344.

United Nations. (2013). Human Development Report. Statistical Annex, 2012.

Warschauer, M. (2003). Dissecting the "digital divide": A case study in Egypt. *The Information Society, 19*, 297–304.

WEF. (2013). *The Global Information Technology Report, 2013*. Geneva, Switzerland: World Economic Forum.

Warf, B. (2013). *Global Geographies of the Internet*. Heidelberg: Springer.

Wheeler, D. (2007). Empowerment zones? Women, internet cafes, and life transformations in Egypt. *Information Technologies & International Development, 4*(2), 89–104.

World Bank. (2012). *World Development Indicators 2012*. Washington, DC: The World Bank.

Chapter 10
Comparisons of Countries

10.1 Introduction

Nations have varying dimensions and spatial patterns of digital usage. This chapter focuses on the comparison of findings between the nations of China, India, Japan, and the United States. It examines how regression results are similar or different as well as how much geographic patterns of ICT uses resemble each other or differ across the four countries. A core of common determinants of ICT uses is arrived at, based on the general model in Chap. 3, while outlier findings specific to certain nations are examined with the goal to interpret the outlier results. For spatial comparisons, the results of cluster analyses for the four nations are compared and common patterns or differences noted.

The goal of this chapter is to seek answers on the extent of common ground in the findings at the state/provincial/prefectural level for the four nations. What factors turn out to be common in predicting ICT levels for states and provinces? What are some unique features of nations related to the state-level divides within the nations? Why are geographic patterns similar or different? The chapter does not seek to generalize results beyond the four nations, but rather to suggest for the future large-scale comparative research studies of state-level or of smaller geographic units that can be generalized over large sets of nations. However, since the book's four study countries account for nearly half of the world's population and GDP, the comparative findings are meaningful and important. Questions of which policies can be derived from the book's findings appear in Chap. 11 on the roles and policies of governments.

The chapter's comparative framework is first assessed in terms of approaches and methods of comparative research (Hantrais 1999; Mills et al. 2006). Comparisons between nations have sometimes sought a universal framework and at other times have attempted to be culturally bound to each nation (Hantrais 1999). The former usually involves analytic, deductive analysis while the latter tends to utilize case study or other qualitative methods to dig deeply into the cultures and

J.B. Pick and A. Sarkar, *The Global Digital Divides*, Progress in IS,
DOI 10.1007/978-3-662-46602-5_10

values of each nation and identify important features for comparison. The present comparison leans towards the former approach but contains some of the latter in interpretation of findings. Also, the models for each nation are by necessity somewhat different from the general model due to differences in the data available for each nation at the state/provincial/prefectural level.

The choice of unit of analysis and the selection criteria for cases are essential to formulating comparisons (Hantrais 1999; Mills et al. 2006). The state, province, or prefecture is chosen as the unit of analysis because (1) the larger and bulkier nation unit might prevent micro effects to be uncovered, (2) the state unit of analysis has rarely been adopted in the digital divide literature, particularly for comparing two or more nations, and (3) from nation to nation, states, provinces, or prefectures are distinctive in their administrative and political structures. Case selection is critical to comparative studies (Ebbinghaus 2005; Mills et al. 2006). The selection of the present small sample of nations is deliberate but not random (Mills et al. 2006) and is based on countries with large GDP and/or population. Also, although this convenience sample is small, the subsamples, i.e. the samples of states/provinces/prefectures within each nation vary from 20 to 47 cases, hence are medium-sized and can be analyzed with positivist, quantitative methods.

Inconsistency of constructs is another challenge in comparative research (Hantrais 1999; Mills et al. 2006). Although some commentators argue for exact operational construct equivalence between cases, i.e. between states/provinces/prefectures (Mills et al. 2006), that is not possible presently due to limitations in the government and nonprofit data available at the state level within the four nations. For instance infrastructure, a construct in the general model, is operationalized as electrical capacity or electrical consumption. Although it is available at the state level for China, India, and Japan, the definitions for the first three nations vary somewhat, ranging from electrical output to electrical capacity to electrical power consumption. Others argue that comparisons can be done without exact equivalence, but rather involve pragmatic choice of factors, with recognition of cultural differences in meanings (Hantrais 1999; Mills et al. 2006).

Our comparison divides state/provincial/prefectural constructs into those that are generic between nations, such as urban, economic prosperity, education, and infrastructure, and those constructs unique to a nation, such as prefectural-level patents for Japan. Accordingly, this approach seeks to "separate patterns that are more general and isolate regularities from the context-laden environment" (Mills et al. 2006). The latter step for our comparison identifies culturally-unique features, such as Indian cooperative society members or overall freedom index of states in the U.S.; and features dependent on data collection limitations, such as ethnic categories in the U.S. and export commodities value in China. Although our approach has advantages and disadvantages discussed later under Limitations and Future Research, any empirical approach for comparative research on nations encompasses a mix of pluses and minuses (Mills et al. 2006).

The subsequent chapter section reviews prior comparative studies of the digital divide, followed by comparison of similarities and specific differences in our findings on determinants of ICT use for the sub-national units within each nation.

Next, the spatial patterns of clusters for the four nations are compared, followed by a section on limitations and future research. The ensuing section concerns the efforts of international organizations, the UN and World Economic Forum, to foster agendas, initiatives, and calls to action regarding the world's digital divide. Lastly, the Lenovo case study, based on primary and secondary data, examines how the firm progressed, first gaining dominance throughout China and moving to provide its products throughout the world. The case is then related to the chapter's comparative findings.

10.2 Review of Prior Comparative Studies of the Digital Divide

Although multiple studies have analyzed national samples (for instance, Robison and Crenshaw 2002; Baliamoune-Lutz 2003; Guillén and Suárez 2004; Gulati and Yates 2012) and some studies have intensively examined a single country (Compaine 2001; Doron and Jeffrey 2013), few have focused on comparison between nations and none on comparisons between state-level findings for nations.

Prior comparisons at the nation unit of analysis use varied approaches that span the range mentioned in the prior section. A cross-national study of e-commerce adoption posited a conceptual framework and tested it using sample of 10 nations chosen to span the socioeconomic spectrum and also be manageable as a large project (Gibbs et al. 2003). Its conceptual framework posited that the global environment (multinational strategies and supply chains, trade liberalization) influences the national environment (demographic, economic financial infrastructure, organizational, consumer factors), which in turn affects national policy, which lastly influences e-commerce adoption. The case studies were analyzed to determine, for all these factors, which ones helped or hindered e-commerce adoption, and a comparison was made across cases to identify common factors and exceptions. A key study finding was that differences in business-to-business (B2B) e-commerce are more based on global competition, while business-to-consumer (B2C) e-commerce is more locally based (Gibbs et al. 2003).

Among the differences of the Gibbs et al. study from book's analysis is that the former is more qualitative; does not include sub-national units; and selects sample nations to cover a broad social-economic-political range, rather than selecting them by size of economy.

A more intensive, follow-on study was based on a huge survey of over 2100 companies in three industries in the same 10 nations (Kraemer et al. 2005). A conceptual model was posited of 3-way relationships between the factors of firm globalization, e-commerce use, and firm performance. The empirical test used regression analysis, with the countries' comparative differences identified through dummy variables. Study findings confirmed the key difference in B2B versus B2C e-commerce from the earlier investigation, but also considered B2C to be both local and multi-domestic i.e. occurring differently in diverse parts of nations.

The Kraemer et al. (2005) study applied equivalent constructs to thousands of firms, with the advantage of more statistical power to build larger models. However, such a survey would have been difficult to accomplish presently because of issues of access to respondents, security, and potential missing data at the state/provincial/prefectural level. By contrast, our approach assures consistent state data within a single nation, but has more limited statistical power as well as some inconsistency in constructs between countries.

A study of five highly developed nations (United States, Sweden, Japan, South Korea, and Singapore), based a very large set of available surveys called "Cyber Life Observations" of individuals for 1997 and 2000, applied regression analysis to each nation to examine the influences of gender, age, education, income, and year on PC ownership at home, PC use at home, Internet use at any location, and Internet use at home (Ono and Zavodny 2007). Moreover, the timing of first computer use was estimated by survival analysis. Findings indicated each country had unique differences in determinants of IT usage and that comparative differences were greater between the three Asian nations than between Sweden and the U.S.

This study's sample exceeded 5000 respondents, yet only included independent variables that are common to all the nations, such as gender, age, and secondary education, avoiding culturally-unique variables. Regression analysis was applied for each nation, with exact equivalence of the survey and constructs. Although large-scale and rigorous, the study is constrained in the scope of its cross-cultural comparison, so findings are of less practical benefit for policymaking and action. Furthermore, such a large-scale survey approach might not allow access to state/provincial/prefectural respondents who would be able and willing to answer questions, whereas the secondary, and largely official data sources in our study provide a breadth of variables which have complete data.

Another type of approach does qualitative comparison of digital divide situations between nations. In an edited book of country-region and country narrative analysis of digital divides, including Japan, U.S. and the Americas, India, South Korea, and Thailand (Kagami et al. 2004), the editors arrived at comparative observations in the conclusion chapter, including that ICT use by socioeconomic groups with countries is increasing; education, income, and ethnicities influence usage; the developed/developing nation gaps is growing; and urban and rural areas have a wide gap. Relative to our research framework, this comparative volume is qualitative and unstructured, and does not examine the state/provincial/prefectural level.

Lastly, in a recent analysis of policies and progress in ICT utilization between China and India, a 2-way case-study comparison informs many aspects of their differences not possible within the present book's research framework (Chaudhuri 2012). Chaudhuri's analysis is helpful in understanding deeply how the two nations diverged greatly in the ICT realm.

For China, the emphasis has been on manufacturing and export of hardware, in concert with the present Chap. 5. This success is ascribed to a supply chain that emphasizes and encourages domestic suppliers, with stress on high volume of production at low cost (Chaudhuri 2012). The total value of China's technology hardware exports is five times that of India's exported outsourcing services (Chaudhuri 2012). Chinese small businesses also benefit by support and connectivity to the larger firms.

In India by contrast the large ICT providers often have supply chains based in China, or European supply chains that include nameplate firms. Hence India lacks the strong domestic supply chains of China and accordingly has higher costs (Chaudhuri 2012). Another distinction of India is that its outsourcing firms provide services predominantly outside the country, services that began as low-level ones and have gradually moved upwards (Chaudhuri 2012). Although presently worldwide outsourcing leaders, the firms confront higher costs if they move up-market, and further face the looming threat of Chinese outsourcing firms. The latter have primarily served the burgeoning China domestic market but are beginning to look outwards as well and threaten to compete with India on cost (Chaudhuri 2012). Additionally, although India has benefitted by English proficiency of its educated population, in its much larger consumer base has multiple languages compared to the rival nation's Chinese language as a broad, nationwide standard.

This study by Choudhuri (2012) informs deeply on the important China–India comparison in ICT. It yields a richness of comparison not possible with our quantitative approach, but does not delve into the quantitative intra-country or geographic differences.

10.3 Comparison of Correlates of ICT for China, India, Japan and the United States

China, India, Japan, and the United States constitute arguably the world's largest grouping of nations on the basis of population and GDP. The three most populous ones of China, India, and the U.S. constitute over 40 % of the planet's population, while the four nations together account for 44 % of world GDP. This implies that many of their state-level populations are very large, in fact some larger than nearly all nations. For instance China's Guangdong Province in 2010 had population of 104.3 million, which would be at 12th rank worldwide as a country, while India's Uttar Pradesh, Maharashtra, and Bihar had 199.5, 112.4, and 103.8 million people respectively, placing them at 5th, 11th, and 12th rank among nations.

The differences in the four nations are seen in a summary of their demographic educational, economic, infrastructural, societal openness, and governmental ICT priorities (Table 10.1). The countries contrast in percent of aged population (65+), led by Japan at almost a quarter of its people, versus 13 % for the U.S., 10 % for China, and 5 % for India. For ICT use, age can be influential as occurred somewhat in our study.

On education and literacy, the Japan and the U.S. are very high while China is moderate and India quite low having a 62.8 % literacy rate and secondary school

Table 10.1 Demographic and economic variables for China, India, Japan, and United States

	China	India	Japan	United States	World	Percent of world
Population, in millions, 2012	1350	1260	128	314	7058	43.2
Population rank worldwide, 2012	1	2	20	3		
Projected population, in millions, 2025	1635	1458	120	351	8082	44.1
Percent of pop. 65 years+, 2012	10	5	24	13		8
Percent urban, 2012	51	31	86	79		51
Adult literacy rate, 2005–2010	94.0	62.8	99.0	99.0		84.1
Secondary education, gross enrollment ratio, 2012, percent	80.1	60.2	105.1	96.5		
Gross domestic product (GDP), in billions of US$, 2010	5926.6	1727.1	5458.8	14,586.7	63,242.1	43.8
GDP per capita, in US$	4389	1371	42,781	46,469	8960	
Merchandise Exports, in billions of US$, 2010	1577.8	220.0	769.8	1278.3	15,211.3	25.3
Electricity production, kWh/capita, 2012	2602.6	697.1	8215.8	13,524.2	2,912.3*	
Judicial independence, on 1-to-7 (best) scale, 2012	3.9	4.3	5.8	4.9		
Intellectual property protection, on 1-to-7 (best) scale, 2012	4.0	3.5	5.3	5.0		
Government prioritization of ICT, on 1-to-7 (best) scale, 2012	5.6	5.1	5.1	5.3		

Sources Population Reference Bureau (2012), World Bank (2012), Dutta and Bilbao-Osorio (2012)
*2009

enrollment percentage of only 60.2 %. As seen in Chap. 6, India's educational challenge underscores and helps account for its large digital divide. Unsurprisingly, education is consistently related to technology use in its states.

GDP per capita varies hugely with the U.S. and Japan at around $45,000, compared to $4389 for China and only $1371 for India. These differences inform our comparisons. For instance, if mobile phones are to succeed in India, pricing needs to be very low, a factor which might limit use and prevent wide diffusion of 3-G mobile devices. On the other hand, Japanese and U.S. users can afford multiple high-end devices, with few cost barriers to widespread use of 3-G and 4-G devices.

Comparison of technology indicators for the four countries, seen in Table 10.2, reflects the educational and economic differences. Japan and the U.S. have similar high levels of mobile phone subscriptions, Internet access and use, PCs, and mobile broadband subscriptions. Fixed broadband subscriptions are less prevalent and have stabilized in growth, but far exceed levels for India and China (see Fig. 2.2). Use of virtual social networks is much higher in the U.S. than the other nations. However, the language and culture may be restraining its use in Japan and India. China has its

Table 10.2 Internet usage for China, India, Japan, and United States

	China	India	Japan	United States
Mobile phone subscriptions/100 pop.	64.0	61.4	95.4	89.9
Internet use, percent by individuals	34.3	7.5	78.2	74.0
Internet access, percent of households	23.7	4.2	81.3	71.6
Personal computer, percent of households with PC	35.4	6.1	83.4	75.5
Broadband internet subscriptions/100 pop.	9.4	0.9	26.9	27.6
Mobile broadband subscriptions/100 pop.	1.8	0.1	64.6	50.6
Use of virtual social networks, by 1-to-7(best) scale	4.9	4.9	5.2	6.2

Source ITU (2013)

own very popular native-language social networks such as QZone and Renren, while Facebook is banned. India appears to have the greatest challenge in Internet usage.

Considering the educational differences already noted with secondary education, another dimension of educational comparison concerns science and engineering education and at the highest level, STEM (Science, Technology, Engineering, and Math) education and R&D dimensions that depend on scientific workforce. Engineering workforce contributes to domestic technology expertise, useful for infrastructure development, build-up of domestic ICT companies, maintenance of technology, and advanced use. As seen in Table 10.3, engineering graduates are much more plentiful in China and India than in Japan or the U.S., a gap that has been widening. In 2006, China and India combined had almost a million engineering graduates, compared to 165,000 for Japan and the U.S., or a ratio of nearly 5 to 1. Although it can be argued that the latter degrees are of higher quality, the disproportionate ratio points to a shifting balance in national engineering capacities, which could become very influential in the future. Recently-hired Indian engineering graduates at seen in training at the large, modern Infosys Training Center in Mysore, India, in Fig. 10.1.

At the doctoral level, science and engineering doctorates remain more numerous in Japan and the U.S. compared to India and China, but not by much, and the balance is likely fairly soon to tilt the other way. Based on the scientific high-end indicators of patents and scientific/engineering articles, the U.S. clearly dominates, which might reflect both its eminence in graduate education and in corporate scientific, engineering, and technology R&D. This is evident in this chapter's case

Table 10.3 Number of engineering graduates, China, India, Japan, U.S., 2000 and 2006–7

	Number of engineering graduates			
	2000	2006	2008	Growth rate (%) 2000–2008
China	212,905	575,634	704,764	14.96
India		350,000[a]		
Japan	103,156	96,675	95,216	−1.00
United States	59,487	68,227	69,908	2.02

[a]2007 data. *Source* Baerjee and Muley (2008), NSF (2012, Chap. 2)

Fig. 10.1 Recently-hired engineering graduates hired by Infosys at its training center, Mysore, India

study, in which a Chinese PC firm seeks to break into global markets by acquiring major foreign technology firms and even actually shifting its headquarters to the U.S. For its population size, Japan also shows R&D leadership as reflected in articles and patents.

National ambition may alter even these balances in the future, at least for China. For instance the latest list of the most powerful supercomputers in the world (Top500 2014), places a Chinese NUDT in first place, followed by two in the U.S. and one in Japan (see Table 10.4). In considering future comparisons between these leading nations, these and other leading-edge R&D variables need to be considered, as early indicators of change.

Having done some general cross-national comparison across many dimensions relevant to the book's research, next the book's regression analysis findings on correlates of ICT use for states, provinces, and prefectures are summarized and compared. This is done by considering variables or variable groups in common across the studies in Chaps. 5–8, as well as the variables or variable groups that are unique to nations. This reflects the comparative approach described earlier in the chapter of distinguish some common attributes and others that are culturally unique.

The summary in Table 10.5 presents only findings for variables which are statistically significant, given in the left-hand column. For statistically insignificant variables, the reader can consult the regression tables in Chaps. 5–8, which provide

Table 10.4 World's largest supercomputers, 2014

Rank	Name	Computer	Site	Manufacturer	Country	Total cores	Rmax	Power	Processor	Operating system
1	Tianhe-2 (MilkyWay-2)	NUDT. Xeon E5-2692_Xeon Phi 31S1P, TH Express-2	National Super Computer Center in Guangzhou	NUDT	China	3,120,000	33,862,700	17,808	Intel Xeon E5-2692v2 12C 2.2 GHz	Kylin Linux
2	Titan	Cray XK7. Opteron 6274 + Tesla K20X, Cray, Genimi Interconnect	Oak Ridge National Laboratory, U.S.	Cray Inc.	United States	560,640	17,590,000	8209	Opteron 6274 16C 2.2 GHz	Cray Linux
3	Sequoia	Bue Gene/Q. Power PC A2, Custom	Lawrence Livermore National Laboratory, U.S.	IBM	United States	1,572,864	17,173,224	7890	Power BQC 16C 1.6 GHz	Linux
4		Riken. SPARC64 VIIIfx, Tofu	RIKEN, Japan	Fujitsu	Japan	705,024	10,510,000	12,660	SPARC64 VIIIfx 8C 2 GHz	Linux
5	Mira	Blue Gene/Q. Power PC A2, Custom	Argonne National Laboratory, U.S.	IBM	United States	786,432	8,586,612	3945	Power BQC 16C 1.6 GHz	Linux
6	Piz Daint	Cray XC30. Xeon E5-2670 + Tesla K20X, Aries	Swiss National Supercomputing Centre, Switzerland	Cray Inc.	Switzerland	115,984	6,271,000	2325	Xeon E5-2670 8C 2.6 GHz	Cray Linux
7	Stampede	PowerEdge C8220. Xeon E5-2680 + Xeon Phi, Infiniband	Texas Advanced Computing Ctr., Univ. of Texas, U.S.	Dell	United States	462,462	5,168,110	4510	Xeon E5-2680 8C 2.7 GHz	Linux
8	JUQUEEN	Blue Gene/Q. PowerPC A2, Custom	Forschungszentrum Juelich, Germany	IBM	Germany	458,752	5,008,857	2301	Power BQC 16C 1.6 GHz	Linux
9	Vulcan	Bue Gene/Q. Power PC A2, Custom	Lawrence Livermore National Laboratory, U.S.	IBM	United States	393,216	4,293,306	1972	Power BQC 16C 1.6 GHz	Linux
10		Cray XC30. Xeon E5-2697v2, Aries	United States	Cray Inc.	United States	225,984	3,143,520		Intel Xeon E5-2697v2 12C 2.7 GHz	Cray Linux
52		iDataPlex DX360M4, Xeon E5-2670 + Infiniband	Indian Institute of Tropical Meteorology, India	IBM	India	38,016	719,220	790	Xeon E5-2670 8C 2.6 GHz	Linux

Source Top500 (2014)

Table 10.5 Summary of significant findings at state/provincial/prefecture levels for model, China, India, Japan, and United States

Variable	General Model Category	Personal Computers/100 Pop.				Mobile Phone Subscribers/100 pop.				Internet Users or Subscribers/100 pop.				Broadband Subscribers or Adoption/100				Landline Subscribers/100 pop.			
		China	India	Japan	U.S.	China	India	Japan	U.S.*	China	India	Japan	U.S.**	China	India	Japan	U.S.***	China	India	Japan	U.S.****
Generic Variable																					
Urban	Demographic		NA	+	+								+			+	+			+	
Economic Prosperity	Economic	+	NA	+		+										+		+			
Secondary, Tertiary Education	Education		NA		+		+				+		+		+		+			-	
Infrastructure	Infrastructure		NA					+		+	+	+							+		
Specific to Countries																					
Export Commodities Value	Economic	+				+				+				+				+			
No. of Published Books	Innovation	+												+							
Innovation Funds	Innovation																	+			
Newspapers and Periodicals	Innovation		NA				+				+				+				+		
Cooperative Society Member	Social Capital		NA								+								+		
Age Structure (young, middle)	Demographic							+				+				-				-	
Technical Workers	Economic											+									
Patents	Innovation							+								+					
Newspaper Circulation	Innovation											+				+					
Asian	Demographic								+												
Hispanic	Demographic												-								+
Overall Freedom Index	Societal Openness																				+
R&D	Innovation																				-
Putnam's Social Capital Index	Social Capital				+								+				+				

Note: China results are given for samples both with and without Beijing and Shanghai. India results are given for samples with and without Delhi

* For U.S., this variable's proxy is mobile wireless high-speed devices/household

** For U.S., this variable's proxy is internet access at home

*** For U.S., this variable's proxy is broadband adoption in household

**** For U.S., this variable's proxy is persons in fixed-phones-only household

the entire results. The second column "general model category" lists the model factor from the book's general theory that applies (see Fig. 3.8). In the rows, under "Generic Variable" are the variables or variable groups which applied to all four nations, while under "Specific to Countries" are the unique variables that did not apply to all four countries. In the table, the specific-result cells for the each of the four nations are highlighted in light grey.

For the generic variables or variable groups, urban location is seen as influential in Japan and the U.S. for PCs and broadband, in the U.S. for Internet, and in Japan for landline subscribers. Urban areas in the two nations dominate and would have improved infrastructure relative to the rural regions (Grubesic 2006; Arai and Naganuma 2010). Its absence for mobile phone subscribers points to geographically more widespread use of mobile phones across these highly developed nations, which reflects phone costs that are quite low relative their high GDP/capita. Economic prosperity variables have relatively small influence at the state level, compared to their greater prevalence in national level research. They are influential only for China and Japan. For China, greater provincial prosperity would make any of the three technologies more affordable, which is important given China's still moderate-low GDP/capita. The influence in Japan of prosperous provinces for PCs and broadband may relate to a high quality level and concomitant high cost of PCs, making prosperous prefectures more PC-intensive, while the effect for broadband is unexplained since its cost is affordable.

ICT use in India and the U.S. is impacted by education. Education's influence for India was explained earlier, while its effects in the U.S. for PCs, Internet, and broadband use might be due to United States' greater range and complexity of ICT uses than for other nations, which a state with lower education level could try to improve by training and support. The State of Florida, for example, had a 2010 college attainment rate, age 25–34, of 36.2 %, 3 % below the U.S. average (U.S. Census Bureau 2014)—Florida college graduates are seen working on laptops at a conference in Fig. 10.2. The absence of education as a factor for China and Japan may point to gaps in technology training in their educational systems.

Infrastructure influences are mostly present for Internet users/subscribers, and for all nations except the U.S. In China and India, infrastructure for electricity continues to be built extensively and would influence Internet use through its impacts on the extension especially of cable and fiber networks throughout the countries. In Japan, the often formidable mountain topography and separation of more remote islands makes infrastructure for the Internet more important. The U.S. registers no infrastructure influences, which points to an exceptional set of varied networks including cable, fiber-optic, cell-tower, and traditional copper.

The findings unique to each country are now discussed. For China, export commodities value dominates, as was explicated in Chap. 5, and is further examined in this chapter's case. Also published books are important for PCs and broadband and are explained as a proxy for web content production and emerging e-publishing. For India, newspapers and periodicals likewise serve as a proxy for generation of web content and emerging e-publishing.

Fig. 10.2 College graduates working on laptops at a conference, Tampa, Florida

The culturally unique variable of cooperative society membership, discussed in Chap. 6, relates to the social capital advantages of belonging to such a society.

For Japan, demographic factors of young and working age population are correlated with increased mobile phone subscribers and Internet users, but reduced fixed phone and broadband subscribers. The first three are explainable by greater young and working age Internet usage due to generational learning and experience, compared to the old population. The broadband effect is unexplained. Innovation, as reflected in patents and the proxy variable of newspaper circulation (related to e-news and web content development), is correlated with Internet use and mobile phone and broadband subscriptions. As seen in Table 10.6, Japan with a population 41 % of the US, has per capita patents roughly equivalent to the U.S. The prefectures with high innovation create an environment of greater ICT use.

For the U.S., the unique variable of Putnam's Social Capital Index influences PCs, Internet access, and broadband adoption. As mentioned in Chap. 8, this index broadly assesses the social interconnections and networking of people. This result

Table 10.6 STEM and R&D variables for China, India, Japan, and United States

	China	India	Japan	United States
Science and engineering (S&E) doctoral degrees, 2010 or most recent year	48,987	7982	15,872	57,405
S&E Articles, all fields combined, 2011	89,894	22,480	47,106	212,394
S&E Articles in computer sciences, 2011	1492	165	162	2254
Patents granted from U.S. patent and trademark office	3717	1277	50,708	107,962

Table 10.7 Summary of significant findings for social media at state/provincial/prefecture levels for model, Japan, and United States

Variable	General Model Category	Facebook[a]		Twitter[a]	
		Japan	U.S.	Japan	U.S.
Generic Variable					
Urban	Demographic	+		+	
Economic Prosperity	Economic				
Secondary, Tertiary Education	Education	+		+	-
Infrastructure	Infrastructure				
Specific to Countries					
Young Dependency Ratio	Demographic			-	
Patents	Innovation	+		+	
Asian	Demographic				
Hispanic	Demographic		-		
Putnam's Social Capital Index	Social Capital		+		
Immigrant Population	Social Capital		+		+

[a]For Japan Facebook or Twitter subscribers/capita. For U.S. Facebook or Twitter users/capita

can be viewed as an individuals' broad, informal exposure to others, which would likely include informal learning about personal uses of technology. As a prominent example, Silicon Valley would foster this advantage of learning by informal networking and socializing with others. Some start-ups have even come from meetings in the Valley's coffee shops. The two ethnicity findings, i.e. Asian population having higher mobile device access, and Hispanic population having reduced Internet access along with higher landline access are plausible, as pointed out in Chap. 8. Ethnicity is a unique U.S. variable since government and nonprofit data sources in the other nations do not break out ICT uses by ethnicity at the state level, nor even nationally.

Overall, this book's detailed findings point to general factors of urban, economic prosperity, education, and infrastructure for states, provinces, and prefectures, which are associated with selective ICTs factors. On the other hand, there are varied culturally-unique factors, which depend on unique definitions of variables of interest in a nation that do not carry over to all four countries.

Both the general and unique findings can be of use in formulating policies and plans for developing and expanding utilization of technologies, topics examined in the next chapter. The findings for social media variables are again divided into generic and country-specific ones (See Table 10.7). Since China and India did not have state/provincial-level data available for social media, the findings are only for

Japan and the U.S. The generic results apply mainly to Japan, for which urban location and education (students and pupils per capita) are related to both Facebook and Twitter use. The importance of urban location is consistent with Japanese results for PCs, broadband, and landline phones. It points strongly to an urban-rural digital divide in the country. For the U.S., the inverse impact of education on Twitter use is unexplained and suggests need for further research.

For country-specific findings, innovation as connoted by patents has positive influence on Facebook and Twitter use and is consistent with innovation correlations in Japan of mobile phone and broadband subscriptions with Internet use. That young dependency ratio reduces Twitter use is unexplained, as youth would logically be a factor increasing use.

For the U.S., social capital variables are related to Facebook and Twitter use. This corresponds to the already discussed influence of social capital on PCs, Internet, and broadband. States with people who have more social and networking connections on the ground are more likely to have higher social media adoption. The association of higher Hispanic proportion with reduced Facebook use resembles the association with Internet use, and points to the need for more detailed study of Hispanic population to confirm and explain these inverse effects. In summary, these state and prefectural findings for social media are exploratory and point to socio-economic correlates that are distinctive for the two countries.

10.4 Comparison of Spatial Patterns for Four Nations

The state/provincial/prefectural geographical patterns of use of ICTs have provided new insights on the digital divide, which were examined for the four nations in Chaps. 5–8 and provided answers about patterns for specific ICTs. A further question is how do the four nations match up on their spatial patterning? This section seeks to answer that question by comparing the results of cluster analyses conducted for the state-units in each nation on its full set of dependent variables. Consequently, comparing the cluster patterns is comparing summarized results for all the ICTs examined for each nation.

The cluster maps for the four nations are shown together in Figs. 10.3 and 10.4. The cluster color sequence is similar for each country, with red representing the clusters with the highest overall ICT level and grey representing those with the lowest ICT level with the exception of the U.S. map in Fig. 10.4 in which blue represents the clusters with the highest overall ICT level.

China. For China in 2009, the two clusters of provinces containing the megacities of Shanghai and Beijing dominate all other clusters in ICTs. Beijing is China's political capital and is regarded also as its intellectual capital, since it has many universities including the prestigious ones of Beijing University and Tsinghua University, and Renmin University. Beijing also contains the city area of Zhongguancun, which has among China's largest concentration of high-tech firm headquarters and R&D centers. Another top ICT cluster is Shanghai municipality containing the megacity of Shanghai and regarded as mainland China's most important business city, with strong presence of international corporations, and many excellent universities. It is not surprising these two giant cities have exceptionally high ICT levels, as their populations are among the nation's most educated and there are strong concentrations of high technology, with spillover to their populations, as well as web content and e-publishing production through their concentrations of publishers and universities. They, like Delhi in India and Tokyo in Japan, are among the world's largest megacities, a topic brought up later.

The cluster of the provinces of Tianjin and Zhejiang rivals the southeastern cluster of Fujian and Guangdong for second place in ICT usage. In the former cluster, as seen in Fig. 10.3, Tianjin borders Beijing to the northwest and Zhejiang borders Shanghai to the south. As mentioned in Chap. 5, China's explicit governmental policy of the "low-end industry transfer model" sought specifically to transfer industries including ICT from Shanghai to Zhejiang. In a parallel way, transfers of ICT knowledge and capability took place from the megacity of Beijing to that of Tianjin, only 70 miles away. Zhejiang also benefits from ICT spillovers from the high-tech cluster of Fujian and Guangdong to the South. China's other clusters spreading out to the north, west, and southwest have moderate to very low ICT levels.

India. India's unique cluster with very high ICT level is Delhi, comprising one of the world's largest megacities. It is the political capital of India and one of the top higher education centers in the country, as well as being the nation's publishing center including e-publishing. With these features and elevated income levels, it has the highest ICT usage among states by far. The cluster of next importance in ICT comprises the neighboring states of Punjab and Himachal Pradesh to the north and northwest of Delhi and the neighboring states of Kerala and Tamil Nadu in the southern tip of India (see Fig. 10.3). A possible reason for high ICT levels in the two northern states is proximity to Delhi, analogous to the Chinese examples. The high technology states in the South relate to Kerala's high average level of education among Indian states, while the high ICT level in Tamil Nadu may be due to the high-tech industry in its largest city of Chennai. Broad reaches of the West and Southwest of India have moderate levels, while the northeast parts have the lowest levels.

Japan. Tokyo, as the world's largest megacity comprises 30 % of Japan's population. It is much higher in all the technology attributes except PCs which is slightly lower, compared to the next cluster group that includes Tokyo's immediately surrounding prefectures of Satiama, Chiba, and Kanagawa. These prefectures are posited to receive a spillover from Tokyo, an effect that needs further research to confirm (see Fig. 10.4). The other three clusters vary from moderate to low and include prefectures in the mountainous areas in the central and northern islands of Honshu and Hokkaido and more peripheral rural areas of the central and southern islands.

United States. The lack of one or two dominant states in the U.S. reflects a cluster pattern that is more evenly balanced than in the other three countries. As mentioned in Chap. 8, one of the two highest clusters, comprising 13 states, is centered in New England and mid-Atlantic ones, plus California, Hawaii, and Alaska. This large and diverse region includes the megacities of New York-New Jersey and Los Angeles, the former being a national center for media, advertising,

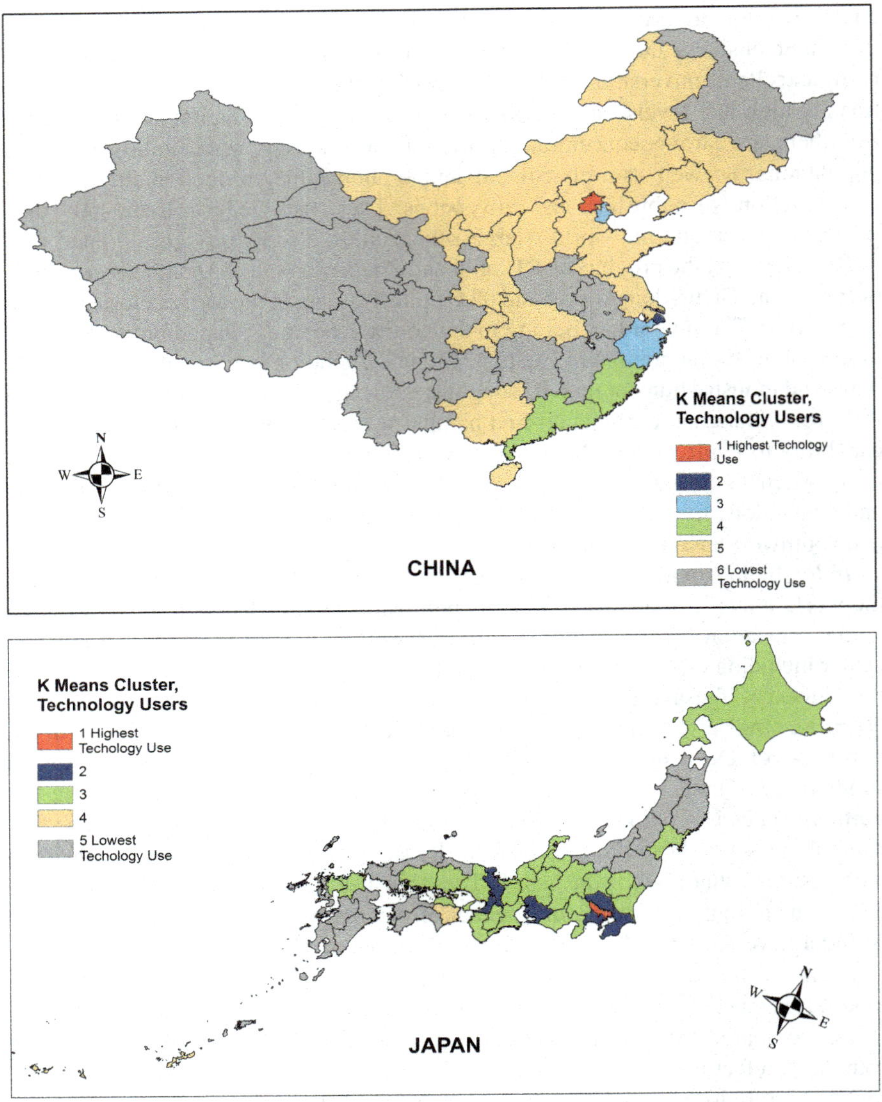

Fig. 10.3 Cluster maps, China 2009 and Japan 2010

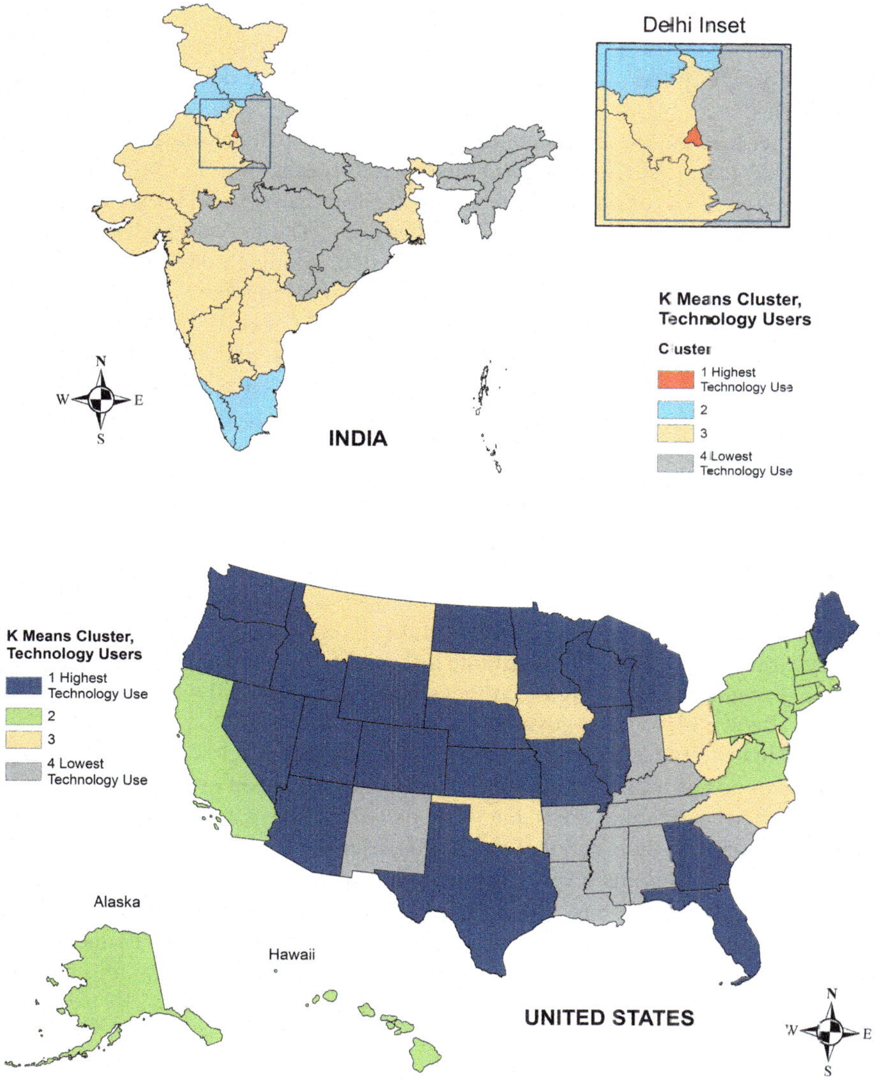

Fig. 10.4 Cluster maps, India 2010 and United States 2010

and web content generation and the latter a technology, engineering and entertainment hub. Many of its states have populations with high average education and income. The US differs from the other nations in lack of a dominant megacity in technologies.

The other leading ICT cluster comprises 21 states which are mostly in the upper Midwest and Rocky Mountain states plus Texas, Georgia, Florida, and Maine. This sprawling area of high tech is unique among the four nations and is especially high in phone attributes (mobile, fixed) and Twitter users. The lowest usage region is in the upper to lower central South plus New Mexico. Although deprived in ICT use relative to the U.S. averages, this "low tech" cluster region varies much less from the national norm than for the Indian and Chinese ICT-deprived clusters.

One of the common spatial features for three of the nations is the high-tech dominance of one or two megacities. A megacity is defined as a city of population more than 10 million, of which there were 28 worldwide in 2014. On this list, Tokyo is the largest with a population of 38 million, while Delhi is 2nd with 25 million, Shanghai 3rd with 23 million, and Beijing 8th at 20 million (UN 2014). Tokyo is also a primate city, defined as one that is twice or more as big as the next largest city.

In summary, the prefecture, provinces and state that contain these four megacities are seen by this comparison to stand out by far as the leaders in technology for three of the study nations. Also, for each of them, there are either directly adjoining or very close nearby units that constitute a second tier of technology use. We posit that the proximity to these high-tech megacities leads to a spillover effect, one that actually aligns with explicit government policy in China.

For China and India, there are broad ICT hinterlands, mostly contiguous clusters of dozens of provinces or states with low levels of ICT. Japan also has its lowest 18 clusters in peripheral, rural areas, but those prefectures have much less extreme low values compared to their national norm than counterparts in China and India. By contrast, the U.S. presents a very different spatial pattern that does not have unique ICT-dominant states, but has a more balanced distribution with differences of the lowest-tech cluster even closer to the national norm than for Japan. The megacities of New York-New Jersey and Los Angeles bear more detailed study for spillover effects, which could be done by utilizing county data.

10.5 Limitations and Future Research

The chapter's comparative analysis has several limitations that suggest future avenues of research. The dependence on national government and nonprofit data sources for state/provincial/prefectural datadata implies that overall equivalence of variables definitions is not attainable, except for very basic ones such as population, dependency ratio, and internet users per capita. Mostly, the variables are differently defined from country to country, or uniquely defined for one nation, such as staff and workers in non-state-owned units in China or cooperative society membership in India. Our approach is to include all such variables as long as they can be well justified conceptually.

Future research could be restricted only to attributes that are uniformly defined at the state level across all nations. However, use of today's government and nonprofit data sources would severely restrict study depth. A recommended way around this problem would be to design and administer a standardized survey of individuals in all the state/provincial/prefectural units within each nation in the study sample. Such a study would require significant funding and resources.

Another challenge to the comparisons is in interpretation of comparative findings. The researcher needs to be careful to not overstretch cultural insights for a variety of nations and to be aware of his/her own biases and perspective, as noted by Hantrais (1999, pp. 103–104). This issue was resolved for one literature study by fielding a research team that included a native investigator for each nation under study, in particular for a 10-country National Science Foundation (NSF)-funded project described in Kraemer et al. (2005). In the present project, native-born investigators participated for three of the four countries under study, while for the fourth country, one of the book co-authors took two sabbatical leaves there. For a study of state-level units for dozens of nations, this challenge would need to be considered.

There are methodological limitations as well. For example with state-level samples that vary from 20 to 47, methods that allow for more complex relationships between variables such as structural equation modeling and geographically-weighted regression, are not applicable. To provide sufficiently large samples for these methods, future research projects might adopt geographical units below the state level, such as counties in the U.S. or districts in India, but there would likely be fewer variables available. This challenge bears also on the complexity of conceptual models that can be formulated and operationalized. Another approach entirely would be to perform case-study or other qualitative analysis of multiple states or provinces throughout a country, but that might require considerable resources.

10.6 Examples of Cooperation of Nations on Digital Divide: United Nations ICT Task Force and World Economic Forum

The results of the comparison between nations in this book's studies lead to the question of implications of the comparison. Whether the findings are general or unique for nations, how can they be applied to inform international collaborative policy applications? The next chapter focuses on policies more specifically for nations, but the questions here are: what type of organizations can apply the comparative results? What are some examples of attempts to do this?

Worldwide nonprofit organizations would be potentially well suited for benefitting by comparative analysis, since many would include the area of technology, such as the World Bank, United Nations, World Economic Forum, World Health

Organization, and International Telecommunications Union. Although they all consider technology to be important, two examples demonstrate how comparative state/provincial/prefectural data can be useful in formulating policy, namely the UN and the World Economic Forum.

The UN from 2001 to 2005 formed the ICT Task Force, which had the goal to support and promote discussions and provide policy advice on ICT to national governments and other international organizations. The group was unusual for the UN since it was the first group with UN inter-governmental approval to have members from national governments, business sectors, other global organizations, and civil society in which each member had an equal voice (Khan 2005). Representatives were chosen from developed and developing nations, with particular focus on helping poor nations. The task force included members from leading ICT firms such as Intel, HP, IBM, and Cisco. The mission was to help to spread the benefits of ICT worldwide and to recommend how to make the benefits available to every global citizen.

The Task Force met regularly, engaged in discussions about how to leverage ICT's benefits globally, produced many publications, and held workshops. In doing so, it often used comparative case studies and data. It also worked closely also with the Information Society, a global group that sponsors the World Summit on the Information Society (WSIS). As the Task Force neared its completion date of 2005, it collaborated with the Information Society in co-presenting pre-workshops and forums, and the full WSIS in Tunis in 2005 (Khan 2005). Although it disbanded in 2005, the task force's publications, ideas, and discussions of global collaboration to reduce the digital divide have continued to inform the UN.

The second example organization, the World Economic Forum (WEF), formed in 1971 and headquartered in Geneva, Switzerland, has the goal to "improve the state of the world through public-private cooperation" (WEF 2014a). It comprises business, nonprofit, and academic stakeholders who meet regularly worldwide and regionally to outline agenda and steps to solve global information technology issues. Starting in 2001 it formally recognized technology as an issue by starting the Global Information Technology annual report series, which has been a leading source for national quantitative data-sets, comparative studies, as well as sub-national case studies, which over time have constituted an excellent qualitative comparative set, several of which are cited in this volume. Recently, the WEF opened up three additional issue initiatives related to ICT (WEF 2014b). "ICT for Growth" focuses on special meetings and reports to broaden inclusion of the world's poor with technology. "Delivering on Digital Infrastructure" seeks to foster an improved environment for innovation and investment in digital infrastructure through a multi-industry collaboration in order to satisfy rapidly expanding demand for ICT in the developing world. "Hyperconnected World" aspires to look forward a decade a world with over 50 billion devices that are connected together, and to determine what the opportunities are, as well as what the threats are, such as security risks and destruction of meaning in human relationships (WEF 2014b).

In short, international organizations have been able to address digital divide issues, to produce worldwide discussions, agendae for action, and initiatives to reduce the digital divide especially for poor nations. The widely recognized work of the two examples has included making use of comparative ICT data and insights for sub-national and national units.

10.7 Case Study: Lenovo

This case study illustrates some of the chapter's comparative findings as well as demonstrating distinctive settings for ICT markets and usage in three nations. Lenovo is a Chinese firm founded under the Legend name in 1984 by a scientist Liu Chuanzhi who left the Chinese Academy of Sciences with its support to start up a business enterprise in computing. Three decades later this small firm has become the world's number one manufacturer of PCs, with a dominant position in China and products in PCs, mobile devices, cloud services, and enterprise systems that are marketed worldwide. Lenovo's story interweaves a variety of collaborative and competitive roles for especially for China and the U.S. but also for India which relate to the focus of this chapter.

Following its founding, Legend was successful in growing its PC business in China to become a leading domestic competitor within China. At first, it worked collaboratively with leading U.S. PC vendors such as IBM and Hewlett Packard that sought entry to Chinese markets. One innovation by Legend (which became Lenovo in 2003) was to develop the best accepted Chinese digital character set. This, along with aggressive management, enabled Legend by the mid-1990s to become the largest PC manufacturer in China with its own brand and occupy one third of the national market (Liu 2007).

In a 2003 month-long strategy session, Liu and Lenovo made the bold decision to shift its goal from domestic focus to global focus. Shortly later Lenovo took the giant step of acquiring IBM's PC division, which was losing money. Liu referred to this as an instance of "as snake swallowing an elephant" (Liu 2007) Fortunately the acquisition worked out well, including moving many key former IBM managers and technical experts to Beijing, many located in the Lenovo Innovation Center as seen in Fig. 10.5, while placing Lenovo's global headquarters in the U.S.

The company demonstrated it intended to become a multinational company by populating its top executive ranks with 1/3 from China, 1/3 from Hong Kong, and 1/3 from Europe and the U.S.

Since this acquisition, Lenovo has expanded rapidly worldwide while retaining its dominance of the Chinese PC market. A key part of its strategy in China has been to market simple, low-end products to the vast West and North regions. It continued strategic, high quality acquisitions by acquiring Stoneware (U.S. cloud software), CCE (leading consumer Brazilian electronics firm), MEDION (German consumer electronics firm with broad products of PCs, notebooks digital cameras, TVs, etc.), Compal (Taiwan original equipment manufacturer) and Motorola Mobility (U.S. mobile phone firm with famous brand name). Lenovo felt that mergers and acquisitions would accelerate its globalization (Liu and Buck 2009).

Fig. 10.5 Lenovo Innovation center, Beijing, China

The Motorola acquisition strengthened Lenovo's existing mobile phone products and it moved its mobile phone sales to 3rd place worldwide.

Also, there were impacts on India, which unlike China has strengths in outsourcing software and services externally to India but has limited manufacturing capability for PCs or mobile devices. The acquisition of Motorola Mobility allowed the combined Lenovo/Motorola India to move to 3rd place in India in mobile phones, while expanding its prominent PC market share (Press Trust of India 2014).

Lenovo's story relates to this chapter and the book in a number of ways. It demonstrates the stabilizing of growth in PC volumes, in all 3 nations, while demand for mobile phones has continued to rise rapidly in China and India (see Fig. 1.2). It demonstrates our finding on the importance of export commodities value for China, since the Lenovo's 2003 turn towards globalization greatly accelerated export from China of the firm's product line. Even the acquired companies' manufacturing was largely done in China for export. Lenovo's marketing throughout all the reaches of China's provinces supports our spatial finding for China of ICT deficits in China's Western and Northern regions.

Lenovo's leaders have been smart in realizing many of the differences shown by the comparative study, for example that education is a driver in India and the U.S., so products need to be higher-level ones, whereas products being sold in India also depend on content generation. For Japan, higher-end PCs and 3-G cell phones are marketed corresponding to regions of economic prosperity as a key determinant.

In summary, Lenovo's upward curve demonstrates ability to adjust to comparative differences in ICT adoption factors between nations, which are often distinctive ones. It succeeded initially by its cultural sensitivity within China, and has carried that further by learning to apply distinctive marketing approaches in contrasting nations and their state-level units. Because of its success as a corporation it is providing major ICTs to over 160 nations around the world (Lenovo 2014), and helping to fill in supply for the developing nations, while maintaining strength as a now multinational company.

10.8 Conclusion

This chapter uses a mixed comparative approach which is explained. Some general comparisons are done of national features of the four diverse nations, which together comprise over two fifths of the planet's population and GDP. A comparative analysis brought findings on the book' four in-depth state/province/prefectural studies together into a framework that indicates a core of general factors as well as unique country-specific ones. The factors are further interpreted based on the comparisons.

Next, comparison is done of the spatial patterns of ICT clusters among the four nations. For China, India, and Japan, there is a similar dominant status of the provinces, state, and prefecture, containing megacities of Shanghai, Beijing, Delhi, and Tokyo. The spillover effect from the Chinese megacities to adjacent provinces is considered, and similar patterns are identified for Delhi and Tokyo on particular surrounding or nearby units. Broad rural expanses of low ICT use are compared for China and India, whereas Japan's rural prefectures are closer in technology levels to the country average than for the former nations. Regarding the U.S., its spatial arrangement is quite different, lacking a unique dominant high-end state but grouping two leading clusters consisting of dozens of states. The low-end region in the middle to lower South is closer yet to the country average than for Japan.

The limitations of our comparative approach are discussed, with suggestions for future research that with appropriate resources could achieve more robust findings, although some cultural-based differences will continue to make comparison challenging.

Two examples are given on how international organizations have been able to adopt an international approach to the digital divide that often has made use of comparative studies and insights. The chapter ends with the remarkable story of Lenovo, which has been able to grow domestically and worldwide by being sensitive to differences in state-level, as well as national-level determinants of ICT adoption and use. At the same time, this company's approach has helped to increase supply of technology products worldwide, and in some cases alleviate shortages in developing nations.

References

Arai, Y., & Naganuma, S. (2010). The geographical digital divide in broadband access and governmental policies in Japan: Three case studies. *Networks and Communication Studies, 24* (1–2), 7–26.

Baerjee, R., & Muley, V. (2008). *Engineering education in India.* Delhi: India Observer Research Foundation

Baliamoune-Lutz, M. (2003). An analysis of the determinants and effects of ICT diffusion in developing countries. *Information Technology for Development, 10*(3), 151–169.

Chaudhuri, A. (2012). Creeping tiger, souring dragon: India, China and competition in information technologies. *China and World Economy, 20*(6), 1–28.

Compaine, B. M. (2001). *The digital divide: Facing a crisis or creating a myth?*. Cambridge, MA: The MIT Press.

Doron, A., & Jeffrey, R. (2013). *The great Indian phone book*. Cambridge, MA: Harvard University Press.

Dutta, S., & Bilbao-Osorio, B. (Eds.). (2012). *The global information technology report 2012*. Geneva: World Economic Forum and INSEAD.

Ebbinghaus, (2005). When less is more: Selection problems in large-N and small-N cross-national comparisons. *International Sociology, 20*(2), 133–152.

Gibbs, J., Kraemer, K. L., & Dedrick, J. (2003). Environment and policy factors shaping global e-commerce diffusion: A cross-country comparison. *The Information Society, 19*, 5–18.

Grubesic, T. H. (2006). A spatial taxonomy of broadband regions in the United States. *Information Economics and Policy, 18*, 423–448.

Guillén, M. F., & Suárez, S. I. (2004). Explaining the global digital divide: Economic, political and sociological drivers of cross-national internet use. *Social Forces, 84*(2), 681–708.

Gulati, G. J., & Yates, D. J. (2012). Different paths to universal access: The impact of policy and regulation on broadband diffusion in the developed and developing worlds. *Telecommunications Policy, 36*, 749–761.

Hantrais, L. (1999). Contextualization in cross-national comparative research. *International Journal of Social Research Methodology, 2*(2), 93–108.

ITU. (2013). *Measuring the information society*. Geneva, Switzerland: International Telecommunications Union.

Kagami, M., Tsuji, M., & Giovannetti, E. (2004). Conclusion (Chap. 15). In K. Kagami, M. Tsuji, & E. Giovannetti (Eds.), *Information technology and the digital divide* (pp. 303–307). Cheltenham, JK: Edward Elgar.

Khan, S. (2005). WSIS and the United Nations ICT task force. In D. Stauffacher & W. Kleinwachter (Eds.), *The world summit on the information society: Moving from the past into the future* (pp. 41–43). New York, NY: United Nations ICT Task Force.

Kraemer, K. L., Gibbs, J., & Dedrick, J. (2005). Impacts of globalization on e-commerce use and firm performance: A cross-country investigation. *The Information Society, 21*, 323–340.

Lenovo Group Ltd. (2014). *Lenovo 2013/14 annual report*. Morrisville, NC: Lenovo Ltd.

Liu, C. Z. (2007). An executive's note: Lenovo: An example of globalization of Chinese enterprises. *Journal of International Business Studies, 38*, 573–577.

Liu, X., & Buck, T. (2009). The internationalization strategies of Chinese firms: Lenovo and BOE. *Journal of Chinese Economic and Business Studies, 7*(2), 167–181.

Mills, M., van de Bunt, G. G., & de Bruijn, J. (2006). Comparative research: Persistent problems and promising solutions. *International Sociology, 21*(5), 619–931.

NSF Science and Education Indicators (2012). *Chap 2: Higher education in science and engineering*. Washington, D.C.: National Science Foundation.

Ono, H., & Zavodny, M. (2007). Digital inequality: A five country comparison using microdata. *Social Science Research, 36*, 1135–1155.

Population Reference Bureau. (2012). *2012 World population data sheet*. Washington, D.C.: Population Reference Bureau.

Press Trust of India (2014). *Motorola and Lenovo brands to co-exist in India*. New Delhi: Press Trust of India. 31 Oct.

Robison, K. K., & Crenshaw, E. M. (2002). Post-industrial transformations and cyber-space: A cross-national analysis of internet development. *Social Science Research, 31*, 334–363.

Top500 (2014). *Top 500 supercomputer sites*. Waibstadt-Daisbach, Germany: Top500 supercomputer sites. http://www.top500.org/lists/2014/06. Available Sept 20, 2014.

U.S. Census Bureau. (2014). American FactFinder. Washington, D.C.: United States Census Bureau.

United Nations (2014). *World urbanization prospects: 2014 revision*. New York, NY: Department of Economic and Social Affairs, United Nations.

World Bank. (2012). *World development indicators*. Washington, D.C.: World Bank.

World Economic Forum. (2014a). *Annual Report 2013–2014*. Geneva, Switzerland: World Economic Forum.

World Economic Forum (2014b). *Issues, the world economic forum*. Geneva, Switzerland World Economic Forum. Available at www.weforum.org.

Chapter 11
The Roles and Policies of Governments

11.1 Introduction

The goal in this chapter is to try to bring together the diversity of government policies into a general approach that stems from the findings of applying our general model to many parts of the globe. In accomplishing this goal, a number of questions arise. Is ICT more than policy-determined but a basic human right? How important has ICT policy been for nations up to this point? What general approach to ICT policy stems from our research? What are examples of nations having ICT policies that have succeeded over time? How does net neutrality alter government policy? How do ICT policies influence states, provinces, and cities? What are the limitations on positive benefits from ICT use and how do/should the limits constrain ICT policies? The chapter attempts to address these questions and more. Its goal is not to provide a clear roadmap to every government seeking to enhance IT use within its jurisdiction, but rather to inform about best practices and pitfalls, and how the book's research and its approach can sharpen ICT planning.

Two starting points are to consider institutional factors in policy setting and to discuss whether ICT should be considered a fundamental human right. Governments are institutions that play a crucial role in realizing policies for use of ICT (King et al. 1994). Several factors we have included in our general model are institutional. *Government prioritization of ICT* stems directly from the institution of government. *Education* is linked to government since in nearly all nations it is wholly or largely funded by it, with the remainder funded by private or nonprofit sources. Likewise, *innovation* stems largely from government in most nations. *Societal openness* is based partly on government since societal laws and regulations impact it (King et al. 1994).

The institutional perspective suggests how ICT policy influence can be exerted, i.e. through supporting knowledge creation and by directives for diffusion of innovations, subsidies, and standards (King et al. 1994). The institutional perspective also results in six "conjectures," several of which inform this chapter.

J.B. Pick and A. Sarkar, *The Global Digital Divides*, Progress in IS,
DOI 10.1007/978-3-662-46602-5_11

Conjecture 1 proposes that institutional intervention is essential to continuing production of knowledge but not required for to successfully diffuse its use (King et al. 1994). It implies that diffusion can take place by movement through networking of users such as the Internet rather than by government stimulation. Conjecture 6 in (King et al. 1994) indicates that "top-down directives" to use technology can be useful in crises but are very limited as a means for sustainable use of innovations in ICT (King et al. 1994). For instance, a case example of the Egyptian government policy mandating implementing state-of-the art labs in secondary schools in the 1990s resulted in implementation of hardware, but the implementation eventually was ineffective due to lack of training in how to run the equipment and how to use it pedagogically (Warschauer 2003).

Another background aspect for the chapter is the increasing recognition of ICT access and use as a human right. In 2012, the United Nations supported this by ruling that access to the Internet is a basic human right which nations need to guarantee and protect, a ruling supported by the entire General Assembly. In particular, the Human Rights Council of the UN affirmed "that the same rights that people have offline must also be protected online, in particular freedom of expression, which is applicable regardless of frontiers and through any media of one's choice, in accordance with the Universal Declaration of Human Rights and the International Covenant on Civil and Political Rights" (UN 2012). The Council also called upon nations to support and facilitate access to the Internet and international cooperation in developing ICT (UN 2011, 2012).

Although not attracting widespread attention at the time, the human right to the Internet for the world's population suggests that government policies should stress an open, free, and affordable Internet for the world's citizens, regardless of a citizen's political environment or social and economic status. Although today the world falls way short of this ideal, as seen by the large variations in ICT access and usage from the book's empirical chapters, this ultimate ideal promulgated by the UN should be kept in mind.

The chapter now turns to examining the ICT policies of national governments, first considering policies of the advanced nations of Japan, Sweden, and the U.S. They are chosen because two of them are study countries in this book and Sweden has been regarded as exemplary in its ICT policies. Next, the distinction between policies of ICT supply and demand is examined. The challenge of the potential loss of net neutrality is discussed, a loss that would inhibit access to the Internet for segments of users, narrowing the scope of application and increasing consumer cost. This chapter proposes a unified set of national internet policies, reasoned from the book's findings and including goals that can be objectively evaluated. Finally, Rwanda is examined as a nation exemplifying the realization, over 15 years, of ambitious ICT policies in a low-level developing nation, while challenges remain.

11.2 National Government Policies

In this section three countries' ICT policies are reviewed, after which two relevant studies are discussed, one concerning policies of supply and demand of information technologies (Pepper and Garrity 2013) and a second that evaluates the relative importance of ICT policy compared to all the other factors influencing use of ICT. This section serves as background for the later section on policies emerging from our research.

11.2.1 Sweden's Policies in ICT

Sweden had policies that early on put it in a leadership position to develop among the fastest and most ubiquitous use of information technologies worldwide. Sweden is a country that historically has been advanced in telecommunications. In 2012, Sweden's broadband internet subscriptions per 100 persons was 31.8 while mobile broadband subscriptions per 100 persons was 88.6, in global 1st place (Dutta and Bilbao-Osorio 2012).

Sweden arrived at this high level through government policies and planning that included its tradition of excellence in technology and engineering, liberalization of the telecommunications market early on in 1993, and collaboration between government and private sector. Government support included information systems and computing in the schools, provision for access to personal computers, and the goal of "an information society for all" leading to legislation in 1999 supporting it (Kim et al. 2010; Hall 2012). The specific policies were directed at both ICT's demand side and its supply side. For instance, on the supply side, the government took responsibility to assure that ICT infrastructure would be available to all citizens throughout the nation, even extending into the Arctic regions (Hall 2012). The central government also supported municipal governments through tax credits and federal grants to develop their own high-speed networks, in advance of when market forces alone would have enabled building them (Kim et al. 2010).

In 2007 the Swedish government introduced a public-private cooperative agreement in which the government encouraged market competition to provide broadband services through grants, subsidies, and regulation (Kim et al. 2010). The most recent policy initiative seeks in the medium term to achieve nationwide digitization at the top world level (Government Offices of Sweden 2011). Its four major areas of policy focus are: (1) ease and safety of ICT use, (2) services to create benefit, (3) continuing development of infrastructure, and (4) providing ICT for societal development.

For the latter, the intent is for ICT policies to develop "rules intended to protect personal privacy, secrecy, copyright etc. often needing to be adapted in order to respond to the changed circumstances to which technological development gives rise" (Government Offices of Sweden 2011). Some of this effort would be directed within the country and some towards global development, to help people elsewhere benefit from web freedom, democracy, and transparent development assistance.

The government in 2013 also put forward a policy to "enhance Internet freedom and human rights through responsible business practices" (Government Offices of Sweden 2011). This policy was stimulated both by the role of Internet technology in the Arab Spring of 2011 and by government surveillance issues that have arisen in large countries including China and the U.S., frequently abetted by private companies. The policy encourages efforts by businesses to utilize ICT to enable human rights and to be responsible when faced with challenges such as revealing user identities, breaching privacy, or participating in transmitting messages inflammatory to violence (Government Offices of Sweden 2011).

Given its remarkable early and prescient government policies over several decades with the necessary legislative follow-through, Sweden in 2012 was rated 1st out of 142 nations in Networked Readiness Index and 2nd in its Political and Regulatory Environment (Dutta and Bilbao-Osorio 2012). The nation has been forward looking with the latest and fastest technologies, while also being on the forefront of providing the benefits to all citizens and striving for recognition of human rights to partake in the benefits of information technology.

11.2.2 Japan's Policies in ICT

In 2012, Japan's broadband internet subscriptions per 100 persons was 26.9 while mobile broadband subscriptions per 100 persons was 64.6, in global 3st place (Dutta and Bilbao-Osorio 2012). The Japanese government has undergone long-term and intensive policy and planning for ICT. One of the backbones of its national policies has been the establishment of a national fiber optic backbone, with compulsory shared access to it (Kim et al. 2010). Even with this network implemented, today the country's infrastructure and digital content are very good but not the world's best, rated 22nd among 142 nations (Dutta and Bilbao-Osorio 2012).

In the late 1990s, the government deregulated its telecommunications industry through privatization and competitive polices, so the incumbent firm Nippon Telegraph and Telephone (NTT) was reduced in dominance and new market entrants offered competitive alternatives (OECD 1999).

The government has had a number of ICT policies, many leading to legislation during the 2000s. Strong government policy leadership led to the Basic Law on the Formation of an Advanced Information and Telecommunications Network Society in 2000, which sought by 2010 to provide a high-speed network access to its citizens and eliminate urban-rural, age-related, and other gaps. In the 2000s, government policies have sought to have ubiquitous computing throughout the country and to strengthen digital literacy. Part of this initiative has been to provide funding to municipal governments to in turn develop regional and local high-speed networks giving access to government services.

Japan has also set policies to become a world leader in wireless broadband. The central government eased regulations making it easier for providers to obtain spectrum for this wireless broadband (Kim et al. 2010). Consequently, Japan has moved to world leadership with a high proportion of its population having mobile broadband access.

The central government's policy emphasis for the period 2010–2020 (IT Strategic Headquarters 2010) goes beyond achieving ICT ubiquity and widespread broadband to emphasize quality of citizen uses of ICT, including e-government, transparency of government information, provision of citizen medical/healthcare information and efficient medical services (termed "My Hospital Everywhere"), providing information on health and monitoring of seniors, developing ICT interfaces for seniors, promotion of telework, creating new IT-related markets, becoming competitive in cloud computing, and developing a global presence and standards in ICT for international export, global supply chains, procurement, and consortia. These policies originated at the President's cabinet level and the government used a thorough, detailed approach that included dozens of detailed modified Gantt charts detailing the decade-long schedule of tasks and responsible government units.

Although Japan's policy has been ambitious and precisely planned, its execution has resulted overall in very good ICT level, but not among the world's best. Although it excelled in business usage of ICT and utilization by individuals, it has not achieved leadership status in government usage of, and social impacts from ICT (Dutta and Bilbao-Osorio 2012). The policies promulgated for the 2010s focus on these wider areas.

11.2.3 United States

The United States broadband internet subscriptions per 100 persons was 27.6 in 2012 while mobile broadband subscriptions per 100 persons was 50.6, in global 8th place (Dutta and Bilbao-Osorio 2012). The United States, although it developed most of the key ICT innovations, is not the world's number one ICT usage nation according to the WEF, which rates it 8th out of 142 (Dutta and Bilbao-Osorio 2012).

The U.S. achieved its high status, in spite of lack of overarching national policies for ICT development. Rather, its status derives more from the nation's leadership in higher education, private sector competitiveness, venture capital, ICT patents, new organizational models, and innovation capacity (Dutta and Bilbao-Osorio 2012).

In achieving remarkable broadband advances, several historical policy factors come into play. First, the U.S. government in the mid-20th century provided the policies and regulatory framework for private firms to develop the world's largest fixed phone network. As broadband appeared in the late 1990s, this national network was poised and ready for providers to offer DSL high speed enhancements. Second, the U.S. pioneered in cable television networks beginning in the late 1940s which served to grow to over half of American homes by 2005. The large national cable network also could be relatively easily converted into high-speed internet services. Third, telecommunications providers installed great amounts of long distance fiber-optic cable so that by the time of the Dot Com Bust in 2001, the U.S. had a surfeit of fiber cable. Recently service providers have put more emphasis on fiber-to-the-home internet access. These internet and broadband advances were led by competitive market forces between firms to provide DSL, cable, wireless and emerging fiber services. The role of government policy was secondary, and was less to plan a national "superhighway" and more to provide standards and regulations to enable these network services to thrive competitively.

This model of deregulation and competitive market leadership has the weakness that sometimes short-term profit motives result in low quality infrastructure and solutions (Kim et al. 2010).

The Obama Administration since 2009 has provided more national policy and planning than previously. In particular, responding to the recession of 2007–2010, the Obama economic stimulus plan of 2008–2009 provided for expansion of broadband infrastructure. The strategy was formalized in 2010 in "Connecting America: The National Broadband Plan" (FCC 2010). This policy called for the federal government to expand the broadband "ecosystem" by continuing robust competition, allocating and managing efficiently ICT resources controlled by the government, increasing the affordability of broadband and voice services for low-income Americans especially in high-cost regions, and reform of policies, laws, and standards to "maximize the benefits of broadband in sectors government influences significantly, such as public education, health care, and government operations" (FCC 2010).

The long-term policy goals of "Connecting America" are for every citizen to have ubiquitous, very high speed, and affordable internet access, corresponding skills to use it in order to lead the world in mobile technology innovation; and to harness its world-leading platform to support clean energy, public safety, schools, hospitals and government buildings (FCC 2010). Essentially this policy, if fully implemented, is intended to restore the U.S. to the top worldwide position in ICT utilization. Results so far fall short of this ideal, but some steps have been taken.

The U.S. government ICT policy has until recently taken a back seat. In contrast, the highly competitive ICT markets in the U.S. have been the drivers of the country's high worldwide standing, with government policies as enablers, and with progress that draws on the country's leading positions in higher education and innovation.

It is helpful to consider the supply side as well as the demand side of ICT policy. Supply side policy concerns policies that provide for supply of software, hardware, services, and infrastructure, whereas demand side policies concern how individuals, businesses or government make use of and expand demand for ICT (see Table 11.1). A study that analyzed demand-side and supply-side for broadband sorted existing national broadband policies into four categories: supply-driven, demand-driven, broad based (consists of both demand-driven and supply driven) and emergent, which have fewer recommendations across a wide range of policy options (Pepper and Garrity 2013). The matrix of supply side versus demand side policies is shown in Table 11.2.

Examples of supply-side and demand-side policies across a dozen nations are given in Table 11.3. It is evident that Sweden, the U.S. and India are broad based in their ICT policy. Emphasizing policy on the supply side are Germany, New Zealand, and the Slovak Republic. This reflects the desire by these advanced economies to provide an infrastructure base on which to build. Weighted heavily on the demand

Table 11.1 Examples of supply-side and demand-side broadband

Supply-side policies	Demand-side policies
I. Competition and investment	I. Affordability of devices and access
II. Spectrum allocation and assignment	II. Government leadership in broadband use and online activity
III. Reducing infrastructure deployment costs	III. ICT skills development
IV. Core network expansion: market led, government led, or a mix	IV. Online and local content, applications new technologies, and services
V. Inclusive broadband availability (e.g., with universal services obligations or universal service fund)	V. Consumer protection and empowerment

Source Pepper and Garrity (2013)

side are Japan, Pakistan, and Singapore, which have policies to enhance skills, provide online content, and encourage consumer protection (Pepper and Garrity 2013).

This research informs our proposed ICT policy issues, which appear later in the chapter. First, it points to our model's independent factors being factors that represent both supply and demand. Referring to the book's conceptual model (Fig. 3.8), *government prioritization of ICT* is on the supply-side; *innovation* and *social capital* are broad-based (both demand side and supply side); and the other factors are demand-side. In policies based on these factors policymakers can vary which ones to emphasize according to needs of overall situation at the time. As is seen in the nation sketches earlier in this section and in case studies such as Azerbaijan and South Korea (Chap. 2) and Rwanda later in this chapter, at times government concern is on such supply challenges as educating underserved populations in technology use, having more affordable subscription charges, or increasing university graduation in ICT subjects, while other times government is more concerned with demand such as putting in fiber-optic networks, de-regulating ICT industries, or agreeing to a higher priority for ICT in its mix of initiatives. The choice of supply-side, demand-side or the broad-based approach might also depend on the government's capacity to achieve policy goals. The same government might change policy orientation over time, as for example has occurred with Japan in recent years in shifting from heavy supply orientation to mixed demand-supply orientation.

Another study bears on the question of national policy concerns, weighting the importance of government policy and regulation versus other factors in determining broadband subscriptions. Based on a sample of data from 113 nations, the impact of

Table 11.2 Matrix of supply-side and demand-side ICT policies (the numbers in parentheses indicate number of countries categorized)

		Demand side	
		Limited	Extensive
Supply side	Limited	Emergent (2)	Demand driven (8)
	Extensive	Supply driven (9)	Broad based (9)

Modified from Pepper and Garrity (2013)

Table 11.3 Classification of broadband/ICT policies into supply-side and demand-side

	India	Sweden	U.S.	Germany	New Zealand	Slovak Republic	Japan	Pakistan	Singapore	Canada	Chile
Supply-side policies									X	X	
1. Competition and investment	X	X	X	X	X	X		X			
2. Spectrum allocation and assignment	X	X	X	X		X	X			X	
3. Reducing infrastructure deployment costs	X	X	X	X	X	X		X	X		
4. Core network expansion: market-led and/or govt-led	X	X	X	X	X	X			X	X	
5. Inclusive broadband availability (universal service or service-funds	X	X	X	X	X	X					X
Demand-side policies											
6. Affordability of devices and access	X		X					X	X		
7. Govt leadership in broadband use and online activity	X	X	X	X			X	X	X	X	X
8. ICT skills development	X	X	X		X	X	X	X	X	X	X
9. Online and local content, apps, technologies, services	X	X	X				X	X	X		X
10. Consumer protection and empowerment	X	X	X				X	X	X	X	

Based on Pepper and Garrity (2013)

a variety regulatory, competition, democracy, governance, wealth, education, and urbanization factors on broadband subscriptions per capita was studied. The major finding was that the traditional factors of income, education, and urbanization dominate as factors for broadband, while regulatory and planning factors are secondary or of no influence (Gulati and Yates 2012). In particular, the following three policy/regulatory factors: national government regulatory reach, adoption of a national broadband plan, and governance index of regulatory quality, had no influence on broadband. For developed nations, government effectiveness in supporting policy implementation increased broadband, while for mid-level developing nations, having a national telecommunications regulatory agency reduced broadband. The latter surprising result might be due to the inhibitory influence of regulatory bureaucracy in developing nations.

This study can suggest for the book's model that adding more policy and regulation factors is not necessary. It also might be that policy's effects are hard to capture in a quantitative model, but rather are easier to portray in individual and unique case studies, as is done in this chapter and throughout the book.

Box: Net Neutrality

Net neutrality refers to equality in treatment of internet traffic by internet service providers (ISPs) (Ammori 2014). If internet traffic is divided into fast lanes at reduced charges for larger-volume customers, versus slow lanes for less-well-off consumers, then net neutrality has been violated. The U.S. dilemma is portrayed here, but similar net neutrality issues confront many nations. For instance, net neutrality laws were passed by Chile in 2010 and in 2014 the Parliament of the European Union adopted strong rules favoring it.

Net neutrality has been in effect in the U.S. for the Internet for two decades but has recently been legally challenged. In the U.S., equal treatment of all parties' messages on the Internet has been a force for openness of information and for disruptive innovation (Naughton 2014). A result of this even-handed approach is that the Internet has acted as infrastructure for a public forum, but not as an Internet Service Provider (ISP)-directed control mechanism of message flows. Consequently many major ICT companies at their origins were small enterprises that could seize opportunities in the "fast lane" to develop world-class competitive services, for example Facebook and Skype, which benefitted by net neutrality to gain early momentum. Skype was developed originally in Denmark and Estonia but under a similar net neutrality environments.

Net neutrality in recent years has become a challenge and obstacle for large media companies such as Netflix which seek to provide growing customer bases with streaming information at large bandwidth. The Obama Administration at its start in 2009 strongly favored net neutrality following candidate Obama's campaign pledges (Ammori 2014). Accordingly in 2010 the FCC adopted rules that closely adhered to the campaign pledge (Quittner 2014).

In 2014, a Federal Appeals Court ruled that the plaintiff ISP Verizon could charge a fee to its customer Netflix to have a fast lane on its Internet services. The FCC has responded that it would now allow fast lanes as long as the charge for the lanes is "commercially reasonable" (Naughton 2014). In 2014, the FCC opened its stance up to public discussion and received nearly one million messages agreeing or disagreeing with the stance. In 2015, the FCC ruled in favor of net neutrality but court cases may eventually determine the outcome.

Related to the book focus on digital divides, abandonment of net neutrality in the U.S. or elsewhere would tend to lead to increased divide differences within a nation. ISP-favored participants would have their messages expedited versus less advantaged, everyday net users. If the world moved away from net neutrality, there would be much more of a patchwork of net neutrality environments than are present today. The Internet Society, a prominent international organization concerned with Internet issues has taken a neutral stance on costing, but does support a global solution to net neutrality based on user-centered principles of "access, choice, and transparency" (Internet Society 2010). A meeting of the Internet Society is seen in Fig. 11.1.

Regarding our model, factors subject to change by abandonment of net neutrality are innovation and societal openness, both of which would be lowered. The fate of net neutrality in the U.S. or worldwide is a complex issue likely to grow with time. Although it would be helpful to formalize it as part

Fig. 11.1 Meeting of the internet society: a forum to discuss net neutrality. *Source* Daniel Anderson, Elon University

of digital divide models, that is not possible since currently there are no standardized global measures of net neutrality for nations or their states and provinces.

11.3 Exceptional Initiatives of Cities and States/Provinces

ICT policies are not only applied at the national level but also at the state or city levels. Examples already discussed in the book are policies that have fostered development of the high tech city of Bangalore in India (Chap. 6), and in the U.S. the concerted policies of Georgia, leading to an exceptional high-tech state. In those instances the city and state policies were not linked to national policies but were due to local and regional leadership in setting goals and establishing policies to advance ICT.

This section discusses the example of the U-City of Songdo in South Korea. The exemplary leadership and accomplishments of ICT by the South Korean government were discussed as a case in Chap. 2. The U-City concept started in the early 2000s. Dozens of U-Cities have been built or are in construction (Lee 2013) The Ubiquitous City (U-City) is a futuristic city than combines cutting edge and ubiquitous access to information as part of the city built environment of vehicles, roads and bridges, buildings, homes and their appliances and other contents (Jang and Suh 2010). From the citizen's standpoint, such a city provides services such as u-Education, u-Work, and u-Health which offer a highly net-centric environment emphasizing rich virtual presence, ease of information access, and connection with devices, appliances, and the physical environment. The U-cities are usually designed and built with ubiquitous technologies as an inherent part of the construction process.

U-Songdo, an example of a U-city, was built from scratch starting in 2002 at an estimated cost of $40 billion up through 2014 (Arbes and Bethea 2014). Located on a new embankment of the waterfront in the northwestern city of Incheon, Songdo emphasizes high tech, green features, and sustainability (Lobo 2014). There are 100 major buildings, which meet LEED (Leadership in Energy and Environmental Design) sustainable requirements. U-Songdo is seen in Fig. 11.2.

The city has extensive networks of sensors that provide real-time environmental and urban metrics on energy consumption, traffic, water flows, and multiple other dimensions of the city. For instance, Songdo's waste disposal system is highly advanced and, in place of garbage trucks, uses a sucking system so garbage flows through tubes to an automated sorting system for sorting of renewables and treatments, that emphasizes the retention of energy from the wastes (Lobo 2014). For Songdo's schools, video telepresence is a standard feature, in particular through large classroom-sized video screens that connect Korean school children with life-size images of their counterparts in the U.S., enabling the feeling of one classroom. In the development phase are mobile phone control of home appliances and other futuristic technologies.

Fig. 11.2 U-Songdo. *Source* Louis Quah

The city has yet to fill up with businesses and residents, but it is providing incentives such as tax reductions and subsidies for arriving businesses, no property tax for businesses for 10 years, and perks associated with companies have international investment. U-Songdo represents a mixture of national and local policies that have been imaginative in encouraging development of a city designed for 50 years hence, and foreseeing a merger of futuristic technologies with a green and highly productive city. For the chapter, Songdo illustrates that policies addressing the digital divide do not originate only at the national level but also at the level of cities and states/provinces. Although the city level is too detailed to be covered thoroughly in this volume, it should be kept in mind as a source of ICT policies, some exceptional in their vision.

11.4 Proposed Global ICT Policies, Based on Book's Findings

This section pulls together policy findings from Chaps. 4–9 in order to formulate general national policy recommendations. The general recommendations might also be useful at the world level or city level. However, since Chaps. 5–8 have focus at the national level, based on data from states/provinces/prefectures, the principal focus is national or state/provincial.

The common elements recommendations for policy are the following;

- Identify the key 3 or 4 ICT determinants for their implications on policy priorities
- Recognize the tightly interrelated factors of education, income, and urbanization
- Gauge whether the government and citizen capacities can respond better to broad-based, supply-side, or demand-side policies
- Recognize that old technologies usually have opposite ICT impacts
- Recognize ethnic differences are culturally linked to the country
- Adjust policies for the wide geographic diversity of ICT

- Recognize the role of ICT-dominant states/provinces/prefectures
- Test for model factors prior to formulating policy
- Adjust ICT policies periodically to conform to changing ICT determinants, technological levels, and government capacities and viewpoints

Identifying the key 3 or 4 ICT determinants for policy priority

The findings for the four nations indicated that key ICT determinants vary among them. For instance, in China the crucial determinants for technology levels are book publishing, export commodities value, and percent of workforce in non-state enterprises. For India, they are registered newspapers/periodicals, engineering/technical education, cooperative society membership, and electrical capacity. For Japan, they are demographic factors, education, and innovation, while for the U.S. they are social capital, societal openness, education, urbanization, and ethnicity. The country factors encompass all of the book's general model factors except government prioritization, data for which are not currently available at the state and provincial level. However, the determinants are quite distinctive from country to country. For policy formulation of a given country, we suggest determining which are the 3 or 4 most significant factors and examine their implications for policies.

Recognizing the tightly interrelated factors of education, income, and urbanization

In Chaps. 4–9, the factors of education, income, and urbanization are consistently highly inter-correlated. Hence at least one of them and sometimes two had to be dropped from our regression analyses, to avoid multi-collinearity. Their strong inter-relationship has also been noted in the digital divide literature. The mechanism of their close relationship can be reasoned as follows: urban areas and cities tend on average to have higher educational levels than the rural hinterlands. Urban areas are usually where stronger schooling and universities are located. Likewise, urban areas tend to have higher average incomes, which are associated with the more specialized economies of cities and the higher levels of education leading to higher level jobs and consistent pay.

Since these variables are common ones to include for demand-based ICT policies, the policymaker should realize that they work together as a closely knit group.

Gauging whether the government and citizen capacities can respond better to broad-based, supply-side, or demand-side policies

The policymaker has the opportunity to decide how to weigh policies between supply-and demand-based factors in order to respond to constituencies in government and citizenry. For instance, if citizens are demanding greater ICT training to use a broader range of applications, then emphasizing demand-side tenets for ICT policies is warranted. If the concern is about government inadequacy in supporting infrastructure to support broadband, then the policy weighting should be towards supply. Once the weighting between supply and demand is determined, particular model factors can be emphasized.

Recognizing that old technologies tend to have opposite ICT impacts

For the national studies, fixed phones consistently have effects opposite to the other technology factors. For instance, if greater societal openness is associated with higher ICT levels for the other technologies beside fixed phones, then it likely is inversely associated with higher fixed phones. The reason is that fixed phones can be regarded as an outmoded technology, so as cell phones or IP phones are adopted more and more, fixed phones are displaced or abandoned. Hence, this argument signals that the policymaker should be careful about mixing in older technologies in the policy message, since their effects are likely to be opposite to those of contemporary ICTs, albeit there are exceptions such as Azerbaijan. Furthermore, what is "outmoded" will change over time so setting of policy needs to review which technologies are obsolete at the advent of new policies.

Recognizing ethnic differences are culturally linked to the country

In the Chap. 5–8, only Chap. 8 on the U.S. includes ethnic variables. That is mostly due to lack, in the other three nations, of state/provincial/prefectural ethnic variables that can be used systematically in research studies. Moreover, since each nation has a different mix of minority ethnic groups, due to the country's culture and history, the policymaker needs to carefully assess the ethnic mix for his/her nation and not assume ICT ethnic effects carry over from other country studies.

Adjusting policies for the wide geographic diversity of ICT

For all four nations, there are large and important geographic differences, such as the variation in China between the ICT-intensive East/Southeast and the ICT-weak Western parts of the nation. One key to initially gauging the geographic patterns is to consider what parts of the land area are urban or rural. The refined methods of cluster and spatial auto-correlation analyses can provide more detail on geographies, but are not presently used much by ICT policymakers (hopefully this book will motivate more use). With knowledge of the spatial patterns of ICT, the policymaker can avoid too much generalization on the entire nation, and add more geographic dimensions into policies.

Recognition of the role of dominant metropolitan states/provinces/prefectures

For China, India, and Japan, the huge and dominant metropolitan areas of Beijing, Shanghai, Delhi, and Tokyo have outlier importance in affecting national ICT levels, so they should be given extra attention in the policy setting. An exception is the U.S., for which no special or unique metropolitan area stands out.

Adjusting ICT policies periodically with changing ICT determinants, technological levels, and government capacities and viewpoints

As seen with China, longitudinal changes are evident in the geographies and determinants of ICTs. Hence, policies likewise should be viewed as needing to be updated regularly. In some of the best-practice case study successes with ICT policy and planning, such as South Korea and Rwanda, national ICT policies were updated every 5 years or so. A weakness in the U.S. has been the lack of a national ICT plan. It has at times led to competitive business excesses and to lack of consistency.

11.5 Limitations of Government's Role

Although this chapter and the book generally emphasize models, empirical findings, and cases where factors influence greater access to and adoption of ICTs, it is important to mention the limitations to such positively oriented models and then outcomes. For instance, there are often reasons in which policies emphasizing the book's factors for technology usage backfire and actually reduce ICT use. This can happen, for instance, if the users are unprepared for, untrained in, or resistant to technologies. For instance a new high-tech lab for school children might be put in a neighborhood of disadvantaged families, but the community as a whole does not value it and some school children are resistant to using the lab. A quite different example would be that concern for confidentiality might slow down or stop the usage of ICT for healthcare information. In another instance, a programmer or group of programmers might have invested most of their careers in mastering a technology, so that when factors point to its replacement, they are resistant because they feel their jobs are threatened. Such fears can impact whole industries in some cases, such as the replacement in the latter half of the 20th century of manual steel production by robotic methods.

Several studies have revealed the resistance of disadvantaged people to policies that seek to upgrade ICT training and infrastructure (Kvasny 2006; Kvasny and Keil 2006). A question to be asked is: could the factors for ICT be regarded by the disadvantaged as evil rather than benefit? (Kvasny 2006). Based on an ethnographic study of a low-income and mostly African-American community, the answer is sometimes affirmative. For instance in one study, regarding disadvantaged people undergoing training at a community technology center, many trainees "experienced ICT as a type of double-bind oppression that occurred when oppressive forces pushed people into situations for which there are few options, all of which are suboptimal" (Kvasny 2006). A person undergoing the training might leave the base of a low-skilled job to undertake training, but subsequently find that promised technology jobs were scarce or non-existent for him/her after finishing up. In another example, an employee goes through months of ICT training quite successfully, only to find out on return to her company that her boss now wanted computers eliminated from the firm (Kvasny 2006).

In another study of the cities of LaGrange and Atlanta, Georgia, two approaches were taken to computer training for the disadvantaged. In LaGrange, internet access was given at home by installing a free digital cable box, while in Atlanta community technology centers were made available for citizens to learn about computers and the Internet (Kvasny and Keil 2006). However, the learning by both approaches was disappointing. The authors ascribed the problem to absorption by the training providers with a "technology fix," whereas the problem was "deeply rooted in longstanding and systemic patterns of spatial, political, and economic disadvantage" (Kvasny and Keil 2006).

What is the solution? If might be partly to have policies and plans not only for technological change in workers, but for concomitant social change (Kvasny 2006). An even more generic solution would be to involve the disadvantaged persons in the formulation of strategies and policies, a step that would include the well-known steps of leadership, user involvement in design of systems solutions, and care in overseeing and monitoring the entire system solution process (Kvasny 2006).

Another study considered the instances when the developers of innovative systems wish to put constraints on its use (Baskerville and Pries-Heje 1998). A common example involves innovations that impact a person's privacy and security. Although this could be contained by the legal methods of copyright and patenting, often the legal means take too long or are inadequate to providing full control. Thus, another approach is to restrict use such as might occur with access to software for security alarms or medical records. Another example of constraints would be to put barriers on a company's ICT products or services so competitors could not integrate them into customers' systems (Baskerville and Pries-Heje 1998).

This study relates to this chapter since policies that eliminate constraints can lead to faster and wider diffusion of ICT innovations and, numerically at least, serve to reduce the digital divide. However, policymakers need to also consider the costs of maximizing diffusion, including invasions of security and privacy, and widespread worker resistance to ICT change.

11.6 Case Study: Rwanda

Rwanda represents a poor African country with low level of education and one that suffered a terrible genocide during the 1994. Its low level of development can be seen by a GDP per capita in 2012 of only $1500 U.S. dollars, with a rank of 206th worldwide (ITU 2012). 91 % of the population is engaged in agriculture, often at subsistence level, and its population 15 years or older is 29 % illiterate (CIA 2014).

By 2000, the country was on its way to restoring its economy and society. As part of its recovery the Rwandan government proposed in 2000 the ambitious Vision 2020 which emphasized strategies to elevate Rwanda to a middle-income nation and move from its nearly total agricultural economy to much more knowledge-based and ICT-oriented one (Republic of Rwanda 2000; Ntale et al. 2013). The plan included investment in infrastructure and human capital related to ICT. In the same year, the government adopted a National Information Communications Infrastructure Plan (NICI I). This plan sought to set a base environment for use of ICTs to develop the nation that would include control and management processes and have legal, regulatory, and institutional structure to support an expansion in the ICT sector (Republic of Rwanda 2014). The government was consistent in following in 2005 with NICI II, with the overall objective to "establish critical national ICT infrastructure" (Republic of Rwanda 2014).

This plan did succeed by 2010 in creating an infrastructure base and furthermore building e-services on top of it. The infrastructure accomplishments included to build a high-speed fiber-optic backbone network to connect all the nation's districts and border stations, advance mobile phone coverage to reach 96 % of the national land area by 2011, instal a Tier 3 data center which has very high reliability and cloud services, implement a digital terrestrial television system throughout the nation, and implement community telecenters, public information kiosks, and a national innovation center to foster ICT start-ups. A crew laying fiber optic cable in Africa is seen in Fig. 11.3. There were even ICT buses that traveled the country providing training and awareness (Republic of Rwanda 2014).

Also as part of NICI II, the government improved e-services available for business, agriculture, and healthcare. The agricultural service E-Soko was initiated as an information system enabling farmers and other consumers to access agricultural market information. For e-business, government services provide online tax calculators, business registration, legal case management, and land administration, while e-health offers a national medical records system and central collection and storage of clinical health statistics (Ntale et al. 2013). NICI II represents a giant step, particularly for an impoverished African nation.

Not stopping there, Rwanda in 2010 instituted NICI III, which focuses on development of ICT skills, creation of a dynamic domestic ICT sector, community development through ICTs, further development of e-government, and adding a

Fig. 11.3 Fiber optic cable being installed in Africa. *Source* Tony Hoffman

strong cybersecurity layer to the national systems (Ntale et al. 2014). Rwanda's progress has even enabled it to attract Carnegie Mellon University to have a permanent campus in its new Kigali Free Trade Zone, the university's highly valued role being to educate Rwanda's own world-class workforce of technology experts and innovators (Republic of Rwanda 2014). NICI III also includes other initiatives in ICT education and research, such as teacher training, STEM scholarships, and a new institute at Kigali Institute of Science and Technology, all of which build up an ICT private sector in Rwanda (Ntale et al. 2013).

The proof of the exceptional government ICT leadership and polices is seen in Rwanda moving up to rank 88th out of 144 nations in Networked Readiness Index in 2013, a remarkable achievement considering its earlier bottom level globally of income and literacy (Bilbao-Osorio et al. 2013). In "importance of ICTs to government vision" Rwanda was ranked 10th worldwide, an astounding level. Nevertheless, government policy challenges remain, which include shortage of highly skilled ICT personnel, low average levels of individual usage, low broadband penetration, limited FDI and other foreign financing, and limited energy supply with high costs (Bilbao-Osorio et al. 2013; Ntale et al. 2013). These challenges point to issues larger than ICT alone, which involve development of the nation's economy and standards of living. The Rwandan success so far reinforces the importance of factors in the book's model including education, economic infrastructure, social capital, and especially government prioritization of ICT, the latter factor leading to ambitious policies and programs.

11.7 Conclusion

Governments have responsibility to formulate ICT policies and support advances that reduce the digital divide. This chapter has provided instances of exemplary policy success at the national level for Japan, Sweden, and the U.S. The approaches and sequences of policies differ, but lead to similar outcomes in ICT access and use. An important aspect of an ICT policy is to consider whether it is demand-driven, supply-driven or balanced. On this basis, policies can be focused better for the needs of constituencies. Quantitative models of ICT usage might not be able to account well for policies and regulations, since they vary greatly and are not measured systematically.

An example of the complexity of ICT policymaking is seen with net neutrality, which represents keeping the Internet usage on an equal footing for all its users, versus creating fast lanes and slow lanes based on users' economic status. The U.S., for decades a country of net neutrality, is at a crossroads in whether to retain or abandon the policy.

Drawing on the findings for four nations in Chaps. 5–8, a set of nine general recommendations are given for setting policies at the national level, based on information and analysis of states, provinces, and prefectures. These recommendations can also inform the setting of ICT policy for states/provinces/prefectures or for cities. Limitations are given on ICT policy setting based on our model of ICT usage, which stems from issues with populations that are underserved and/or resistant to ICT change or from groups that have privacy or security reasons to restrict the diffusion of ICT innovations.

The case study of Rwanda illustrates that even a very poor country lacking in education and literacy can have the resolve to set ambitious government ICT policies and make remarkable progress in achieving success, given the circumstances. The ambitions are constrained by the challenges of the nation's social and economic level and lack of skilled ICT workforce. The case illustrates factors in the book's model and the chapter's policy recommendations.

References

Ammori, M. (2014). The case for net neutrality. *Foreign Affairs, 93*(4), 62–73.

Arbes, R., & Bethea, C. (2014). Songdo, South Korea: City of the future. The Atlantic, Sept 27, 2014. Available October 1, 2014, at www.theatlantic.com.

Baskerville, R., & Pries-Heje, J. (1998). Information technology diffusion. *European Journal of Information Systems, 7*, 17–28.

Bilbao-Osorio, B., Dutta, S., & Lanvin, B. (2013). *The global information technology report 2013*. Geneva, Switzerland: World Economic Forum and INSEAD.

CIA. (2014). *World Factbook*. Washington, DC: Central Intelligence Agency.

Dutta, S., & Bilbao-Osorio, B. (Eds.). (2012). *The global information technology report 2012*. Geneva: World Economic Forum and INSEAD.

FCC. (2010). *Connecting America: The national broadband plan*. Washington, DC: Federal Communications Commission.

Government Offices of Sweden. (2011). *ICT for everyone—A digital Agenda for Sweden*. Stockholm, Sweden: Ministry of Enterprise, Energy, and Communications, Government Offices of Sweden. Available November 1, 2014, at http://www.government.se/content/1/c6/18/19/14/70f489cb.pdf.

Gulati, G. J., & Yates, D. J. (2012). Different paths to universal access: The impact of policy and regulation on broadband diffusion in the developed and developing worlds. *Telecommunications Policy, 36*, 749–761.

Hall, M. (2012). *Swedish IT policy*. Stockholm: Government Offices of Sweden.

Internet Society. (2010). *Internet society international chapter toolkit*. Reston, VA: Internet Society. Available November 4, 2014, at http://InternetSociety.org.

IT Strategic Headquarters. (2010). *The new strategy in information and communications technology (IT): Roadmaps*. Tokyo, Japan: The strategic headquarters for the promotion of an advanced information and telecommunications network society, Government of Japan.

ITU. (2012). *Measuring the information society* (Report). Geneva, Switzerland: International Telecommunication Union.

Jang, M., & Suh, S.-T. (2010). U-city: New trends of urban planning in Korea based on pervasive and ubiquitous geotechnology and geoinformation. In *Proceedings of Internctional Conference on Computational Science and Applications* (Part I, pp. 262–270). Heidelberg: Springer.

Kim, Y., Kelly, T., & Siddhrtha, R. (2010). *Building broadband: Strategies and policies for the developing world*. Washington, DC: The World Bank.

King, J. L. (1994). Institutional factors in information technology innovation. *Information Systems Research, 5*(2), 139–169.

Kvasny, L. (2006). The existential significance of the digital divide for America's historically underserved populations. In S. Hongladaram & C. Ess (Eds.), *Information technology ethics: Cultural perspectives* (pp. 200–212). Hershey, PA: IGI Global.

Kvasny, L., & Keil, M. (2006). The challenges of redressing the digital divide: A tale of two US cities. *Information Systems Journal, 16*, 23–53.

Lee, Y. W. (2013). Ubiquitous [smart] city. Presentation at European Union Parliament Seminar. May 13, 2013.

Lobo, R. (2014). Could Songdo be the world's smartest city? World Finance, January 21. Available October 1, 2014, at www.worldfinance.com.

Naughton, J. (2014). Net neutrality is dead—Welcome to the age of digital discrimination. The Guardian, July 26. Available September 2, 2014, at www.theguardian.com.

Ntale, A., Yamanaka, A., & Nkurikiymfura, D. (2013). The metamorphosis to a knowledge-based society: Rwanda, Chap. 2.2. In B. Bilbao-Osorio, S. Dutta & B. Lanvin (Eds.), *The global information technology report 2013* (pp. 119–125). Geneva, Switzerland: World Economic Forum.

OECD. (1999). *Regulatory reform in the telecommunications industry*. Paris, France: Organisation for Economic Co-operation and Development.

Pepper, R., & Garrity, J. (2013). Convergent objectives, divergent strategies: A taxonomy of national broad and ICT plans, Chap. 1.3. In B. Bilbao-Osorio, S. Dutta & B. Lanvin (Eds.), *The Global Information Technology Report 2013* (pp. 43–50). Geneva, Switzerland: World Economic Forum.

Quittner, J. (2014). The net neutrality debate is far from over (And It Could Go Either Way), July 30. Harlan, Iowa: Inc. Magazine. Available September 10, 2014, at http://www.inc.com.

Republic of Rwanda. (2000). *Rwanda Vision 2020*. Kigali: Republic of Rwanda.

Republic of Rwanda. (2014). *National ICT Strategy and Plan, NICI – 2015*. Republic of Rwanda: Office of the President, Ministry in charge of ICT.

United Nations. (2011). Report of the special rapporteur on the promotion and protection of the right to freedom of opinion and expression, Frank La Rue. A/HRC/17/27. Human Rights Council, 17th Session, United Nations General Assembly. May 16, 2011. New York, NY: United Nations.

United Nations. (2012). The promotion, protection and enjoyment of human rights on the internet. A/HRC/20.L.13. Human Rights Council, 20th Session, United Nations General Assembly, June 29. New York, NY: United Nations.

Warschauer, M. (2003). Dissecting the "digital divide": A case study in Egypt. *The Information Society, 19*, 297–304.

Chapter 12
The Future of the Digital Divide

12.1 Introduction

The world is becoming more technologically advanced year by year, but digital divides remain between world regions, nations, and state-level units, and metropolitan areas. Although technology has an adverse or injurious side as alluded to in the last chapter, most would argue that overall the impacts of information technologies have been beneficial. Further most would argue it would be beneficial to reduce the digital divide, rather than leave often large differences in access and use of ICT in place or even increase the divide further. This final chapter is written to examine future trends in technologies and how digital divides might be impacted by changes. The goal of the chapter is not to try to prognosticate the future of technology, as that is outside the scope of the book, but rather to examine some trends and offer ideas about potential factors that could impact future divides. The chapter is not based solely on book findings, as Chap. 10 has sought to consolidate those findings, but rather to consider future digital divides, sometimes bringing in our research model and findings.

As a base, this introduction first summarizes briefly the most important findings from our study. Following that, a section examines two important technology trends of hyperconnectivity and big data that appear to characterize this decade as well as directly impact the digital divides throughout the planet. There are other technological trends not covered or emerging ones that other books devoted to technology forecasts cover more thoroughly. A box on Google's Project Loon discusses a creative idea to bring broadband Internet access fairly quickly to the 4 billion people who currently lack this access, and what some of the opportunities and challenges/roadblocks to doing so might be. Next, the chapter considers ideas on the increasing virtual presence of nations (Schmidt and Cohen 2013), and how those ideas might affect the selection of nations or state-level units of analysis for digital divide research.

A subsequent section gives special attention to factors that might affect the paths of information technology development for the developing world including its poorest

J.B. Pick and A. Sarkar, *The Global Digital Divides*, Progress in IS,
DOI 10.1007/978-3-662-46602-5_12

members. A concluding section considers and tries to distill from the whole volume a summary of what are the problems and opportunities for nations and sub-national units in seeking to advance ICT usage. The chapter ends by taking up the critical factor of leadership in advancing technology adoption and usage and doing it well, first analyzing the case example of broadband leadership in Poland, and finishing with a summary of key challenges of leadership in reducing the digital divides.

12.1.1 Summary of Book Findings

The major findings of this volume are as follows.

1. Nations are distinctive in their internal digital divides.

The evolution and challenges of improving access to and utilization of ICTs is found to be distinctive between nations. The reasons relate to unique histories, cultures, and institutions. For instance, since China has an authoritarian government, export economy, and wide socio-economic differences across its 31 mainland provinces, the factors driving its ICT growth and extent of internal digital divide relate to export, innovation as measured by generation of content, while geographically its ICT concentrations are in its two best known and leading megacities of Beijing and Shanghai, yet very large and poor provinces in the West and North trail in information technologies. At the same time, China aspires to provide government priority in achieving world-class status in infrastructure, innovation and business technology. India, on the other hand, has emphasized software services supported by talented domestic programmers and systems engineers, and a strong and English-speaking higher education system, but has only modest manufacturing and export of ICT hardware. Although the government has planned several times to implement broadband to its vast rural population, so far results have had limited outcomes. Spatially, India resembles China with the megacities of Delhi and Bangalore leading in ICT and with large low-tech regions in the central and northeastern states. The Indian government has not had strong long-range ICT planning and implementation.

This and other instances of disputes between study nations support the general point that there are unique digital divide differences among nations. Similarly, states, provinces, and prefectures have their own distinctive histories and cultures, although they tend to be less well known.

2. The major common determinants of ICT usage at the state/provincial/prefecural level for four study nations are urban location, economic prosperity, education, and infrastructure.

These factors are well known in the literature at the national level. Although all were present at the state level, they varied from nation to nation by type of technology and by country, further confirming the uniqueness between nations in sub-national determinants.

3. Factors available in datasets from governments and nonprofit organizations on ICT usage at the state/provincial/prefectural level are often unique to particular countries.

Since state-level variables collected by government and nonprofit agencies differ among nations, our study design includes some state-level factors that are unique for particular nations. For instance cooperative society membership is significant for some technologies in India, while newspaper circulation is significant for Japan, and social capital for the U.S.

4. There are major geographical differences in ICT usage patterns for each nation studied and for the African continent.

The differences are seen in the mapping of individual determinants, as well as of the cluster groupings from k-means cluster analysis. For 3 out of the 4 study nations, the geographical patterns conform roughly to an urban hierarchy, with the largest and most prominent megacities leading in usage of technologies, areas nearby moderate to high, and ICT rural "hinterlands" of low usage. The clusters can be characterized and compared with each other on multiple technology dimensions. For Africa, cluster analysis reveals high technology levels in clusters in North Africa and the far South, but with the landlocked center of Africa at very low levels. The geographic analysis of ICT trends needs to recognize and test for the problem of spatial bias.

5. A conceptual model of ICT utilization is constructed based on literature and the authors' reasoning. The model is tested and empirically validated with international data sets.
6. Case studies can go beyond the quantitative model in exemplifying the evolution and complexity of the digital divide within a country.

The book's case studies, drawn from secondary sources, provide behavioral, leadership, and organizational findings, many longitudinal, that are not possible through the quantitative approach.

7. Government policies are often in place for nations, but because of technological and societal changes are often modified and changing. Chapter 11 has discussed the variety of policy approaches and their relationship to our empirical studies. Common elements are recommended in that chapter for policy development.
8. Leadership is often a critical element in determining the success or failure of initiatives to lessen or eliminate the digital divide.

Leadership is crucial in many of our case studies, including Estonia, State of Georgia, and Rwanda, as well as the Polish case study later in this chapter. Other cases of countries did not fulfill their potential of advancing ICT usage due to deficits in leadership including the U.S., and Mexico. Leadership was not incorporated in our general model, except partially for the factor "government prioritization of ICT," which was included only for the world and for African nations. That factor generally proved unimportant based on the quantitative analysis of the world, world regions, and Africa. However, that analysis is not able to capture the critical strategic influence of key leaders. A leadership factor was excluded from our general model due to lack of data available consistently for it at the state/provincial/prefectural level.

12.2 Hyperconnectivity and Big Data

Two trends are highlighted in this section as essential for the trajectories of the digital divides for the rest of this decade—hyperconnectivity and Big Data. Hyperconnectivity is defined as very fast, continual connectivity that roams through space and time, and can seamlessly move from network to network (Biggs et al. 2012). Hyperconnectivity broadens the networking concept from connecting people to connecting people and things. For instance, a citizen may be internet-connected to other citizens, home appliances, cars, merchandise in stores, cameras, locks, and other physical objects. A factory manager may be continually internet-connected to workers, machines, inventory items, security devices, and testing devices. Some sensor devices would be miniaturized, even based on nanotechnology and connected by wireless communications. Hyperconnectivity enlarges the size and richness of information of the Internet, with estimates that 15 billion hyper-connected devices would be present by 2019 (Google estimate, cited in Biggs et al. 2012). Many of the physical items would be RFID-enabled (Radio Frequency Identification-enabled), to allow tracking and geospatial analysis.

Hyperconnectivity is economically possible through falling prices for high-speed fixed and mobile broadband. As seen in Fig. 12.1, fixed broadband pricing, as a percent of gross national income (GNI) per capita, decreased worldwide from 115.1 in 2008 to 22.1 in 2012, and is continuing to decline. What had been limited to rich users in 2008 now becomes widely affordable albeit still expensive in 2012, and can be expected to become quite affordable in the next several years. The lowering price trend provides the economic basis for extending the high-speed Internet to billions of physical items and devices.

It is also evident that a sharp pricing differential is present between developed and developing nations. Developed nations' broadband has been highly affordable at 1.7–2.5 % of Gross National Income (GNI) per capita from 2008–2012, while that for developing ones dropped from an unaffordable level of 164.6 % of GNI/capita in 2008 to an expensive 31 % in 2012. Accordingly, the initial rollout of hyperconnectivity has occurred in the advanced economies, as exemplified by robotic factories in Japan, the hyperconnected U-city of Songdo in South Korea discussed in the last chapter, and U.S. military which has implemented an RFID-enabled supply chain with Internet connectivity for its inventory. As seen in Fig. 12.2, the U.S. and Japan have very low fixed broadband pricing compared to affordable pricing in India, China and the Commonwealth of Independent States i.e. former Soviet satellites, while the Americas are expensive and even more so for Asia-Pacific countries. Africa at 64.3 GNI/capita is largely unaffordable for broadband. In this book, the four study nations have affordable broadband and have

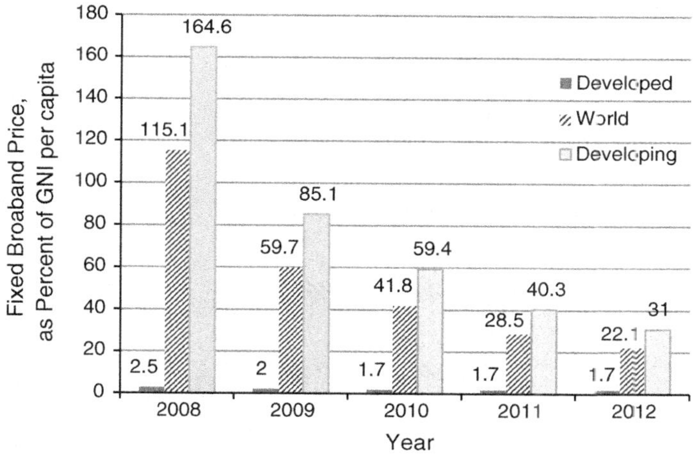

Fig. 12.1 Fixed broadband prices, as percent of gross national income, 2008–2012. *Source* ITU (2013)

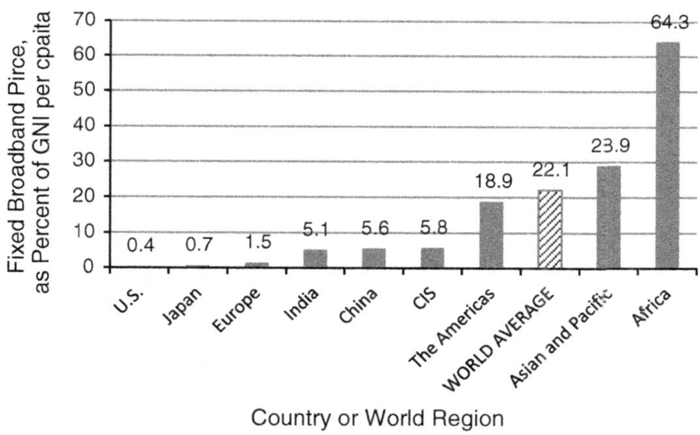

Fig. 12.2 Fixed broadband price, as percent of gross national income per capita, by country or world region, 2012. *Source* ITU (2013)

entered at least partly into the hyperconnected era, while African nations remain predominantly not hyperconnected.

Similarly, broadband pricing and world-region differentials are present for mobile broadband, which consists mostly of 3-G and 4-G cell phones. As seen in Fig. 12.3, developed nations' mobile broadband pricing is low, approximating the level of fixed broadband. In the developing world, fixed broadband pricing is unaffordable at 17.7 % GNI/capita. Handset-based mobile broadband for the developing world offers the lowest percent pricing at only 8.3 % for postpaid

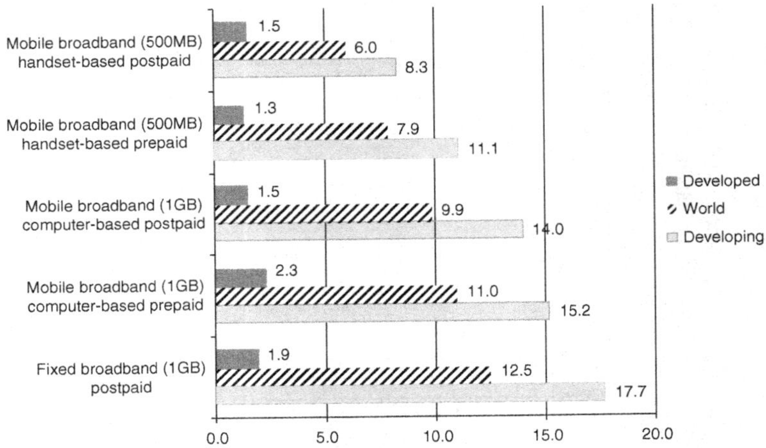

Fig. 12.3 Mobile-broadband and Fixed-broadband prices, as percentage of gross national income per capita, 2012. *Source* ITU (2013)

handset-based broadband. This helps explain why 3-G cellphones have become the primary Internet access method for the poor countries, for instance those in most of Sub-Saharan Africa. Another implication is that hyperconnectivity in the developing world would be led by inexpensive mobile devices for people and inexpensive sensors and other physical device connectors.

Big Data refers to massive data storage now being collected that exceeds the traditional data-base and requires new methods to process, manage, and make use of the data. It has been caused by the huge expansion in the amount of data being collected and utilized globally, as well as the exponential growth in processing capability of microprocessor-driven devices (doubling every 18 months), which was foretold by Intel scientist and top executive Gordon Moore, originator of Moore's Law, seen in Fig. 12.4.

Large companies and governments have been storing very large amounts of data for a decade or more, but much of the data was "locked up" and used slightly. A government example is the huge amount of NASA satellite imagery that was collected since 1972, the date of the first Landsat satellite. However, the data are so massive that only a tiny portion has been analyzed.

The flood of information is today stimulated by exploding Internet use, sensor deployments, satellite imagery, government and corporate data collection, and social media. The volume growth is reflected in the expansion of Internet users from 16 million in 1995 to over 3 billion today (Maney 2014), and in yearly global IP traffic in 2012 estimated by Cisco at 1 zettabyte (a zettabyte equals one trillion gigabytes) (cited in Biggs et al. 2012). This massive store is growing—International Data Corporation (IDC) estimates that the amount of digital information doubles every 2 years (Biggs et al., citing IDC). An estimated 80 % of the information being presently produced is unstructured, which refers to being unorganized and un-indexed (Maney 2014). As a result conventional data-base software is being replaced by software with the

Fig. 12.4 Gordon Moore, inventor of Moore's Law, in his Intel Office, Santa Clara, California. *Source* Intel Free Press

capability to store and access unstructured data, such as Apache Hadoop, an open source software service that can store and handle massive amounts of information on thousands of systems. Hadoop also has the Map Reduce program to efficiently combine information, and connects with related software that can perform Big Data search and analytics such as R on Hadoop, Watson Explorer, and text analytics programs (IBM 2014; Maney 2014). The impact on an individual user is that much more data on him/her will be collected, and he/she will find more detailed information is being applied to daily life, yet he/she will potentially be assisted more thoroughly in answering many questions and making decisions (Maney 2014).

Hyperconnectivity and Big Data are raising regulatory, security, and standards questions that regulatory agencies and courts will have to contend with for many years. They include questions of how much open access will be available to networks, shortage of addresses to identify devices, net neutrality (covered in the last chapter), standards for networks to communicate with each other, and privacy invasion of individuals' personal data. "More and more of the everyday real world around us [will become] reflected online" (Biggs et al. 2012), so concepts of privacy, security provisions, and the role of regulators will need to change.

Hyperconnectivity and Big Data will potentially influence our general model of technology utilization. In our present general model (Fig. 3.8) they impact the factor of infrastructure, which will change in ways just described and other unforeseen ways, and on the dependent technology factors, which will likely need to evolve into new or modified forms keyed to these trends. For instance, social media has been changing rapidly and will be impacted by hyperconnectivity and Big Data in unknown ways to become richer in information and more immediate. This evolution may impact the independent model factor of social capital, which may evolve to more of a mixture of physical and virtual social capital. The traditional landline phone, lacking broadband hyperconnectivity, will likely accelerate to its decline.

Innovative new ways to access and use technology are likely in a hyperconnected world. At the same time, parts of the model should remain robust in the new environment including demographic/ethnic, education, economic, societal openness, government ICT prioritization, and innovation factors.

Geographic aspects of the general model may alter as more detailed geo-referenced information is produced by the flood of Big Data, yielding robust and systematic data-sets for geographic units smaller than state-level, for instance at county or urban-area levels.

Box: Google's Project Loon

Project Loon from Google illustrates a risky technology that might or might not succeed, and have greater or lesser impacts on the global digital divide. Whatever its fate is, it is a reminder that global digital divides are likely to be influenced by breakthrough technologies, little known or unknown at present.

Project Loon was started by Google-X in June of 2013 with the goal to serve the estimated 4.4 billion people who were not yet online in 2013 (ITU 2013). Google-X is Google's secretive, special projects unit located near its headquarters in Mountain View, California. The plan is to put a ring of high altitude balloons (termed "loons") that have capability to receive and transmit wireless broadband around the world in a controlled and coordinated loon network. Because the developing world and remote, rural parts of the developed world have potential for wireless but often not wired broadband, the network could provide the potential to make broadband accessible to the preponderance of these broadband underserved people.

A loon is a balloon about 50 feet wide and 40 feet high which is currently helium filled, although Google might switch to hydrogen (Butler 2013). With a skin of polyethylene fabric, it carries a payload with WiFi electronics, special custom Google-X circuit boards, satellite receiver, and radio equipment. A loon has solar collectors that store energy in batteries during the day, using the energy at night to power the loon air vents and electronic equipment. After a loon launch, such as is seen in Fig. 12.5, the loon ascends in the stratosphere to about 60,000 feet feet of altitude.

The loon floats in the stratosphere above the earth, moving by finding and taking advantage of known wind currents in the stratosphere, under the control of a combination of mostly automatic pre-programmed controls as well as human ground-based flight controllers. It can gain or lower altitude through dual automatic air vents. Hundreds or thousands of loons currently circle the

Fig. 12.5 Google loon launch, New Zealand. *Source* Carlos Fernandez

globe along its 40th parallel of latitude in the southern hemisphere (Butler 2013; Barr 2014), a fleet that can communicate and coordinate with each other.

By 2014, Project Loon should have a ring of operational loons stretching around the world. The project was mostly successfully pre-tested in New Zealand, northern California, Nevada, and Brazil (Barr 2014). Although a balloon crashed in Nevada, and adjustments and changes have been made to the design and systems, user testing in Brazil confirms that medium wireless broadband speeds for earth users can be achieved by the network of loons (Barr 2014).

Project Loon faces regulatory and legal issues that involve the U.S. and possibly any nation, the territory over which loons fly (Butler 2013). For instance, in the U.S. it is uncertain if the loon falls under "unmanned free balloon" regulation of the U.S. federal code or under Federal Aviation Administration (FAA) regulation for non-military drones (Butler 2013). Regardless, for commercial purposes Google will need to follow federal guidelines. The unresolved question for other nations is whether balloon flight in the stratosphere is subject to laws of the country underneath.

Google's motivations might be both self-serving and philanthropic. If Project Loon works, Google would gain enlarged markets of up to potentially a billion or more users for most of its product lines; at the same time the lives of huge numbers of underserved people could benefit. Even in the U.S. some rural and remote areas still lack broadband, which the Project could remedy (Morphy 2014). This type of atmospheric network has been pioneered by Google-X, but there might be other market entrants, such as Facebook, which announced a project somewhat like Project Loon but using large drones instead, with plans to be operational by 2018 (Lapowsky 2014; Griffiths 2014).

For the book's model, Project Loon would be a leapfrog step representing both innovation and infrastructure that would yield step changes in some of the dependent variables, with particular impact on developing nations. Accordingly, it would contribute to the narrowing of the digital divide both at the national level, as well as at the state, provincial, and prefectural levels. As was seen in Chaps. 5–9, rural, remote areas are often the ones with lower levels of ICTs. As explained earlier in this chapter, the cost of such services in different nations would be influential on its extent of loon usage, and regulations and laws could limit or restrict access and use.

12.3 Future Technology, Nations, and States

The geographic units of analysis in this book are the nation, and states, provinces, and prefectures. These units have provided a strong basis in our study and most of the literature for understanding the correlates of ICT usage and geographies of use. These units are administratively very long-term, almost permanent governmental units and provide a stable geographic basis for studies of the digital divide and potential for longitudinal studies as consistent state-level data series build up over time.

Recently the idea of nation has been viewed from the ICT perspective as a combined physical ground concept coupled with a virtual nation concept (Schmidt and Cohen 2013). This section briefly describes the idea of both a physical and virtual nation, and considers how it is useful in understanding and possibly studying digital divides.

Nations have been classified as containers, regulators, competitors, and collaborators (Dicken 2011). Container refers to the nation as containing distinctive cultures, history, institutions, and values, yet not a sealed-off container but one that has connections outside. Regulator refers broadly to a nations' regulations, laws, and practices. Nations have often competed on multiple dimensions that include economies, education, innovation, and trade, while collaboration represents world economic and political alliances such as the European Union (EU) and North American Free Trade Agreement (NAFTA), to less formal good neighbor relationships.

Many nations today are also forming powerful virtual forms, such as the Chinese internet, which is huge and expanding rapidly but tightly controlled and censored or the Indian internet, which is multi-lingual, fractionated, with reduced government regulation along with federal government web presence that reflects the nation's complex bureaucracy. By contrast, many developing nations such as Bolivia (http://www.bolivia.gob.bo) and Rwanda (http://www.gov.rw) have fairly weak virtual presence, although it is difficult to judge the strength of their internal federal ICTs.

Since nations have regulatory control of the physical structure of the national Internets, the national government can step back and support net neutrality for the country, as did Chile through legislation in 2010, while some other nations at the opposite extreme can exert extensive control, filtering, and censoring of information, with North Korea being an extreme example. Highly controlling nations can put restrictions on the gateway routers for all traffic entering or leaving the nation and/or put restrictions on the servers of the country's domain name system (DNS), which designate address names for web access (Schmidt and Cohen 2013).

The intensity of national government control of its domestic Internet has been classified as (1) politically and culturally acceptable filtering, (2) sheepish Internet filterers, and (3) blatant filtering (Schmidt and Cohen 2013). The first category covers nations that do limited filtering often considered for worthy purposes such as eliminating access to pornography or filtering out propaganda from an enemy, as South Korea does for North Korean content. In the second category, many sites and connections are filtered but usually without knowledge that filtering is taking place, with Turkey as an example. The third category involves strong and visible censoring of information, both internal and external, with China as an example.

In response to differences in virtual presence of nations, regulation, competition, and collaboration can take place (Dicken 2011). The collaboration, termed "virtual multilateralism" would bring together nations of similar political forms such as several democratic nations, or a group of autocratic nations, to share virtual resources and even presence in alliances. An example of a virtual alliance of autocratic nations is that North Korea's national website is 3/4 run by Egypt's Orascom telecommunications company, with the other quarter run by North Korea (Schmidt and Cohen 2013). An instance of collaboration of democracies is the European Union's Digital Agenda for Europe, discussed as part of this chapter's Poland case. National virtual collaboration could also involve a nation with a strong virtual presence influencing the virtual presence of a small and weaker neighbor.

Virtual competition between nations is already taking place, as seen in government-sponsored Internet spying and information theft or at greater extreme by cyber-attacks, perhaps the most well-known of which was the Stuxnet Worm's successful attack by Israel and the United States on the physical infrastructure of largest Iranian nuclear facility.

Regulation of virtual control of national Internets is beginning to happen in certain countries, although it is done under varying extents of secrecy. The virtual assets and sophistication of nations is likely to grow, with their virtual multilateralism, isolation, and collaboration, and completion becoming more influential and important for nations.

The implications of the virtual manifestation of nations for the book first of all relate to geography. If strong multilateral collaboration of national virtual assets takes place, models of the digital divide might need to recognize and include new geographical constructs for such alliances in the future. Another example is that nations which legislate national adherence to net neutrality such as Chile might become destinations for "internet refugees" escaping nations with strong censoring and control of the Internet. That would impact the book model's societal openness factor. A further example is that poor and weak nations highly dependent on large ones' virtual assets could be recognized in case studies and controlled for in digital divide models. In summary, conceptual modelers of digital divides in the future will need to assess the evolution in the virtual forms of nations and states and consider adding constructs to account for them.

12.4 Future Paths in Information Technology for the Developing World

Countries in the developing world have choices on their paths to advancement on technologies and the closing of their digital divides. Customized pathways to improvements align with the book theme mentioned earlier in this chapter and in Chap. 11 that our study tends to show unique national patterns of ICT utilization. This section offers some observations partly based on the work of Kim et al. (2010) that suggest steps poor nations can take to advance their technology levels; then it discusses a well-known example of failure of introducing laptops in developing nations the One Laptop Per Family initiative, linking this failure to this section's recommended success steps.

The recommended steps for advancing ICTs in a developing nation start with *flexibility*. The planning of ICT is unlikely to be able to be controlled precisely, as developing nations often undergo economic and political ups and downs. For instance, with the Azerbaijan case in Chap. 10, the rich petroleum resources discovered gave sufficient steady funding support to carry out ambitious large steps in network strengthening and expansion, education, and collaboration with near neighbors and even Europe and Asian ones. For Rwanda, the coincidence of external funding for its fiber backbone with arrival of an overseas campus of Carnegie Mellon University required flexibility to take full advantage of opportunities. Another aspect of flexibility is to realize that ICT maturation is taking place and adapt policy to conform to maturation steps along the way. South Korea, elucidated as a case in Chap. 11, demonstrated flexible adjustments as it moved on a long trajectory from developing-world digital status to a global leader.

A second step is to *seek financial partnerships with caution to lower risk*. Since each developing nation has its own strengths and weaknesses, financing should be sought that addresses areas of weakness, a step seeking a balance in capabilities. Public-private partnerships can tap into new funding sources. For instance, Azerbaijan leveraged off of an existing nationwide fixed phone network in urban areas by collaborating with those providers to add to the network DSL capabilities. Such a step presents opportunities in India, which has providers of cable networks that cover most of the nation (Kim et al. 2010). Another way to lower risk is to leverage limited government financial resources based on market supply and demand (Kim et al. 2010). A further essential aspect of this step that can lower risk is to strengthen the incentives, subsidies, and environment for private investment (Kim et al. 2010). Azerbaijan did this by encouraging and creating a welcoming environment for Microsoft and Hewlett Packard to provide PCs nationally at discounted prices focused on school children.

A third step is *to support education and training*, with careful attention to language and computing skills. The education and training factor has been shown repeatedly in the academic literature and case studies, including in this volume, to be essential for developing nations. Children in a poor nation in Africa are seen collaborating in Fig. 12.6 to learn about a notebook's functioning and its knowledge content.

Fig. 12.6 Children learning collaboratively using a notebook, Africa. *Source* Alikem Tamakloe, Ho Regional Library in Ghana

A developing nation's federal government can incentivize and support all educational levels from primary through graduate education. For instance, Mexico's strong education systems for a middle-level nation have been helpful in its moderate success with ICT, while Estonia consistently put emphasis nationally on technology in its schools and universities.

In summary, recommended steps for developing nations to raise their position on the digital divide are to emphasize flexibility, seek financial partnerships while lowering risk, and support education and training.

An example of a partly failed attempt to strengthen usage of ICTs in the developing world was the One Laptop per Child (OLPC) initiative initiated in 2005 by Nicholas Negroponte, MIT professor and former director of its Media Lab. He proposed that US$100 laptops could be provided to hundreds of millions of school children in impoverished nations. Negroponte had received commitments of $20 million in start-up funds from investments made by important technology industry companies.

The OLPC's XO laptop was unique at the time for its small size, ruggedness, multi-lingual keyboard, free Linux open source operating system, and low cost (Kraemer et al. 2009). The XO was intended for ease of use in teaching and learning.

Although promising and well-intended to address digital deficits for children in poor countries, the project failed for a number of reasons. Perhaps the two largest issues were misunderstandings of cultural and social values and customs in the intended countries and not realizing the need for a complex of support and maintenance processes in the recipient countries. Regarding cultural misunderstanding, Negroponte and other team members of the OLPC nonprofit did not foresee that strong resistance to the laptop technology, educational change, and new learning methods would be prevalent among users and their organizations. Also the countries lacked a network of service, maintenance, support, content production, and trained teachers for the XO laptops (Kraemer et al. 2009). This gave users difficult experiences that lowered diffusion of the new technology.

Another unanticipated issue was that the computer industry, some members of which had provided start-up funds, saw market potential for themselves in the XO-type product and became increasingly competitive to claim a large market share from OLPC. Hence during 2008, the industry ramped up netbook production from a small pilot market to over 10 million units of low-cost netbooks in global sales. The aforementioned problems plus this competition toppled the plans for OLPC, caused staff defections and disillusionment, and eventually led to OLPC's failure to reach its goals. At the end of 2009, sales from OLPC total only several hundred thousand XO's and the financing went away.

The case illustrates why the three steps recommended earlier were not working. OLPC was not flexible by not immediately seeing the cultural and system support/maintenance/training issues and addressing them. In its initial financing, investors' conflicts of interest were present and were destined to arise later, albeit hard to discern ahead of time. However, once those conflicts were evident, OLPC should have sought other financing. Finally, OLPC did not realize the recipient countries' deficits in education and training, nor the technology resistances built into their school systems. Although OLPC was an entity headquartered external to the developing world, the lessons apply just as well for the governments of developing countries.

12.5 Digital Divides: Problems and Prospects

Overall, the book has put forward a conceptual model to understand the determinants of the digital divide, geographic aspects of ICT usage, and resultant policies and implications of the findings.

The problems issuing from this study are many. First, from a research standpoint, although substantial government and nonprofit data are available for nations worldwide, there is limited data available for sub-national units within nations, and the data are not standardized nor is there effort afoot in international organizations to accomplish this standardization. This is a barrier for major expansion of research at the sub-national level, unless researchers can secure sufficient funding to conduct their own state-level surveys across multiple nations. Another challenge for gaining broader understanding of sub-national digital divides around the world is the common place problem with international comparative research of not having sufficient language, cultural, historical, and values understanding of each nation.

From the perspective of challenges to policymakers and implementers of change in countries and states/provinces around the world, there is a challenge of communication of the importance of the digital divide. For many nations, complacency may have snuck its way in, especially in developed nations which look at upward curves of technology use across nations and convince themselves the curves will inexorably increase and saturate markets with ICT use everywhere (James 2011). However, others argue that this view can be deceptive because of the wide dispersion of ICT levels and challenges in the developing world, compared to a much narrower dispersion of ICT levels and great similarity of environments in the developed world (James 2011). At the high end of use in the developing world, with good underlying starting base of infrastructure, the transition to moderate or high ICT level is often swift and fairly easy, but for the ICT-poor nations at the lower end, the challenge is much greater and their divide may actually enlarge with time (James 2011).

Among the opportunities available for nations, states, and provinces to improve their ICT access and usage are the following:

1. Some advanced nations such as Sweden and South Korea not only recognize the persistence of digital divides, but are proactively forming government-sponsored teams that provide direct expertise, training, and assistance to other nations that request it.
2. Public-private partnering has often proved beneficial, and offers opportunities for motivated governments to develop cooperative/collaborative relationships with technology companies that have specific goals, provided there is expectation of performance on both sides of the table.
3. For nations that have tight control and regulation of digital information of its inhabitants, there is opportunity to reduce those controls, leading to a more open flows of information and communications. This is known to be stimulus for improved usage of ICTs with accompanying benefits.
4. Development of leadership in ICT should be cultivated and encouraged. The last two sections of the chapter focus on this issue as a final point in the book, first with a case study of Poland and then some final comments related to leadership.

12.6 Leadership Case Study: Poland

The adoption and implementation of Poland's National Broadband Plan involved major preliminary steps forward, followed by collaboration with the European Union and stakeholders in Poland to realize approval in late 2013. Poland in 2005 was way underneath the EU averages in ICT. It had 19.3 % of PC users versus 42.1 % for the EU and only 3.3 % broadband users, versus 13.5 % for the EU (World Bank 2007). A number of preliminary steps were taken. In 2008, the "Strategy for the Development of the Information Society in Poland 2008–2013" was adopted by the Polish central government, recognizing its deficiencies relative to the EU. This was followed in 2011 by "Poland 2030," a twenty year strategic vision of growing ICT (Biggs and Polomsha, 2013).

Even before this period, Magdalena Gaj, a lawyer and expert in telecommunications law, joined the Polish government in the Office of Telecommunications Regulation and worked her way up to become by 2005 the director of the legal department in the Office of Electronic Communications, the key office for implementing the Internet and telecommunications in the country. Next, as Undersecretary of State in the Ministry of Infrastructure she started the "Broadband Round Table for Poland," a forum that brought together key stakeholders from business, national and local government to discuss ways to expand broadband substantially throughout Poland. She helped to originate "digital lamplighters," teachers who educated others in the communities about ICTs and the need to be inclusive. She became an international representative for Poland service as a member of the ITU Council, in charge of the EU Council work on telecommunications during the Polish Presidency of the Council. In 2012, she was appointed as President of the Office of Electronic Communications.

All this led up to the proposal of Poland's Broadband Plan by the Office of Electronic Communications (UKE), under President Gaj's leadership; the Plan was approved by the Polish government in 2013. The plan meets the requirements of the European Commission's "Digital Agenda for Europe," but goes beyond that plan's goals. In particular the Polish plan calls for all of the 38.5 million inhabitants of Poland to have access by 2020 to a broadband downlink of 30 Mbps (megabits per second) or higher, with over half the population having access of 100 Mbps or higher. This plan also reduces bureaucracy by putting much of the EU funding and the responsibilities for implementation in the hands of local governments (EU 2014).

For instance, in Nowosolna, a small set rural of villages on the outskirts of the city of Lodz, a fiber network for education, funded by Lodz City and EU funding, is implementing fiber cable network by sharing conduit digging costs with the network operating company Toya, a shared arrangement referred to as "piggybacking" (Van Blommestein 2013).

Significant funds were raised from the nation and EU to implement the Plan in amount of US $2.82 billion for the period 2014–2020, while other related funding over the period include US$1.3 billion for e-administration and open government (Telecompaper 2014).

Also the Broadband Plan includes "Broadband Mapping Poland," which provides interactive maps and accompanying data at the national, and sub-national levels that display Poland's Internet, mobile phone coverage, and internet hotspots, and points to deficient zones. This is innovative and useful to officials, government planners, and the general public.

Besides President Gaj's leadership on the National Broadband Plans, she has led and made progress in multiple areas. For instance, she motivated and encouraged telecommunications firms to invest in contemporary infrastructure, improved the experience of consumers and their service quality, and worked at changing regulations to satisfy operations, investors, and consumers, while retaining competitive markets (UKE 2014).

For her long-time, varied, energetic, and productive contributions, now as President of UKE and the key champion in moving Poland's ICT level from deficient for the EU to above average and moving towards a high EU level, President Gaj personifies ICT leadership. Besides exemplifying a successful national leader in ICT, her story relates to the chapter and book in a number of ways. It demonstrates the concept of a regional grouping of nations in IT, namely the 27 EU nations, which are cooperating towards common Union-wide broadband goals. It exemplifies the prior chapter concept of positive cooperative arrangements between a group of democratic nations.

The case illustrates a government that recognizes in its formal planning and budgeting the importance of local geography and local decision-making on broadband implementation. This resonates with the book's goal to emphasize sub-national geographies of ICT as an important research and practical topic. Poland also stands as a country that is striving to reach hyperconnectivity by year 2020 or earlier. Regarding the book's general model in Table 3.8, the Polish case illustrates the importance of education, infrastructure, government prioritization of ICT, and societal openness (rule of law and open communications). Poland's excellence in education has been built up over many years following its departure from the Soviet Bloc, and has constituted a base strength for ICT development.

12.7 Conclusion: Challenges of Leadership

As seen in the book, there are many positive factors illustrated and confirmed that influence the usage of information and communication technologies. Leadership may be at the top of the list. Readiness factors might be present due to strengths of the country, its firms, and its workforce, but the exceptional results in closing and bridging the digital divides generally depend on one or several leaders, who motivate and galvanize governmental endeavors, corporate involvement, and real-world projects to understand and try to stimulate high levels of ICT understanding and use. Such leaders show the ability to involve all internal and external stakeholders, motivate employees through their enthusiasm, circulate widely, reflect the latest trends and applications, communicate well, and relate to the average citizen, while having the flexibility and motivation to learn new knowledge and skills in a rapidly changing field. We hope this volume will be of use to academic researchers, professionals in relevant fields, business analysts and planners and importantly to existing and incipient leaders in ICT who can help to achieve broad technology utilization and narrowing of the digital divides.

References

Barr, A. (2014). Google's moonshot chief claims progress on project loon balloons. September 24. *Wall Street Journal Digits*. Available on October 16, 2014, at http://blogs.wsj.com/digits/2014/09/24/googles-moonshot-chief-claims-progress-on-project-loon-balloons/.

Biggs, P., Johnson, T., Lozanova, Y., & Sundberg, N. (2012). Emerging issues for our hyperconnected world. In S. Dutta & B. Bilbao-Osorio (Eds.), *The global technology report 2012* (pp. 47–56). World Economic Forum: Geneva, Switzerland.

Biggs, P., & Polomska, A. (2013). The importance of national policy leadership. In B. Bilbao-Osorio, S. Dutta, & B. Lanvin (Eds.), *The global technology report 2013* (pp. 53–65). World Economic Forum: Geneva, Switzerland.

Butler, W. (2013). Can we trust google with the stratosphere? *The Atlantic*. Available September 20, 2014, at www.theatlantic.com.

Dicken, P. (2011). *Global shift*. (6th ed.). New York: The Guilford Press.

EU. (2014). Country information—Poland, digital agenda for Europe. Brussels, Belgium: European Commission. Available October 15, 2014, at https://ec.europa.eu/digital-agenda.

Griffiths, S. (2014). Facebook's internet-providing drones will be as large as jumbo jets—and could be use by 2018. *Daily mail*. September 29, 2014. Available on November 10, 2014, at www.dailymail.co.uk.

IBM. (2014). *Hadoop*. Armonk, NY: IBM Corporation. Available on October 28, 2014, at http://www-01.ibm.com/software/data/infosphere/hadoop.

ITU. (2013). *Measuring the information society*. Geneva, Switzerland: ITU.

James, J. (2011). Are changes in the digital divide consistent with global equality or inequality? *The Information Society, 27*, 121–128.

Kim, Y., Kelly, T., & Raja, S. (2010). *Building broadband: Strategies and policies for the developing world*. Washington, D.C.: The World Bank.

Kraemer, K. L., Dedrick, J., & Sharma, P. (2009). One laptop per child: Vision vs. reality. *Communications of the ACM, 52*(6), 66–73.

Lapowsky, I. (2014). Facebook lays out its roadmap for creating internet-connected drones. *Wired magazine*. September 23. Available November 10, 2014, at http://www.wired.com.

Maney, K. (2014). Running with numbers. *Newsweek, 163*(5), 43–45.

Morphy, E. (2014). Why I am more excited about Google's Project Loon than Project Wing. *Forbes*, August 30, 2014.

Schmidt, E., & Cohen, J. (2013). *The new digital age: Reshaping the future of people, nations, and business*. New York: Alfred A. Knopf.

Telecompaper. (2014). Poland to spend EUR 1 billion on improving broadband access. *Telecompaper*, January 9, 2014. Houten, The Netherlands: Telecompaper. Available October 15, 2014, at http://www.telecompaper.com/news/poland-to-spend-eur-1-billion-on-improving-broadband-access–989310.

UKE. (2014). The President of UKE presents the implementation status of the strategy for 2012–2015. Warsaw: Republic of Poland, Office of Electronic Communications (UKE). Available on November 25, 2014, at http://www.en.uke.gov.pl/the-president-of-uke-presents-the-implementation-status-of-the-strategy-for-2012-2015-14939.

Van Blommestein, M. (2013). Fibre piggybacking; how Poland found a way to get the fastest broadband to the remotest regions. October 16. Zdnet.com. Available September 28, 2014, at www.zdnet.com.

World Bank. (2007). *World development indicators*. Geneva, Switzerland: World Bank.

Index

J.B. Pick and A. Sarkar, *The Global Digital Divides*, Progress in IS,
DOI 10.1007/978-3-662-46602-5

E

CPSIA information can be obtained
at www.ICGtesting.com
Printed in the USA
LVOW05*0018230716
497468LV00001B/19/P

9 783662 466018